Titanium Carbide MXenes

Titanium Carbide MXenes

Synthesis, Characterization, Energy, and
Environmental Applications

Edited by Muhammad Tahir

WILEY-VCH

Editor

Muhammad Tahir
United Arab Emirates University
Chemical and Petroleum Engineering
Department
P.O. Box 15551
Al Ain, Abu Dhabi
United Arab Emirates

Library of Congress Card No.: applied for

British Library Cataloguing-in-Publication Data
A catalogue record for this book is available from the British Library.

Bibliographic information published by the Deutsche Nationalbibliothek
The Deutsche Nationalbibliothek lists this publication in the Deutsche Nationalbibliografie; detailed bibliographic data are available on the Internet at <http://dnb.d-nb.de>.

Print ISBN: 978-3-527-35074-2
ePDF ISBN: 978-3-527-83868-4
ePub ISBN: 978-3-527-83869-1
oBook ISBN: 978-3-527-83870-7

Cover Image: © THOM LEACH/SCIENCE PHOTO LIBRARY/Getty Images

Typesetting Straive, Chennai, India
Printing CPI Group (UK) Ltd, Croydon CR0 4YY
C9783527350742_211223

Contents

Preface

Nanotechnology is one of the most important developments in science and technology in recent years. Since the 1960s, MAX phases, layered carbides, and nitride materials have been under development with their high thermal resistance, chemical stability, hardness, water resistance, and conductive properties. However, there has been only a small increment in the increase in surface area and the energy storage and conversion performance of these metal carbides and nitrides. A breakthrough in the field of MAX-phase materials occurred in 2011 with the discovery of MXene materials. MXenes are a class of two-dimensional inorganic compounds that consist of a few atom-thick layers of transition metal carbides, nitrides, or carbonitrides and combine the metallic conductivity of transition metal carbides with a hydrophilic nature due to their hydroxyl- or oxygen-terminated surfaces. There is growing interest in the use of MXenes as single and composite materials because of their promising physical, electrical, catalytic, and optical properties. Among the MXene materials, titanium carbide ($Ti_3C_2T_x$) MXenes, due to their low cost and promising optical and electrical characteristics, have been widely explored in energy storage, conversion, and environmental applications.

The book, *Titanium Carbide MXenes: Synthesis, Characterization, Energy, and Environmental Applications*, is intended to present state-of-the-art developments in different synthesis techniques of $Ti_3C_2T_x$ MXenes as a single material and their composites, as well as their detailed energy and environmental applications. In general, the book is structured into three parts – Fundamentals, Synthesis, Energy, and Environmental Applications – offering a broad introduction to 2D Titanium Carbide MXene nanomaterials for photocatalytic applications, which are divided into 10 chapters. Every chapter begins with an introduction, followed by sections that provide comprehensive information on current technological development, including comparative analysis and discussion.

The introduction of titanium carbide (Ti_3C_2) MXene has been discussed in Chapter 1. Chapter 2 describes the fundamentals, properties, and characteristics of titanium carbide MXenes. The synthesis and characterization of titanium carbide MXenes have been discussed in Chapter 3. The synthesis and characterization of Ti_3C_2-MXene-based composites have been presented in Chapter 4. Chapter 5 discusses titanium dioxide composites based on titanium carbide MXenes for solar fuel production and environmental applications. MXene-based graphitic carbon nitride composites for solar fuel production and environmental applications have been discussed in Chapter 6. The titanium carbide MXene-based MOF composites for energy and environmental applications are discussed in Chapter 7. Titanium carbide MXene-based layered double hydride (LDH) composites for solar fuel production, energy, and environmental applications are depicted in Chapter 8. Titanium carbide-MXene-based perovskite composites for energy and environmental applications are discussed in Chapter 9. Finally, titanium carbide-based MXenes for energy storage have been discussed in Chapter 10. In summary, this book attempts to exemplify emerging new MXene materials, synthesis approaches, and the most exciting developments for energy storage, photocatalytic CO_2 reduction, hydrogen production, and degradation applications. Discussions about the reaction mechanisms and future prospects are also presented. We hope that this book will be a resource for future research in the related fields of layered conductive materials for solar energy-assisted fuel production and environmental applications.

Finally, I want to thank Ms. Katherine Wong from Wiley, who helped a lot with a great deal of work to bring it to completion. I am thankful to United Arab Emirates University for the financial support of this work. I am also thankful to all of the authors who contributed their very informative and in-depth chapters that made this book a reality.

December 2022

Muhammad Tahir
United Arab Emirates University (UAEU)

1

Introduction to Titanium Carbide (Ti_3C_2) MXenes for Energy and Environmental Applications

Muhammad Tahir

United Arab Emirates University, Chemical and Petroleum Engineering Department, Sheik Khalifa Bin Zayed Street, Asharej, P.O. Box 15551, Al Ain, United Arab Emirates

1.1 Introduction

Transition metal carbide- and nitride-based materials were explored in the twentieth century as having high thermal resistance, chemically stable, hard, water resistance, and conductive characteristics. Since the 1970s, several designs of nanotextures have been proposed; however, there was a small increase in the surface area and the performance of these metal carbides and nitrides. A breakthrough in this field happened in 2011, with the discovery of 2D transition metal carbide and nitride nanotexture compounds, named titanium carbide ($Ti_3C_2T_x$) MXenes.

The general formula of MXene is $M_{n+1}X_nT_X$, which is derived from MAX materials. M stands for the early transitional metal elements such as Sc, Ti, Hf, Mo, Ta, V, Zr, Nb, and Cr. The "X" constitutes either carbon or nitrogen elements, and the acronym "A" represents the Al or Si layer and n ranges from 1 to 3. The chemical exfoliation of the A element of MAX results in the termination of the M surface with abundant functional groups, T_x, such as –F, –Cl, –OH, and –O [1, 2]. The first MXene material (titanium carbide MXene [$Ti_3C_2T_x$]) was synthesized in 2011 by Naguib et al. via chemical etching with hydrofluoric acid (HF) [3]. In general, 2D sheets have a thickness in the range of 1 nm; however, it can be altered by varying the number n in the MXene formula such as M_2XT, $M_3X_2T_x$, and $M_4X_3T_x$. Most of the MXenes (25 out of 29) are made of MAX materials by selective etching of the Al layer. A recent study showed that instead of Al and Si, transition metals of groups 8–12 of the periodic table (Fe, Cu, Zn, Cd, Ir, and Au) can also become the A layer of MAX phases.

Ti_3C_2 MXene, due to its 2D layered structure, has drawn attention worldwide in the area of energy storage and conversion applications [4–6]. This is due to its promising properties such as its higher specific area and terminal functional groups (OH, O, and –F). In photocatalytic applications, the Ti_3C_2 functional group aids in the formation of Schottky junctions for the trapping of electrons through a strong chemical connection at the semiconductor photocatalyst's interface and prevent the

Titanium Carbide MXenes: Synthesis, Characterization, Energy, and Environmental Applications, First Edition. Edited by Muhammad Tahir.

recombination of charge carriers produced by light. Due to Ti_3C_2 MXenes' better electron conductivity and increased reactivity, the exposed metallic active sites have an advantage. This causes efficient movement of charge carriers and separation of photogenerated electrons from holes [7–9]. Additionally, because of its diverse morphologies in terms of forms and dimensions, its application and photocatalyst activity capabilities have considerably changed and improved, which has expanded its use in photocatalytic energy conversions [10, 11]. The detailed information about the fundamental properties and characteristics of Ti_3C_2 MXenes has been discussed in Chapter 2.

The synthesis of Ti_3C_2 Mxenes is possible by using an etching agent under different operating conditions. Different etching agents are used for the removal of the Al layer in order to produce Ti_3C_2 in a variety of ways. The most often used techniques include electrochemical etching, molten salt etching, halogen etching, hydrofluoric acid etching, and acid F-salt etching. The detailed information related to synthesis and characterization of MXenes has been discussed in Chapter 3. The kind and concentration of the etching agent, reaction duration, intercalation agents, and synthesis conditions are only a few of the variables that have a significant impact on Ti_3C_2 etching [12]. These factors have the ability to change the structure, termination groups, defect morphology, conducting and semiconducting characteristics, and other features of Ti_3C_2 MXene. The features of the formed Ti_3C_2 can affect its quality, environmental stability, and other qualities [13, 14]. The conductive metal cores in the layered structure endow MXene with excellent metallic conductivities, and the modification of properties can lead to the formation of Ti_3C_2 MXene with the required properties for various applications in the field of energy conversion applications.

The single-Ti_3C_2 MXenes have less photocatalytic activity due to conductive characteristics; thus, they are commonly used as a cocatalysts with semiconductor materials. Ti_3C_2 MXenes can be combined with a variety of materials for use in energy conversion applications because of the several benefits such as (i) numerous functional beneficial characteristics for the construction of an intimate contact interface with other semiconductors; and (ii) the bandgap alignment of Ti_3C_2 modulated for tuning the surface chemistry [15–17]. A detailed discussion about the synthesis and characterization of MXenes based heterojunction formation has been given in Chapter 4. The layered Ti_3C_2 MXene structure, which resembles an accordion, may be transformed into different dimensions, including a three-dimensional structure, very thin two-dimensional nanosheets, one-dimensional (1D) nanorods and nanowires, and zero-dimensional (0D) quantum dots [18–21]. A range of 2D, 0D, and 1D photocatalyst materials may also be loaded onto Ti_3C_2 MXene to provide efficient, renewable solar fuel [22, 23]. For the production of Ti_3C_2 MXene-based composite photocatalysts, a number of techniques have been published in the literature. The mechanical mixing, self-assembly, in situ semiconductor decorating over the MXene surface, and the in situ oxidation approach are the most often utilized techniques [24–27].

Titanium dioxide material is considered the most extensively investigated photocatalysis material owing to its photocatalytic activity, good chemical stability, abundancy, and low cost. However, because of its wide bandgap energy (3.2 eV), the TiO_2 photocatalyst is UV-active only and functions poorly under visible light, thus lowering its photoactivity [28, 29]. Recently, titanium carbides (MXenes) have gained significant concern in improving the photocatalytic activity of semiconductors, in particular TiO_2, due to their exceptional chemical and physical properties. In Chapter 5, the properties and characteristics of the benchmark TiO_2 photocatalysts and the highly conductive 2D MXene (Ti_3C_2) materials are discussed. The role of MXene in enhancing the photocatalytic performance of TiO_2 photocatalysts is summarized in three main sections; the formation of Schottky heterojunctions, enhancing light harvesting; and enhancing reactant adsorption. Finally, Chapter 5 discloses the recent developments of MXene/TiO_2 nanocomposites for environmental remediation and energy production applications such as photocatalytic CO_2 reduction, photocatalytic H_2 production, and pollutant photodegradation.

Due to the appropriate construction, non-toxicity, and high reduction capacity, graphitic carbon nitride (g-C_3N_4) is regarded as an ideal element for hydrogen production in the visible light spectrum. However, they have a higher recombination of charged carriers, which significantly limits g-C_3N_4 photocatalytic efficacy. The performance of g-C_3N_4 with fast charge carrier separation can be effectively achieved by constructing composites with MXenes. In this perspective, 2D/2D heterojunction of g-C_3N_4 with Ti_3C_2 can be achieved. Due to its tunable terminal groups, metallic conductivity, layered structure, and simple morphological arrangement, Ti_3C_2 MXene as a cocatalyst could boost photocatalytic performance [16, 17]. The detailed information about the Ti_3C_2 coupled g-C_3N_4 composite has been discussed in Chapter 6.

In recent years, metal-organic frameworks (MOFs) have been explored for photocatalytic applications because of their high specific surface area, tunable bandgap, changeable crystalline structures, adjustable chemistry, and functionality. Furthermore, MOFs have outstanding physical qualities which can be employed for adsorption processes in environmental applications [30, 31]. However, the crystalline architecture of MOFs inherently introduces structural flaws such as electron-hole recombination centers and low electrical conductivity, limiting their efficiency in photocatalytic water splitting [32, 33]. Ti_3C_2 MXenes are permissive materials that can be used with MOFs for energy and environmental applications. Notably, conductive TiC MXenes may improve charge carrier transport efficiency and allow for fine-tuning of catalytic performance in MXene/MOF multicomponent catalyst systems [34]. Chapter 7 discloses detailed information about the Ti_3C_2-based MOF composites for energy and environmental application.

Currently, several two-dimensional (2D) materials have been employed to improve the catalytic performance by virtue of their non-toxic behavior, high durability, and huge surface area. Layered double hydroxides (LDHs) have been widely utilized in energy storage and conversion application due to their unique properties. For example, the Fe-, Co-, and Ni-based LDHs have demonstrated efficient performance in the oxygen evolution reaction (OER) and hydrogen evolution reaction (HER); and the space of the layered structure can be used as storage sites for Li^+ in Li-ion batteries. However, the LDHs still endure the poor performance owing to the shortage of active sites and effective electronic transport. Among these 2D materials, LDH and titanium carbide MXene (TiC MXene) having highly active sites with high density and high uniformity can work as ideal models to modify the important parameters for energy storage and environmental applications. The detailed information about the recent development in LDH-based materials and their Ti_3C_2-based composites has been discussed in Chapter 8.

Perovskites, on the other hand, are an interesting class of photocatalysts since they can be synthesized under benign circumstances and a broad range of catalytic properties can be controlled. They have broad-spectrum activity, enabling their use for a variety of purposes, including the production of value-added compounds for expanding chemical feedstocks and lowering atmospheric CO_2 emissions [35]. However, perovskites have drawbacks such as high electron-hole recombination and limited specific surface area [36]. As a result, a variety of techniques have been devised to increase the photocatalytic efficiency of perovskites. In this regard, MXene, because of their exceptional conductive and photothermal properties, is one of the fastest-growing research domains among all existing 2D materials [37]. Ti_3C_2 MXene-based semiconductors as a mediator or heterojunction have been accepted as a viable strategy for promoting charge separation and enhancing the photostability of the final product. More interestingly, it is possible to increase light absorption with improved photoactivity and stability by combining MXene multilayer materials with perovskite [38]. Chapter 9 discloses detailed information about the application of perovskites-based MXenes for energy and environmental applications.

With the popularization of electric automobiles and handy electronic devices, the development and improvement of electrochemical energy storage devices (EESD) are needed to meet the required power and energy density. Also, harnessing energy from renewable energy sources, such as solar, wind, and hydro energy, greatly depends on weather conditions (tide, wind, and sun), and these conditions are not continuous due to environmental weather conditions. Hence, highly performing energy storage devices are required to save the energy generated. MXenes due to their intrinsic structure and properties have received increasing attention and have the advantages of being employed in particular as an electrode material for electrochemical energy storage devices. Detailed information about energy storage performances of MXenes has been discussed in Chapter 10.

1.2 Layout of the Book

The introduction of titanium carbide (Ti_3C_2) MXenes has been discussed in Chapter 1. Chapter 2 describes the fundamentals, properties, and characteristics of titanium carbide MXenes ($Ti_3C_2T_x$). The synthesis and characterization of titanium carbide (Ti_3C_2) MXenes have been discussed in Chapter 3. The synthesis and characterization of Ti_2C_3-MXene-based composites for energy storage and conversion have been presented in Chapter 4. Titanium carbide MXene-based titanium dioxide composites for energy and environmental applications are discussed in Chapter 5. MXene-based graphitic carbon nitride composites for energy and environmental applications have been discussed in Chapter 6. The titanium carbide MXene-based MOF composites for energy and environmental applications are discussed in Chapter 7. Chapter 8 depicts titanium carbide (TiC) MXene-based layered double hydroxide (LDH) composites for energy and environmental applications. Chapter 9 discusses titanium carbide MXene-based perovskite composites for energy and environmental applications. Finally, titanium carbide-based MXenes for energy storage applications have been discussed in Chapter 10.

Acknowledgment

The authors would like to extend their deepest appreciation to United Arab Emirates University for the financial support of this work under Start-up Research Grant Vote No. 12N097.

References

1 Tahir, M. and Tahir, B. (2022). Constructing S-scheme 2D/0D g-C_3N_4/TiO_2 NPs/MPs heterojunction with 2D-Ti_3AlC_2 MAX cocatalyst for photocatalytic CO_2 reduction to CO/CH_4 in fixed-bed and monolith photoreactors. *J. Mater. Sci. Technol.* 106: 195–210.
2 Tahir, M., Sherryna, A., Mansoor, R. et al. (2022). Titanium carbide MXene nanostructures as catalysts and cocatalysts for photocatalytic fuel production: a review. *ACS Appl. Nano Mater.* 5 (1): 18–54.
3 Naguib, M., Kurtoglu, M., Presser, V. et al. (2011). Two-dimensional nanocrystals produced by exfoliation of Ti_3AlC_2. *Adv. Mater.* 23 (37): 4248–4253.
4 Aghamohammadi, H., Amousa, N., and Eslami-Farsani, R. (2021). Recent advances in developing the MXene/polymer nanocomposites with multiple properties: a review study. *Synth. Met.* 273.
5 Verger, L., Xu, C., Natu, V. et al. (2019). Overview of the synthesis of MXenes and other ultrathin 2D transition metal carbides and nitrides. *Curr. Opin. Solid State Mater. Sci.* 23 (3): 149–163.
6 Xie, X., Zhang, N., Tang, Z.-R. et al. (2018). $Ti_3C_2T_x$ MXene as a Janus cocatalyst for concurrent promoted photoactivity and inhibited photocorrosion. *Appl. Catal. B* 237: 43–49.
7 Xu, Y., Wang, S., Yang, J. et al. (2018). Highly efficient photoelectrocatalytic reduction of CO_2 on the Ti_3C_2/g-C_3N_4 heterojunction with rich $Ti^{3+}$and pyri-N species. *J. Mater. Chem. A* 6 (31): 15213–15220.
8 Li, X., Wang, C., Cao, Y., and Wang, G. (2018). Functional MXene materials: progress of their applications. *Chem. Asian J.* 13 (19): 2742–2757.
9 Pang, J., Mendes, R.G., Bachmatiuk, A. et al. (2019). Applications of 2D MXenes in energy conversion and storage systems. *Chem. Soc. Rev.* 48 (1): 72–133.
10 Dong, Y., Zheng, S., Qin, J. et al. (2018). All-MXene-based integrated electrode constructed by Ti_3C_2 nanoribbon framework host and nanosheet interlayer for high-energy-density Li–S batteries. *ACS Nano* 12 (3): 2381–2388.
11 Xiong, D., Li, X., Bai, Z., and Lu, S. (2018). Recent advances in layered $Ti_3C_2T_x$ MXene for electrochemical energy storage. *Small* 14 (17): 1703419.

12 Tahir, M., Khan, A.A., Tasleem, S. et al. (2023). Recent advances in titanium carbide MXene-based nanotextures with influential effect of synthesis parameters for solar CO_2 reduction and H_2 production: a critical review. *J. Energy Chem.* 76: 295–331.

13 Alhabeb, M., Maleski, K., Anasori, B. et al. (2017). Guidelines for synthesis and processing of two-dimensional titanium carbide ($Ti_3C_2T_x$ MXene). *Chem. Mater.* 29 (18): 7633–7644.

14 Munir, S., Rasheed, A., Rasheed, T. et al. (2020). Exploring the influence of critical parameters for the effective synthesis of high-quality 2D MXene. *ACS Omega* 5 (41): 26845–26854.

15 Tang, R., Xiong, S., Gong, D. et al. (2020). Ti_3C_2 2D MXene: recent progress and perspectives in photocatalysis. *ACS Appl. Mater. Interfaces* 12 (51): 56663–56680.

16 Tahir, M., Ali Khan, A., Tasleem, S. et al. (2021). Titanium carbide (Ti_3C_2) MXene as a promising co-catalyst for photocatalytic CO_2 conversion to energy-efficient fuels: a review. *Energy Fuels* 35 (13): 10374–10404.

17 Tahir, M. (2021). Binary Ni_2P/Ti_3C_2 multilayer cocatalyst anchored TiO_2 nanocomposite with etchant/oxidation grown TiO_2 NPs for enhancing photocatalytic H_2 production. *Energy Fuels* 35 (17): 14197–14211.

18 Liu, J., Xu, H., Xu, Y. et al. (2017). Graphene quantum dots modified mesoporous graphite carbon nitride with significant enhancement of photocatalytic activity. *Appl. Catal., B* 207: 429–437.

19 Li, K., Liang, M., Wang, H. et al. (2020). 3D MXene architectures for efficient energy storage and conversion. *Adv. Funct. Mater.* 30 (47): 2000842.

20 Li, Z., Wang, L., Sun, D. et al. (2015). Synthesis and thermal stability of two-dimensional carbide MXene Ti_3C_2. *Mater. Sci. Eng., B* 191: 33–40.

21 Zeng, Z., Yan, Y., Chen, J. et al. (2019). Boosting the photocatalytic ability of Cu_2O nanowires for CO_2 conversion by MXene quantum dots. *Adv. Funct. Mater.* 29 (2): 1806500.

22 Fu, Y., Li, Z., Liu, Q. et al. (2017). Construction of carbon nitride and MoS_2 quantum dot 2D/0D hybrid photocatalyst: direct Z-scheme mechanism for improved photocatalytic activity. *Chin. J. Catal.* 38 (12): 2160–2170.

23 Feng, A., Yu, Y., Wang, Y. et al. (2017). Two-dimensional MXene Ti_3C_2 produced by exfoliation of Ti_3AlC_2. *Mater. Design* 114: 161–166.

24 Li, Z. and Wu, Y. (2019). 2D early transition metal carbides (MXenes) for catalysis. *Small* 15 (29): 1804736.

25 Yang, C., Tan, Q., Li, Q. et al. (2020). 2D/2D Ti_3C_2 MXene/g-C_3N_4 nanosheets heterojunction for high efficient CO_2 reduction photocatalyst: dual effects of urea. *Appl. Catal., B* 268: 118738.

26 Su, T., Hood, Z.D., Naguib, M. et al. (2019). 2D/2D heterojunction of Ti_3C_2/g-C_3N_4 nanosheets for enhanced photocatalytic hydrogen evolution. *Nanoscale* 11 (17): 8138–8149.

27 Su, T., Peng, R., Hood, Z.D. et al. (2018). One-step synthesis of $Nb_2O_5/C/Nb_2C$ (MXene) composites and their use as photocatalysts for hydrogen evolution. *ChemSusChem* 11 (4): 688–699.

28 Tahir, B., Tahir, M., and Amin, N.S. (2015). Gold–indium modified TiO_2 nanocatalysts for photocatalytic CO_2 reduction with H_2 as reductant in a monolith photoreactor. *Appl. Surf. Sci.* 338: 1–14.

29 Tahir, M. and Tahir, B. (2021). In-situ growth of TiO_2 imbedded $Ti_3C_2T_A$ nanosheets to construct PCN/$Ti_3C_2T_A$ MXenes 2D/3D heterojunction for efficient solar driven photocatalytic CO_2 reduction towards CO and CH_4 production. *J. Colloid Interface Sci.* 591: 20–37.

30 Liu, X., Shan, Y., Zhang, S. et al. (2022). Application of metal organic framework in wastewater treatment. *Green Energy Environ.* 8 (3): 698–721.

31 Lin, S., Song, Z., Che, G. et al. (2014). Adsorption behavior of metal–organic frameworks for methylene blue from aqueous solution. *Microporous Mesoporous Mater.* 193: 27–34.

32 Liu, S., Zhang, C., Sun, Y. et al. (2020). Design of metal-organic framework-based photocatalysts for hydrogen generation. *Coord. Chem. Rev.* 413: 213266.

33 Reddy, C.V., Reddy, K.R., Harish, V.V.N. et al. (2020). Metal-organic frameworks (MOFs)-based efficient heterogeneous photocatalysts: synthesis, properties and its applications in photocatalytic hydrogen generation, CO_2 reduction and photodegradation of organic dyes. *Int. J. Hydrogen Energy* 45 (13): 7656–7679.

34 Sharma, K., Hasija, V., Patial, S. et al. (2022). Recent progress on MXenes and MOFs hybrids: structure, synthetic strategies and catalytic water splitting. *Int. J. Hydrogen Energy*.

35 Tabish, A., Varghese, A.M., Wahab, M.A., and Karanikolos, G.N. (2020). Perovskites in the energy grid and CO_2 conversion: current context and future directions. *Catalysts* 10 (1): 95.

36 Xiang, S., Zhang, Z., Gong, C. et al. (2018). $LaFeO_3$ nanoparticle-coupled TiO_2 nanotube array composite with enhanced visible light photocatalytic activity. *Mater. Lett.* 216: 1–4.

37 Lin, H., Gao, S., Dai, C. et al. (2017). A two-dimensional biodegradable niobium carbide (MXene) for photothermal tumor eradication in NIR-I and NIR-II biowindows. *J. Am. Chem. Soc.* 139 (45): 16235–16247.

38 Tasleem, S. and Tahir, M. (2020). Current trends in strategies to improve photocatalytic performance of perovskites materials for solar to hydrogen production. *Renewable Sustainable Energy Rev.* 132: 110073.

2

Fundamentals, Properties, and Characteristics of Titanium Carbides MXenes ($Ti_3C_2T_x$)

Areen Sherryna[1] *and Muhammad Tahir*[2]

[1] *Universiti Teknologi Malaysia, Faculty of Chemical and Energy Engineering, Faculty of Engineering, 81310, UTM, Johor Bahru, Johor, Malaysia*
[2] *UAE University, Chemical and Petroleum Engineering Department, P.O. Box 15551, Sheik Khalifa Bin Zayed Street, Al Ain, United Arab Emirates*

2.1 Introduction

The energy crisis and the fleeting consumption of fossil fuels detrimentally affect the ecological systems and economic development. Therefore, alternative replacement to fossil fuels is critically required as their dependency has become a growing concern and is expected to increase by 56% by 2040 [1]. Over the years, efforts have been devoted to seek a newly sustainable, environmentally friendly, and economical technology to overcome the energy crisis and find alternative carriers to replace high-demand fuel sources [2]. The technological advancement led to the discovery of photocatalysis, an up-and-coming system for generating low-cost, renewable, energy-efficient, and highly commercial energy carriers. Fundamentally, photocatalysis mainly utilizes natural sources such as solar light to drive the photocatalytic reaction. Nevertheless, primary elements to ensure a thermodynamically stable reaction system can be examined through the efficiency of the photocatalyst. Therefore, developing highly efficient, inexpensive, nanomaterials which manifest excellent photocatalytic properties is significant to provide a maximal solar-to-fuel conversion [3, 4].

MXenes are a type of two-dimensional (2D) nanomaterials obtained through selective etching of MAX phases and have been studied as novel probable photocatalytic compounds. It is a desirable 2D compound due to several benefits, including (a) chemical and mechanical durability, (b) existence as a single- and multilayered form, and (c) synthesis from heavier and lighter transition metals, which assist in adjusting the valence electrons. Furthermore, on the interface of MXenes, other chemical groups can also be designed and synthesized. These distinct qualities distinguish MXenes as effective 2D semiconductor co-catalysts to promote the photocatalytic reaction [5]. Moreover, MXenes are known to have adjustable chemical and physical characteristics due to their stoichiometry and interface activities. Besides, they also contain a changeable bandgap, ion intercalation ability, photonic and electrical

Titanium Carbide MXenes: Synthesis, Characterization, Energy, and Environmental Applications,
First Edition. Edited by Muhammad Tahir.

capabilities, metallic nature, half semiconducting aspects, and aqueous adaptability [6]. Therefore, more research should focus especially on the fundamental aspects of MXenes owing to their relevancy of 2D MXene nanomaterials in materials science and energy conversion. Herein, the fundamentals, properties, and photocatalytic attributes of widely utilized titanium carbide MXenes ($Ti_3C_2T_x$) have been reviewed. First, the structural overview and the fundamentals of the metallic-semiconductor provision are briefly discussed. In the next section, the photocatalytic properties of $Ti_3C_2T_x$ MXenes with the view of their opto-electronic and electrical properties are deeply focused. Next, the photocatalytic attributes of $Ti_3C_2T_x$ MXenes are comprehensively discussed. The insights into the $Ti_3C_2T_x$ MXene functionalization with the views of photocatalytic performances are elucidated. Besides, the intrinsic 2D structure of $Ti_3C_2T_x$ MXenes with their contribution to the construction of multidimensional $Ti_3C_2T_x$ MXene-based catalysts are profoundly scrutinized. Finally, the views on the future progress in $Ti_3C_2T_x$ MXene have been proffered to further explore $Ti_3C_2T_x$ MXenes in future studies.

2.2 Fundamentals of MXene

2.2.1 Structural Overview of MXene

MXenes are a type of two-dimensional (2D) nanomaterials obtained through selective etching of MAX phases. The general formula for MXene is $M_{n+1}X_nT_x$, in which the M stands for the early transitional metal elements such as Sc, Ti, Hf, Mo, Ta, Nb, and Cr. The acronym A represents the Al or Si layer, while X constitutes either C or N elements. The chemical exfoliation of the A element of MAX results in the termination of the M surface with abundant functional groups, T_x, such as –F, –Cl, –OH, and –O. The first MXene material ever synthesized in 2011 was titanium carbide MXene ($Ti_3C_2T_x$) by Naguib and coworkers via chemical etching with hydrofluoric acid (HF) [7]. In view of the chemical bonding between the elements, the M-X constituted of a mixed covalent/metallic/ionic bond while M-A exhibits metallic bonding. Studies found that the mechanical method was not able to exfoliate the $M_{n+1}X_n$ layers due to the strong metallic M-A bonds in the MAX phase. Therefore, selective etching is a highly recognized and broadly adopted method to etch away the A layer of MAX. This is because the A atomic layers are chemically reactive and easier to be etched owing to the weaker interlayer M-A and interatomic A-A bonds [8].

In the configurational formula of MXene, *n* represents the atomic layer in a unit cell. The different compositional formulas of MXene with different atomic layers can be obtained as presented in Figure 2.1, such as M_2X with three atomic layers, M_3X_2 with five atomic layers, and M_4X_3 with seven atomic layers. The difference in the lattice structure is correlated with the *n* value of MAX. The typical structure of MXenes could be categorized as mono-M MXenes and double-M MXenes based on their compositional configurations [9, 10]. In mono-M MXenes, the compositional arrangement is distinguished by the position of an early transitional metal covering (M) the C or N (X) elements. On the other hand, double-M Mxenes are sub-categorized

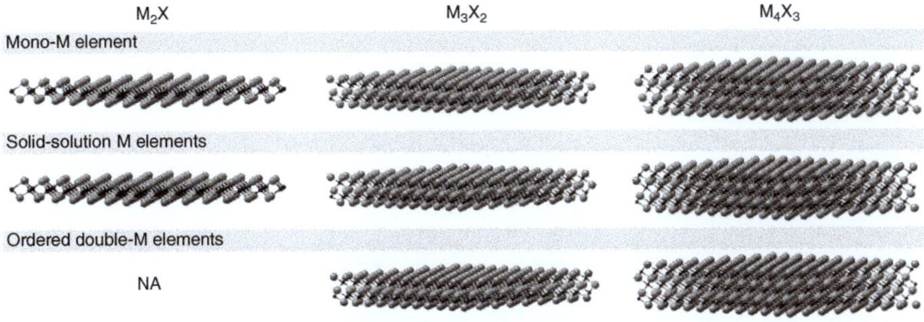

Figure 2.1 Different compositional formulae of $Ti_3C_2T_x$ MXenes with different atomic layers. Source: Anasori et al. [9]/Springer Nature.

into solid solutions and ordered types. In solid-solution double-M Mxenes, two early transitional metals, M1 and M2, are randomly distributed into the MXenes, while for the ordered type, one of the M metals is positioned between the second transition M metal. Generally, MXenes constitute a hexagonal closed-packed crystal structure, but various MXenes show a different sequence of M atoms such as, in the case of M_2X, the hexagonal closed-packed structure is in a sequence of (ABABAB), whereas M_3X_2 and M_4X_3 have a face center cubic sequence (ABCABC) [11]. This atomic ordering is significant as it was theoretically hypothesized to affect the stability of different M-instituted MXenes. In this account, the hexagonal closed-packed structure of Mo_2CT_x MXene is known to be more stable than in their face center cubic sequence ordering. As mentioned, there are various types of Mxenes in relation to the types of M atoms, and the number of atomic layers and C/N incorporated have been explored and reported. For instance, some successfully reported MXenes include Ti_2C, Cr_2TiC_2, Ti_3CN, MO_2C, Ti_3C_2, Zr_3C_2, Ti_4N_3, V_2C, and Mo_2ScC_2 [12]. However, in titanium carbide MXenes, Ti_3C_2 is widely recognized and broadly used in various scientific fields such as photocatalytic water splitting, energy storage field, electrocatalysis, and other environment-related applications.

2.2.2 MXene–Semiconductor Interface Junction

Fundamentally, the charge transfer mechanism in $Ti_3C_2T_x$ MXene-based composites was achieved through the difference in the work function between $Ti_3C_2T_x$ MXene and the semiconductor materials [13]. The built-in internal electric field at the metal–semiconductor junction called the Schottky barrier was formed through the band alignment when the semiconductor and $Ti_3C_2T_x$ MXene are in contact with each other. $Ti_3C_2T_x$ MXenes normally serve as an electron acceptor due to its large metal work function. Typically, the electrons will migrate from higher Fermi level to lower Fermi level materials [14]. Note that the metal work function value (ϕ) is defined as the energy required to transfer an electron from the Fermi level to the vacuum level, where the vacuum level denotes the reference level at which the electrons possess an energy level outside of the metal with zero kinetic energy. It was worth mentioning that the larger the difference in the work function

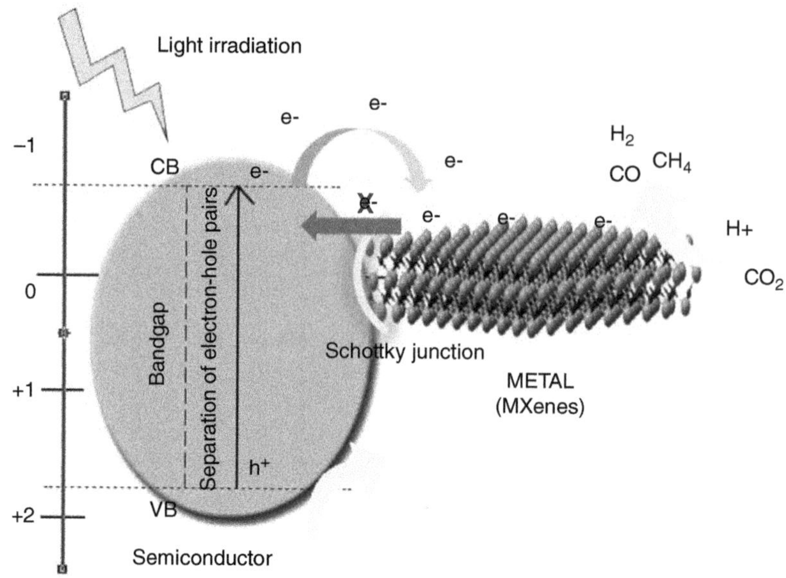

Figure 2.2 Construction of Schottky junction and electron transfer mechanism.

between the hybrid materials, the stronger the Schottky barrier that offers excellent separation of the photocarriers. Therefore, metals with higher work function value such as noble metals and $Ti_3C_2T_x$ MXenes are highly sought after to optimize the solar-to-fuel conversion [15]. Figure 2.2 illustrates the electron transfer mechanism in Schottky junction composite materials. It can be observed that the electrons migrating to the surfaces of the $Ti_3C_2T_x$ MXene are trapped and unable to return to the semiconductor region. Therefore, it reserves more electrons to undergo a redox reaction. The creation of the Schottky junction between semiconductor–$Ti_3C_2T_x$ MXene interfaces prevents the reverse injection of electrons, hastens the electron transfer to the $Ti_3C_2T_x$ MXene surfaces, and inhibits the recombination of photogenerated charges [16].

Given the fact that MXenes are a conductor and can act as co-catalysts, the mechanism of action of the MXene-dependent photocatalytic process is based on rapid charge segregation and prevention of carrier recombination [17]. By absorbing visible light, the photocatalysts activate the CB with electrons created during the absorption process while leaving holes in the valence band (VB). At the contact, stimulated charge carriers are converted to MXenes primarily due to MXenes' greater potential. Without recombination, electrons migrate to MXenes and interact on their surface to form H_2 by reducing H^+ [18], CH_4, and CO by reducing CO_2 [19]. On the other hand, holes are transferred to MXenes and interact to form OH·, which can be used to degrade organics; electrons could also generate OH· for organic degrading [20]. Photocatalysts benefit from increased electron–hole pair separation and reduced charge recombination when the charge is transferred from the photocatalyst to MXenes, which increases the overall photocatalytic activity.

2.2.3 Opto-electronic Properties of MXene

From the view of the electronic conductivity, $Ti_3C_2T_x$ MXenes are primarily known to have a strong metallic conductivity and are theoretically predicted to exhibit non-semiconducting properties [21, 22]. In this account, they are typically employed as an electron mediator to co-assist the photocatalysis process and accelerate the charge transmission [23]. Higher work function is one of the distinct properties of $Ti_3C_2T_x$ MXenes that make them play a significant role as a photocatalytic enhancer and favorable as a noble metal substitution. Work function as high as 6.25 eV could be achieved, which is as high as that of other renowned metal cocatalysts such as Ag (~4.5 eV), Au (~5.38 eV), and Pt (~6.10 eV) [24–26]. In the photocatalysis process, incorporating metals with higher work function could improve the carrier dynamics and induce a stronger redox reaction.

Nevertheless, the work function of $Ti_3C_2T_x$ MXenes can be varied depending on the tailored surface termination groups. This is because their work function is mainly controlled by the induced dipole moments arising from the transfer of charges between the termination group of $Ti_3C_2T_x$ MXenes and the changes in the total surface dipole moments due to surface relaxation. In this context, our findings suggested that $Ti_3C_2T_x$ MXenes with –OH terminal group exhibited a work function of 1.6–2.8 eV. On the other hand, –O terminated $Ti_3C_2T_x$ MXenes displayed a higher work function between 5.75 and 6.25 eV [27]. Furthermore, the termination of –OH groups shows the decrease of work function compared to –O and –F termination, where the work function is shown to either increase or decrease depending on the M metal constituted [28]. However, theoretical analysis through the first-principles calculations ascribes that some types of MXenes can exhibit semiconducting properties depending on the M-metal constitution, surface terminating group, and the number of the MXenes layers such as presented in Figure 2.3. However, the finding suggested that $Ti_3C_2T_x$ manifests only metallic properties regardless of surface functionalization. The same findings were observed on the MXenes with thicker sheets and a large number of transition metals. Therefore, the probability of MXenes undergoing a shift in the electronic properties into semiconducting properties is higher in thinner MXenes such that $n = 2$.

Therefore, not all MXenes are metallic, and some can exhibit semiconducting properties by configuring their termination groups, morphological structure, and type of metal composition. The feasibility of tailoring the electronic and electrical properties is one of the distinct properties of MXenes that made them favorable in many scientific applications. In view of the photocatalytic activity, the bandgaps of MXenes influence the optical property of the material governed by the type of structural property, compositional property, and terminations. Generally, the bandgap of MXenes ranges from 0.05 to 2.87 eV and can be tuned by modifying the presence of terminations. Most of the MXenes constitute indirect bandgap like Sc_2C, having values of 1.8, 1.03, and 0.45 eV with the presence of terminations –O, –F, and –OH, respectively [30]. This opens up the ability to tune the structural properties of MXenes for photocatalytic processes. In the context of titanium-based MXene, the $Ti_3C_2T_x$ having a multilayer structure with metallic nature possesses zero absorption

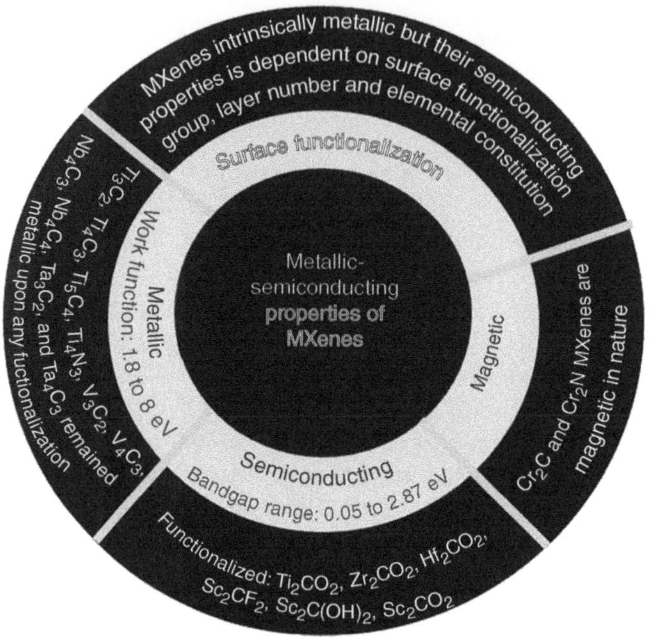

Figure 2.3 Summary of the optical properties of different types of MXenes through first principles calculation. Source: Reproduced with permission from Sherryna and Tahir [29]. Copyright 2021 American Chemical Society.

edge, but the optical properties of Ti_3C_2 are affected by the in-plane transitions in the interband, comprising orbital hybridization. Furthermore, in the $Ti_3C_2T_x$ material, these changes were due to the termination of T_x that improved the visible light consumption. The study of Berdiyorov et al. [31] resulted in the enhanced absorption of light irradiations with an improved reflective property of F- and OH terminated $Ti_3C_2T_x$ under UV and visible light. Similarly, films of $Ti_3C_2T_x$ flakes showed good absorption in the UV–visible region of (300–500 nm) and in the visible light region of (700–800 nm).

Additionally, the difference in the electronic conductivity held by $Ti_3C_2T_x$ can be attributed to the changes in the MXene's flakes d-spacing, termination groups, delamination, the lateral side rise, and defects governed by the amount of the etchant and the time of etching [32]. Their conductivity is similar to that of graphene oxide, but it is higher than that of reduced graphene oxide and other C-based materials [33]. The transmittance of MXenes is also been reported to be dependent on the etchant types and be influenced by the electron-donor surface groups. For instance, MXenes etched by NH_4HF_2 were reported to have higher transmittance. Halim et al. reported that spin-coated MXene films constituted transmittance of 97% per manometer, which is comparable to that of the single-layered GO [34]. In another study, an increase in the transmittance of $Ti_3C_2T_x$ from 74.9% to 92% was reported by using tetramethyl ammonium hydroxide instead of urea, hydrazine, and dimethyl sulfoxide [17, 35]. Moreover, a 5-nm-thick film of $Ti_3C_2T_x$ which is known to show absorption in the UV–visible region i.e. wavelength 300–500 nm and showed 91.2% of transmittance. The film thickness-dependent absorption of irradiation in 700–800 nm wavelength was also reported, and it was recognized as a significant greenish photocatalyst for photothermal applications [36].

2.2.4 Electrical Properties of MXene

$Ti_3C_2T_x$ MXenes are highly sought after due to their higher electrical conductivity, which regards them as a significant cocatalyst to promote the photocatalysis process. The electrical conductivity of $Ti_3C_2T_x$ MXenes is in the range of 6000 to 8000 S cm^{-1}, which is comparable to that of noble metal and graphene materials. MXenes constitute outstanding characteristics like elevated Young's modulus, thermal conductivity, electric conductivity, and flexible band gaps. Their electrical properties can also be tailored by modulating the synthesis conditions, morphology, thickness, and termination properties [37]. Outstanding conductivity was reported by $Ti_3C_2T_x$ MXenes with 1 μm film thickness, by which their electrical conductivity was shown to reach up to 14 000 S cm^{-1}[38]. Besides, it has been suggested that maintaining higher electrical properties can be obtained by etching the $Ti_3C_2T_x$ MXenes in the inert condition to minimize the occurrence of oxidation.

On the other hand, the assistance of nonreactive proton acids such as HCl and $HClO_4$ during the synthesis process was reported to improve the electrical conductivity of the $Ti_3C_2T_x$ MXenes up to 10 400 S cm^{-1} [39]. The improvement in the electrical conductivity is in correlation with the morphological structure of the $Ti_3C_2T_x$ MXene flakes. Such as shown in Figure 2.4a, the proton-acid processing induces aggregation of $Ti_3C_2T_x$ flakes where it was morphologically engineered to follow the planar stacking geometry. In this regard, the proton acid-treated $Ti_3C_2T_x$ MXenes flakes were shown to manifest a thin densely packed sheet-like morphology with lateral size up to millimeter scale and thickness about several micrometers compared to without the proton acid assistance. The highest conductivity of 15 000 S cm^{-1} was reported in the monolayer $Ti_3C_2T_x$ MXene flakes by Zhang et al., with an average lateral size of 10 ± 2.1 μm and flake thickness of ≈1 nm [41]. They elucidate that delaminating the multilayer $Ti_3C_2T_x$ MXenes into a single layer endowed them with higher electrical conductivity. Another study by Lipatov et al. strongly suggested that the thickness, morphology, and synthesis conditions to produce higher quality $Ti_3C_2T_x$ MXenes are crucial to tune their electrical conductivity [42, 43].

The study of defect engineering in the 2D materials shows that monolayer molybdenum disulfide (MoS_2) [44], graphene [45], and hexagonal boron nitride (h-BN) [46] are interconnected with their electrical properties. Interestingly, in $Ti_3C_2T_x$ MXenes, the defect formation was found to be electrically inactive, suggesting their excellent physical stability [40]. Technically, the atomic defects in single-layer $Ti_3C_2T_x$ MXenes can be formed during the etching process or simply by controlling the etchant concentration. The STEM images in Figure 2.4c–e depict the creation of single Ti vacancies (red) and Ti vacancy cluster (white) with varied HF concentration. It was inferred that increasing the concentration of HF acid (7 wt%) was shown to create a more defective structure. However, computational analysis through the

Figure 2.4 (a) SEM images of normal-treated $Ti_3C_2T_x$ MXene flakes and (b) proton acid-treated flakes. Source: (a,b) Chen et al. [39]/John Wiley & Sons; (c) HAADF-STEM images from monolayer $Ti_3C_2T_x$ MXenes with single Ti vacancies (red) at 2.7 wt% HF; (d) single (red) and cluster (white) of Ti vacancies at 5.3 wt% HF; (e) single (red) and cluster (white) of Ti vacancies at 7wt% HF. Source: (c–e) Reproduced with permission from Sang et al. [40]. © 2016, American Chemical Society.

DFT study suggests that higher conductivity retained by defective $Ti_3C_2T_x$ MXenes intrinsically suggests that the formation of defects, regardless of their concentration, does not affect the conductivity. However, mediation of defects on the electrical conductivity of $Ti_3C_2T_x$ MXenes is limited to the theoretical understanding, and their experimental revelation is yet to be studied. Nevertheless, the atomic defects are predicted to affect the morphology and terminaton group of $Ti_3C_2T_x$ MXenes which might practically affect the electrical conductivity. In accordance to this, the finding by Wang et al. suggested that the improvement in the electrical conductivity was found in $Ti_3C_2T_x$ MXene nanosheets prepared by post-treatments of alkalization and calcination via HF etching [47, 48]. In this context, the removal of the functional groups on the surface of $Ti_3C_2T_x$ MXene nanosheets through the post-treatment process induced a three times higher electrical conductivity than the intrinsic $Ti_3C_2T_x$ MXene nanosheets. Moreover, the substitution of –F terminations with halogen groups disclosed by Shi et al. significantly affects the electrical conductivity where it was found to slightly decrease to 1250 $S\,cm^{-1}$ compared to –F induced $Ti_3C_2T_x$ MXenes (1500 $S\,cm^{-1}$) [49, 50]. On that account, the electrical properties in $Ti_3C_2T_x$ MXenes can be governed by critically examining the termination groups, thickness, morphology, and synthesis parameters. Nevertheless, the film thickness or flake sizes were observed as a primary element that can critically govern their electrical conductivity. Even though defect creation was theoretically found to have no impact on the conductivity, more experimental disclosure is required to identify their interconnectivity.

2.3 Photocatalytic Attributes of MXene

2.3.1 $Ti_3C_2T_x$ MXene Functionalization

The etching of the Al layer out of the parent MAX material in the presence of an etchant at normal temperature is the most commonly adopted approach for synthesizing MXenes. Generally, the Al layer from the MAX parent material i.e., Ti_3AlC_2 is etched to obtain $Ti_3C_2T_x$ [51]. The possibility of breaking the flexible M–A bond is higher in comparison to the metallic nature of strong M–X bonds. Generally, hydrofluoric acid (HF) is the most preferred etchant to etch the MAX phase and also offers the synthesized $M_{n+1}AX_n$ to attain terminations (–O, –OH, –F) due to surface functionalization as discussed in Eq. (2.1)–(2.3) and Figure 2.5a. Moreover, the surface functionalization based on the electrophilic property is shown to directly affect the physical and chemical properties of the synthesized MXenes. Furthermore, the exfoliation of multilayered MXenes is reported to be achieved by utilizing polar organic solvents such as isopropanol, dimethyl sulfoxide, and an intercalant. Sonicating the MXenes with the solvents and intercalant will aid the intercalation of the intercalant into the multilayered structure and will lead to the expansion of interlayer spaces to transform the multilayers into single layered, few layered, or flaked structure of MXene [55]. As shown in Figure 2.5a,b, after HF etching, the layered $Ti_3C_2T_x$ partially contains a functional group (–OH, –O, and –F) as termination along with intrinsic defects. Moreover, after annealing in air, the Ti on Ti_3C_2 was oxidized due to the presence of –O, leading to the formation of surface TiO_2 [53].

$$M_{n+1}AX_n + 3HF/AIF_3 + M_{n+1}X_n + 1.5H_2 \tag{2.1}$$

$$M_{n+1}X_n + 2H_2O/M_{n+1}X_n(OH)_2 + H_2 \tag{2.2}$$

$$M_{n+1}X_n + 2HF/M_{n+1}X_nF_2 + H_2 \tag{2.3}$$

Various studies report the use of HF for etching Ti_3AlC_2 along with functional group formation. However, there is still a lack of research on the mechanism study of the functional groups covered $Ti_3C_2T_x$ MXenes for photocatalysis. For instance, a study reported the preparation comparison between the traditional HF etching method and a hydrothermal approach with $NaBF_4$ as an etchant and reported the generation of –OH group termination in both synthesis approaches, whereas the elemental distribution of the $Ti_3C_2T_x$ MXenes prepared by the hydrothermal approach depicted the –F group, resulting from adsorbed F [56]. In another work on photocatalytic H_2 generation, HF was utilized in etching Ti_3AlC_2 into $Ti_3C_2T_x$ with –O and –F functional groups. However, when the prepared $Ti_3C_2T_x$ MXenes were hydrothermally combined with other materials for $CdLa_2S_4$ composite formation, all the –F was transformed into –O [57]. Peng et al. explored the existence of –OH

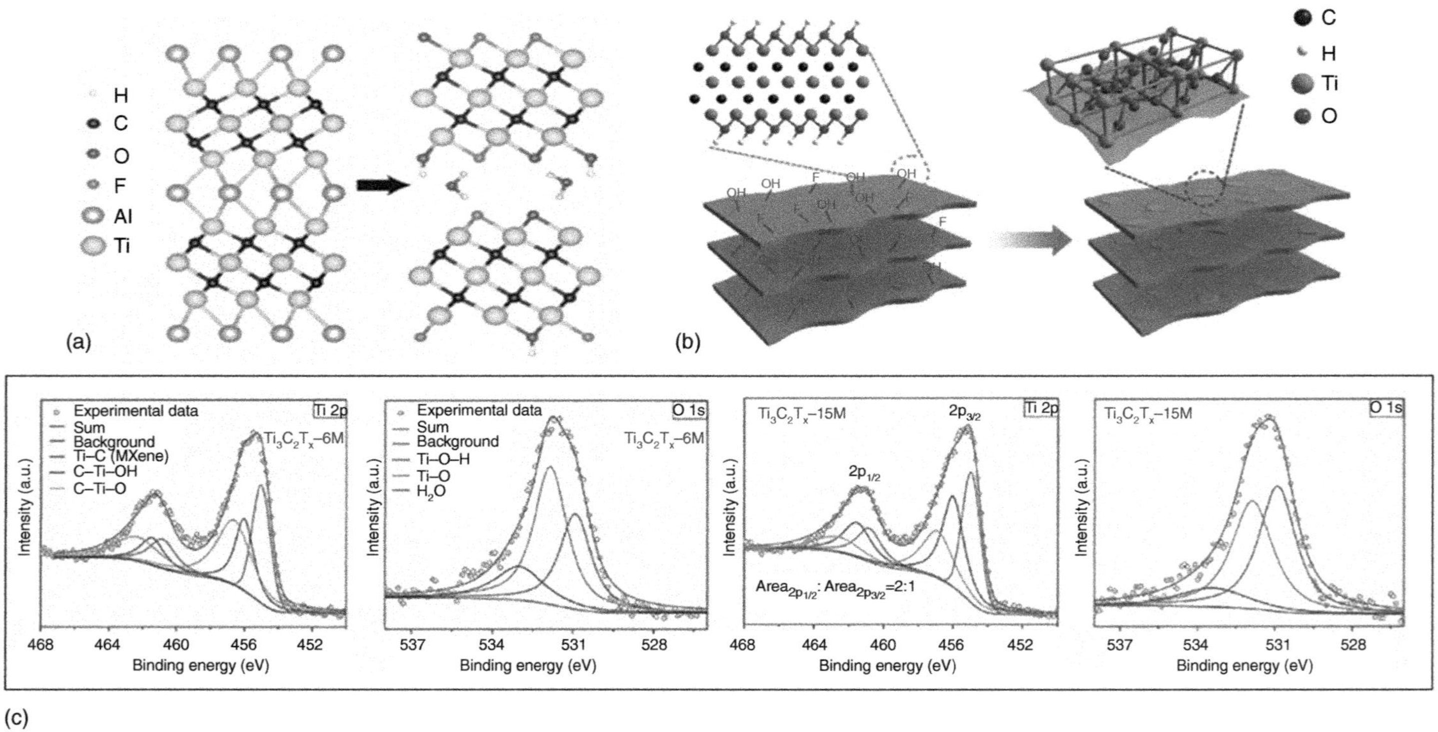

Figure 2.5 (a) Etching from MAX to Ti_3C_2Tx by HF, leading to formation of terminations and existence of interlayer water. Source: Hope et al. [52]/Royal Society of Chemistry/CC BY 3.0.; (b) Etching and annealing of Ti_3C_2, leading to formation of the functional group and TiO_2 on the surface. Source: Reproduced with permission from ref. [53]. Copyright 2016 American Chemical Society; (c) XPS analysis of Ti 2P and O 1s in $Ti_3C_2T_x$ etched with 6 and 15 mol l^{-1} HF. Source: Reproduced with permission from ref. [54]. Copyright 2018 American Chemical Society.

termination for HF-etched $Ti_3C_2T_x$ MXenes, where the –OH terminations offered an ultra-low work function beneficial for efficient charge migration and acted as a hole trapper in photocatalytic degradation [58]. Likewise, another study reported the effect of terminations, especially –OH, on attaining a low work function, i.e. 1.6 eV [59]. Xie et al. reported the etching of Ti_3AlC_2 into $Ti_3C_2T_x$, followed by annealing at a lower temperature. The $Ti_3C_2T_x$ MXenes were found to be covered with –H, –OH, and –F surface terminations in the form of multiple single layers. However, the –F groups were reduced at a lower concentration of HF. The Li storage of the –OH functional group was observed to hold insufficient capacitance, whereas –O terminations were observed to have more capacitance. The –OH functional groups can be reduced to improve the capacitance by annealing [60]. Wastewater treatment was carried out by the heterojunction of $BiPiO_4/Ti_3C_2$ in which the hydrophilic terminations –O and –OH originated, acting as a material to bind the two semiconductors without any other agent [13]. He et al. synthesized $Ti_3C_2T_x$ MXene quantum dots with g-TiO_2/C_3N_4 for photocatalytic CO_2 reduction. The Ti_3C_2 after exfoliation from MAX was treated with dimethyl sulfoxide followed by sonication for 12 hours in an N_2-protected environment along with the use of ammonia to control pH around 9. This led to the generation of –OH, –NH, =O, and –F terminations. The quantum dots offered good dispersion onto g-C_3N_4 due to the binding of the heterojunction aided by the terminations and also acted a pathway for the charge transport [61]. The generation of the functional group –O onto $Ti_3C_2T_x$ MXenes has been studied by using HF with 6 and 15 $mol\,l^{-1}$, as depicted in Figure 2.5c, the –O has been observed to be higher by using 6 $mol\,l^{-1}$ HF than 15 $mol\,l^{-1}$. This causes a higher amount of H_2O molecules in between the layers of $Ti_3C_2T_x$ MXenes, which makes more H^+ reachable to the active sites [54]. In another work, g-C_3N_4/Ti_3C_2 studied for enhancement in the photocatalytic H_2 generation was significantly correlated to the presence of the functional group –F on $Ti_3C_2T_x$ MXenes due to the treatment with HF. However, the –F group was removed due to its unstable nature at high annealing temperature while forming a heterojunction with g-C_3N_4 and instead was converted to –O. The –O terminations were tested to have a lower Femi level than $Ti_3C_2T_x$ and $Ti_3C_2T_x$ with the F-functional group. Moreover, under the annealing process, the Fermi level of $Ti_3C_2T_x$ MXenes achieved was more negative as compared to that of g-C_3N_4, leading to charge separation by electron transfer from g-C_3N_4 to $Ti_3C_2T_x$ MXenes, where more active sites were also present in the form of terminations [62]. $Ti_3C_2T_x$ MXenes were synthesized by etching with NaF and HCl solution, causing the formation of many terminal groups such as –O and –OH. This Ti_3C_2 was coupled with Ag_3PO_4 for photodegradation, which was effective due to the strong termination supported interface between both materials along with electron transport and improved photoactivity [63]. Zhang et al. also recognized the usefulness of functional groups, including –O and –OH, with negative charges on $Ti_3C_2T_x$ for photodegradation based on the enlarged provision of surface area and hydrophilic nature [64]. In another work, the functional groups containing Ti_3C_2 were identified as an alternate for noble metal and cocatalyst due to their metal-like nature and narrow bandgap [65]. The coupling of Ti_3C_2 with ZnS contained –O/–OH functional groups converted from –F after hydrothermal treatment with ZnS as a cocatalyst was

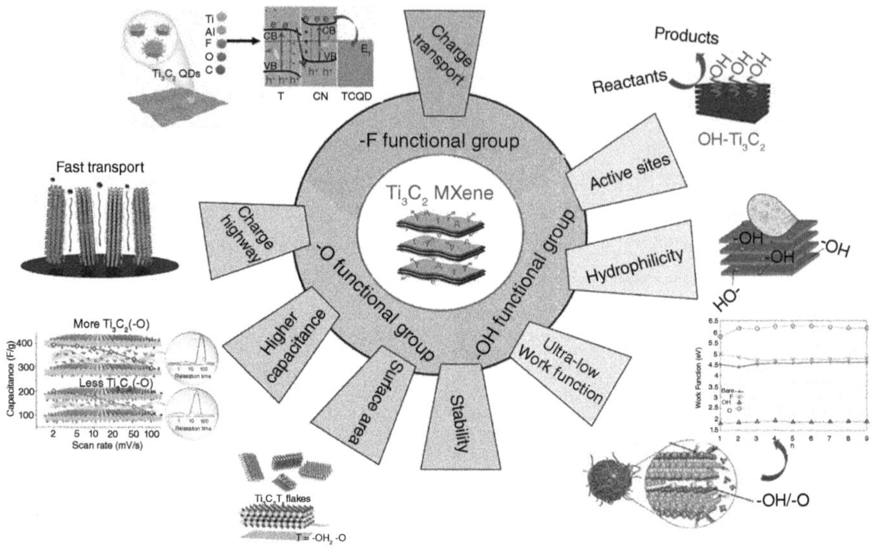

Figure 2.6 Illustration of the function and advantages of functional group terminations Ti_3C_2 MXene.

reported to facilitate the charge transport and separation, leading to an improved photoactivity and stability [66].

Therefore, the existence of terminal groups on $Ti_3C_2T_x$ MXenes is known to be highly photoactive. However, the $Ti_3C_2T_x$ MXenes with functional groups as terminations have not been researched much. On the other hand, researching the location, quantity, and ratio of various terminations can improve the stability of $Ti_3C_2T_x$ MXenes along with its optical properties, ensuring its improved catalytic activity. Figure 2.6 summarizes and displays the advantages of $Ti_3C_2T_x$ MXenes with functional group terminations for photocatalytic applications.

2.3.2 Intrinsic 2D Accordion-like Nanosheets

$Ti_3C_2T_x$ MXenes originally exhibited a two-dimensional (2D) structure with an accordion-like multilayer sheets. However, the morphology of $Ti_3C_2T_x$ MXenes can also be modulated into different dimensional structures such as zero-dimensional (0D) quantum dots [67–70], one-dimensional nanowires, nanoribbons, nanotubes [71–73], and also three-dimensional hierarchical structures [74–76]. The difference in the dimensionality of the $Ti_3C_2T_x$ MXene offers significant contributions to different scientific applications. The sizes and dimensionality of $Ti_3C_2T_x$ MXenes can be mainly controlled during the preparation process. Over the decades, many newly developed techniques have been introduced, which is apparent with the new etchants utilized, intercalating agents, surface groups, reactions systems, and conditions which revolutionized their morphological structures [77, 78]. The multilayered $Ti_3C_2T_x$ MXenes act as a support to grow and finely disperse other photocatalysts onto its surface, whereas the delamination of Ti_3C_2 into nanosheets could result in photocatalysts with lower dimensions such as 0D, 1D, and 2D with a more outstanding interface, larger surface area, and active sites, e.g., 0D Ti_3C_2 QDs [79]. Nevertheless, the intrinsic structure of $Ti_3C_2T_x$ MXenes is shown to provide the highest beneficial attributions, especially in photocatalysis. Therefore, this section will further disseminate the attributions of the intrinsic 2D structure of $Ti_3C_2T_x$ MXene nanosheets regardless of their accordion multilayers and single-layer structures.

The 2D structure nanomaterials are highly coveted as they possess great functionality in promoting the photocatalytic process. Producing the planar structure of materials can be performed either by bottom-up or top-down approaches. Generally, producing the 2D $Ti_3C_2T_x$ MXene sheets is mainly feasible via top-down approaches where the chemical etching of the parent layer of MAX has been utilized. Ti_3AlC_2 MAX is typically a densely packed bulk structure where the Al layer was observed to be intact and attached to the Ti layer. Most of the reported studies suggested that Ti_3AlC_2 MAX exhibits lower catalytic performances than the $Ti_3C_2T_x$ MXenes, which can be due to the inability to maximize the catalytic redox reaction highly reasoned by their packed layered morphology as depicted in Figure 2.8a [49, 80]. On the other hand, wet-chemical etching of the parent Ti_3AlC_2 MAX induced the accordion-like multi-structure of $Ti_3C_2T_x$ MXenes, as reflected in Figure 2.8b [81]. The accordion-like structure of $Ti_3C_2T_x$ MXenes promotes rapid ion diffusion with more ion insertion in multiple channels, inducing higher reaction activity [82]. Therefore, 2D materials with planar orientation, larger surface reactive sites, excellent crystallinity degree, and tunable layer number are indeed significant for producing high-efficient photocatalytic materials [83].

In the view of the photocatalysis process, $Ti_3C_2T_x$ MXenes with a 2D planar structure was shown to provide a universal platform for integrating other dimensional photocatalysts, as illustrated in Figure 2.7. $Ti_3C_2T_x$ MXenes can be observed to be highly feasible for the integration with other 2D materials, such as with Bi_2MoO_6, g-C_3N_4, MoS_2, LDH [84–86], 1D materials such as with CdS, CNT, BN [87–89] and also 0D particulate materials such as ZnCdS, TiO_2 NPs, ZnS, and $CdLa_2S_4$ [57, 90, 91]. In the view of Mxene-based 2D nanomaterials, the intimate contact formation between $Ti_3C_2T_x$ MXenes and 2D-Bi_2MoO_6 fabricated by Zuo et al. was found to boost the visible-light photocatalytic water oxidation to 7.9-fold higher than Bi_2MoO_6 [74]. The 2D planar structure offers efficient attachment sites to tightly integrate Bi_2MoO_6, thereby facilitating charge transportation. Integrating the 2D $Ti_3C_2T_x$ MXenes with other 2D materials offers maximized surface reaction actives sites while inhibiting the sheet aggregation. Correspondingly, the formation of 2D/2D flower-like sheets of $Ti_3C_2T_x/MoS_2$ prepared via hydrothermal treatment was found to augment hydrogen production and photodegradation of methyl orange [85]. The finding suggests that constituting planar sheets of $Ti_3C_2T_x$ MXenes improves the surface area and substantially converts the morphology of MoS_2 from a crouching to a stretching state, as depicted in Figure 2.8c,d. Besides, increasing the amount of $Ti_3C_2T_x$ MXenes could minimize the overlapping of the MoS_2 sheets, which improves the absorption of the visible lights as observed from the FESEM image in Figure 2.8e. Accordingly, the construction of 2D/2D $Ti_3C_2T_x$/NiFe

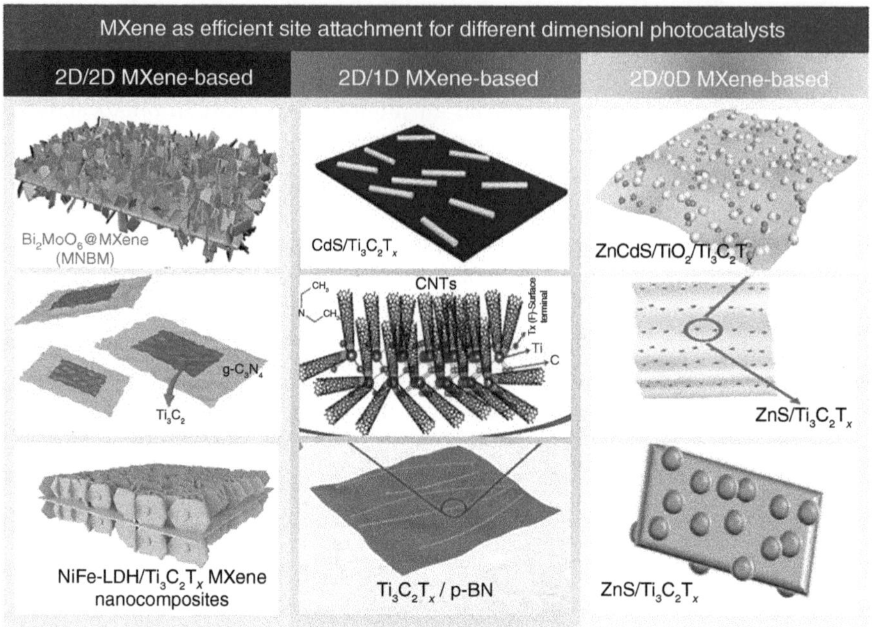

Figure 2.7 MXene as efficient site attachment for different dimensional photocatalysts.

LDH shown was observed to stabilize the LDH sheets with an intact crystal form [84]. The $Ti_3C_2T_x$ MXenes offer an ideal platform to grow the sheets of LDH. As widely known, LDH intrinsically suffered from sheet deactivation, thereby deteriorating their photocatalytic performances [2, 92]. The vertical growth of the LDH nanosheets on the $Ti_3C_2T_x$ MXene platform provides efficient contact between the two sheets with the presence of the negatively charged hydroxyl groups of the $Ti_3C_2T_x$ MXenes assisting the in situ nucleation and crystallization of the LDH through ion capturing.

On the other hand, other multidimensional 1D/2D MXene hybrid catalysts can be constructed by utilizing 2D $Ti_3C_2T_x$ MXene sheets as a growth matrix. Nevertheless, integrating 1D materials on the 2D $Ti_3C_2T_x$ MXenes in the photocatalysis field is very limited. Few studies only have been performed on the 1D/2D MXene hybrid, which might be due to the difficult parameter control for synthesizing 1D materials. However, the construction of 1D/2D $CdS/Ti_3C_2T_x$ through in situ assembling solvothermal treatment by Xiao et al. disclosed an enhancement in the carrier's transmission [89]. The significant improvement in the catalytic reaction is mainly due to the synergistic effects of different morphological dimensions offered by the nanorod-on-sheet structures. An efficient attachment of the CdS nanorods on the $Ti_3C_2T_x$ sheets was found, where the higher length-to-diameter ratio of the CdS with the provision of the Schottky barrier by the $Ti_3C_2T_x$ sheets offered oriented charge transportation with outstanding photocarrier separation. The confinement effects by ultrathin $Ti_3C_2T_x$ induced a well-defined CdS nanorod covered on the sheets, providing beneficial attribution for maximizing solar absorption performances.

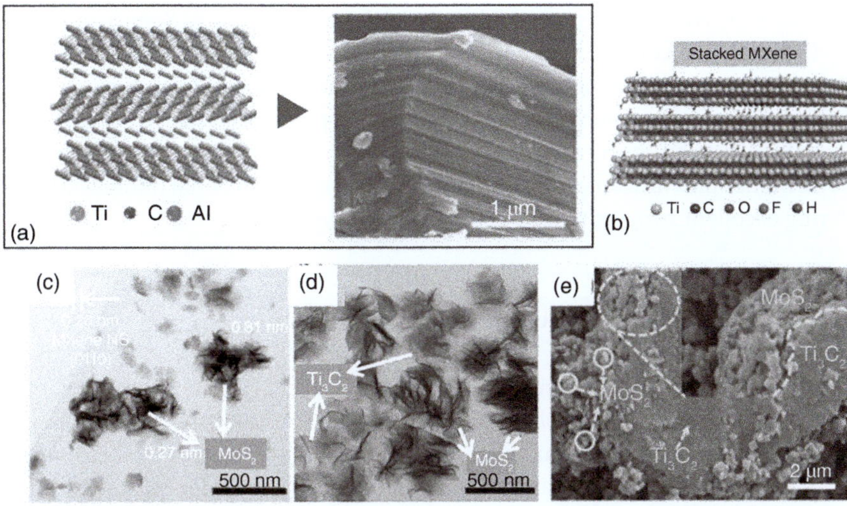

Figure 2.8 (a) Schematic illustration of Ti_3AlC_2 MAX along with the SEM images. Source: Shi et al. [49]/with permission of John Wiley & Sons; (b) schematic of $Ti_3C_2T_x$ MXenes showing accordion-like structures. Source: Wang et al. [81]/with permission of John Wiley & Sons; (c, d) TEM; (e) FESEM image of 2D/2D $Ti_3C_2T_x/MoS_2$. Source: (c–e) Reproduced with permission from Yao et al. [85]. © 2020, Springer Nature/CC BY 4.0.

Accordingly, the formation of 1D/2D hybrid structures through the growth of the CNT on the $Ti_3C_2T_x$ sheets ascribed an increase in the photocatalytic degradation of the Rhodamine B (RhB) [88]. The CNT was observed to be efficiently dispersed into the interlayers of $Ti_3C_2T_x$ MXene flakes, creating a robust contact formation and activating the surface reaction. Therefore, the multidimensional domains such as 1D materials on the 2D surface platform can be a great attribution for aiding the carrier transportation, providing longer inter-transmission of charges.

Additionally, particulate photocatalysts such as TiO_2 nanoparticles offered excellent photocatalytic performances. However, the particulate photocatalyst is typically subjected to particle agglomeration, deactivating the surface reaction [76]. Therefore, an effective platform is required to enable a robust interaction between $Ti_3C_2T_x$ MXenes and the nanoparticles. Interestingly, the 2D structure of $Ti_3C_2T_x$ MXenes provides a confinement effect in offering efficient control of the photocatalytic process [93]. The highly conductive 2D geometry of $Ti_3C_2T_x$ MXenes was found to provide a uniform dispersion of the nanoparticles, which accords an outstanding light absorption potential [90]. Higher surface energy manifested by a low-dimensional photocatalyst intrinsically reduced the photocatalytic activity. Indicatively, the agglomeration in the nanoparticles could block the exposed catalytic active sites, thereby minimizing the solar absorption optimization. Stabilizing the nanoparticles by the merit of 2D planar structures remarkably improves the photostability of the hybrid catalysts. The TiO_2 NP in situ growth on the accordion-like structure of $Ti_3C_2T_x$ MXenes investigated by our group revealed that the uniformed distribution of the TiO_2 NPs into the interlayer spacing of $Ti_3C_2T_x$ MXenes contribute to a maximized surface catalytic reaction [82]. The coverage of TiO_2 NPs on the inner and outer surface of $Ti_3C_2T_x$ MXene accordion layers offers a robust contact with maximal solar harvesting performances, thereby greatly enhancing the photocatalytic hydrogen production. Therefore, hybrid multidimensional composites render a unique morphological structure with its own beneficial attribution toward the photocatalytic reaction. Planar geometry of $Ti_3C_2T_x$ MXenes provides an excellent platform to other dimensional photocatalysts by offering excellent support to fully activate the photocatalysts.

2.4 Conclusion and Future Perspectives

The development of two-dimensional (2D) materials has been encouraged by the fact that they have the following characteristics: (i) mobile electrons and high charge carrier concentration due to changes in electronic and optical features, (ii) large surface-to-mass ratio for rapid charge carrier migration and extended solar spectrum absorption, and (iii) exposed surface atoms that make defects and vacancies that can be used as trapping sites. There are a lot of different types of 2D materials that can be used for different types of energy conversion and storage applications [94]. However, with the provision of electrical conductivity and metallic in nature properties, $Ti_3C_2T_x$ MXenes, which is the class of 2D nanomaterials, have been extensively researched as they proffered a superior photocatalytic enhancement. Their budding advancement in the photocatalytic energy conversion field thereby authenticates their role as a prominent co-catalyst for driving expeditious catalytic reactions.

$Ti_3C_2T_x$ MXenes generally existed in 2D multilayered structures where their morphology was presumed to mimic the accordion-like structure. The 2D geometry of $Ti_3C_2T_x$ MXenes, regardless of their multi- or single-layer structure, was observed to serve as a proficient support to other dimensional photocatalysts, inducing a hybrid composite with multidimensional domains, which has shown to greatly benefit the carriers' dynamics. Generally, sheet-on-sheet hybrid materials were observed to outperform other multidimensional photocatalysts. This is because the synergistic combination of the planar structure of both materials could maximize the surface reaction active sites and optimizes the light harvesting. The universal platform of $Ti_3C_2T_x$ MXenes sheets was disclosed to improve the catalytic performances of lower-dimensional photocatalysts such as 0D nanoparticles. The size dependence surface energy might detrimentally lead to the agglomeration of the nanoparticles. However, good dispersion of the nanoparticles and site attachment offered by $Ti_3C_2T_x$ MXenes were shown to inhibit the deactivation of the photocatalysts, thereby improving the photocatalytic efficiency. On the other hand, integration of 1D semiconductors on the 2D $Ti_3C_2T_x$ MXenes was ascribed to promote the photocarrier's separation, owing to the axial transmission of charges along the 1D heterostructure, which intrinsically minimizes the charge recombination time. Nevertheless, the effectual transmission of the photocarriers in 1D materials requires extensive research to further verify their functionality in promoting the photocatalytic reaction. In view of electrical conductivity, it was revealed that governing their electrical conductivity can be performed mainly by controlling the film/sheet thickness, the synthesis conditions, and terminating groups. On the other hand, $Ti_3C_2T_x$ MXenes were theoretically divulged to manifest metallic properties regardless of any surface functionalization. However, the experimental analysis is shown to be contradicting their theoretical studies. Similar analysis can be observed in the $Ti_3C_2T_x$ MXene-induced defects, where the concentration of defects was theoretically found to be non-significant in governing the electrical conductivity, which was in contrast with the experimental revelation [40, 95].

Therefore, views on the future progress of $Ti_3C_2T_x$ MXenes have been disclosed to give an insight into their future study of the energy conversion fields, such as follows:

- Fundamental analysis on the dynamics of the photocarriers, especially in multidimensional photocatalysts, requires attention to evince the morphology-dependent carrier transmission.
- Theoretical and experimental analysis of the surface functionalization and their interconnection with the morphology of $Ti_3C_2T_x$ MXene sheets should be scrutinized.
- Defects engineering in $Ti_3C_2T_x$ MXenes should be thoroughly researched to understand their roles in governing the catalytic properties of $Ti_3C_2T_x$ MXenes, with comparative studies on the computational and experimental analysis.
- Mixed termination groups found to exist in $Ti_3C_2T_x$ MXenes were affirmed through experimental research. Mixed termination groups might affect the optical properties of $Ti_3C_2T_x$ MXenes, where it might induce semiconducting properties in $Ti_3C_2T_x$ MXenes, thereby needing thorough research attention both theoretically and experimentally.

Acknowledgment

The financial support of this work is provided by United Arab Emirates University (UAEU), UAE under Start up research grant vot no. 12N097.

References

1 Fan, W.K. and Tahir, M. (2022). Recent advances on cobalt metal organic frameworks (MOFs) for photocatalytic CO_2 reduction to renewable energy and fuels: a review on current progress and future directions. *Energy Convers. Manage.* 253: 115180.

2 Sherryna, A., Tahir, M., and Nabgan, W. (2021). Recent advancements of layered double hydroxide heterojunction composites with engineering approach towards photocatalytic hydrogen production: A review. *Int. J. Hydrogen Energy* 47 (2): 862–901.

3 Tahir, M. and Tahir, B. (2020). 2D/2D/2D O-C_3N_4/Bt/$Ti_3C_2T_x$ heterojunction with novel MXene/clay multi-electron mediator for stimulating photo-induced CO_2 reforming to CO and CH_4. *Chem. Eng. J.* 400: 125868.

4 Khan, A.A., Tahir, M., and Mohamed, A.R. (2022). Constructing S-scheme heterojunction of carbon nitride nanorods (g-CNR) assisted trimetallic CoAlLa LDH nanosheets with electron and holes moderation for boosting photocatalytic CO_2 reduction under solar energy. *Chem. Eng. J.* 433: 133693.
5 Jun, B.-M., Kim, S., Heo, J. et al. (2018). Review of MXenes as new nanomaterials for energy storage/delivery and selected environmental applications. *Nano Res.* 12 (3): 471–487.
6 Jiang, X., Kuklin, A.V., Baev, A. et al. (2020). Two-dimensional MXenes: From morphological to optical, electric, and magnetic properties and applications. *Phys. Rep.* 848: 1–58.
7 Naguib, M., Kurtoglu, M., Presser, V. et al. (2011). Two-dimensional nanocrystals produced by exfoliation of Ti_3AlC_2. *Adv. Mater.* 23 (37): 4248–4253.
8 Handoko, A.D., Steinmann, S.N., and Seh, Z.W. (2019). Theory-guided materials design: two-dimensional MXenes in electro- and photocatalysis. *Nanoscale Horiz.* 4 (4): 809–827.
9 Anasori, B., Lukatskaya, M.R., and Gogotsi, Y. (2017). 2D metal carbides and nitrides (MXenes) for energy storage. *Nat. Rev. Mater.* 2 (2): 16098.
10 Li, Z. and Wu, Y. (2019). 2D early transition metal carbides (MXenes) for catalysis. *Small* 15 (29): 1804736.
11 Salim, O., Mahmoud, K., Pant, K., and Joshi, R. (2019). Introduction to MXenes: synthesis and characteristics. *Mater. Today Chem.* 14: 100191.
12 Hong, L.-f., Guo, R.-t., Yuan, Y. et al. (2020). Recent progress of two-dimensional MXenes in photocatalytic applications: a review. *Mater. Today Energy* 18: 100521.
13 Zhang, L., Yang, J., Xie, T. et al. (2020). Boosting visible-light-driven photocatalytic activity of $BiPO_4$ via constructing Schottky junction with Ti_3C_2 MXene. *Mater. Design* 192: 108772.
14 Li, J., Li, J., Wu, C. et al. (2021). Crystalline carbon nitride anchored on MXene as an ordered Schottky heterojunction photocatalyst for enhanced visible-light hydrogen evolution. *Carbon* 179: 387–399.
15 Fajrina, N. and Tahir, M. (2019). A critical review in strategies to improve photocatalytic water splitting towards hydrogen production. *Int. J. Hydrogen Energy* 44 (2): 540–577.
16 Ding, M., Xiao, R., Zhao, C. et al. (2020). Evidencing interfacial charge transfer in 2D CdS/2D MXene Schottky heterojunctions toward high-efficiency photocatalytic hydrogen production. *Solar RRL* 5 (2).
17 Ye, M., Wang, X., Liu, E. et al. (2018). Boosting the photocatalytic activity of P25 for carbon dioxide reduction by using a surface-alkalinized titanium carbide MXene as cocatalyst. *ChemSusChem* 11 (10): 1606–1611.
18 Sun, Y., Jin, D., Sun, Y. et al. (2018). Wang, g-C_3N_4/$Ti_3C_2T_x$ (MXenes) composite with oxidized surface groups for efficient photocatalytic hydrogen evolution. *J. Mater. Chem. A* 6 (19): 9124–9131.
19 Yang, C., Tan, Q., Li, Q. et al. (2020). 2D/2D Ti_3C_2 MXene/g-C_3N_4 nanosheets heterojunction for high efficient CO_2 reduction photocatalyst: dual effects of urea. *Appl. Catal., B* 268: 118738.

20 Peng, C., Wang, H., Yu, H., and Peng, F. (2017). (111) TiO_{2-x}/Ti_3C_2: synergy of active facets, interfacial charge transfer and Ti^{3+} doping for enhance photocatalytic activity. *Mater. Res. Bull.* 89: 16–25.

21 Wang, W.-T., Batool, N., Zhang, T.-H. et al. (2021). When MOFs meet MXenes: superior ORR performance in both alkaline and acidic solutions. *J. Mater. Chem. A* 9 (7): 3952–3960.

22 Khazaei, M., Arai, M., Sasaki, T. et al. (2013). Novel electronic and magnetic properties of two-dimensional transition metal carbides and nitrides. *Adv. Funct. Mater.* 23 (17): 2185–2192.

23 K. Wang, X.Li, N.Wang, Q.Shen, M.Liu, J.Zhou, N.Li, Z-Scheme core–shell meso-TiO_2@$ZnIn_2S_4/Ti_3C_2$ MXene enhances visible light-driven CO_2-to-CH_4 selectivity, *Ind. Eng. Chem. Res.*60(24) (2021) 8720-8732.

24 Uda, M., Nakamura, A., Yamamoto, T., and Fujirnoto, Y. (1998). Work function of polycrystalline Ag, Au and Al. *J. Electron. Spectrosc. Relat. Phenom.* 88-91: 643–648.

25 Agresti, A., Pazniak, A., Pescetelli, S. et al. (2019). Titanium-carbide MXenes for work function and interface engineering in perovskite solar cells. *Nat. Mater.* 18 (11): 1228–1234.

26 Derry, G.N. and Ji-Zhong, Z. (1989). Work function of Pt(111). *Phys. Rev. B: Condens. Matter* 39 (3): 1940–1941.

27 Chertopalov, S. and Mochalin, V.N. (2018). Environment-sensitive photoresponse of spontaneously partially oxidized Ti_3C_2 MXene thin films. *ACS Nano* 12 (6): 6109–6116.

28 Khazaei, M., Ranjbar, A., Arai, M. et al. (2017). Electronic properties and applications of MXenes: a theoretical review. *J. Mater. Chem. C* 5 (10): 2488–2503.

29 Sherryna, A. and Tahir, M. (2021). Role of Ti_3C_2 MXene as prominent Schottky barriers in driving hydrogen production through photoinduced water splitting: a comprehensive review. *ACS Appl. Energy Mater.* 4 (11): 11982–12006.

30 Khazaei, M., Arai, M., Sasaki, T. et al. (2013). Novel electronic and magnetic properties of two-dimensional transition metal carbides and nitrides. *Adv. Funct. Mater.* 23 (17): 2185–2192.

31 Berdiyorov, G. (2016). Optical properties of functionalized $Ti_3C_2T_2$ (T = F, O, OH) MXene: First-principles calculations. *AIP Adv.* 6 (5): 055105.

32 Shabalin, I.L. (2019). *Ultra-High Temperature Materials II: Refractory Carbides I (Ta, Hf, Nb and Zr Carbides)*. Springer.

33 M. Naguib, V.Mochalin, M.W.Barsoum, Y.G.Gogotsi, *MXenes: A New Family of Two-Dimensional Materials--25th Anniversary Article*,(2014).

34 Halim, J., Lukatskaya, M.R., Cook, K.M. et al. (2014). Transparent conductive two-dimensional titanium carbide epitaxial thin films. *Chem. Mater.* 26 (7): 2374–2381.

35 Huang, K., Li, Z., Lin, J. et al. (2018). Two-dimensional transition metal carbides and nitrides (MXenes) for biomedical applications. *Chem. Soc. Rev.* 47 (14): 5109–5124.

36 Hantanasirisakul, K., Zhao, M.Q., Urbankowski, P. et al. (2016). Fabrication of Ti3C2Tx MXene transparent thin films with tunable optoelectronic properties. *Adv. Electron. Mater.* 2 (6): 1600050.

37 Tahir, M., Khan, A.A., Tasleem, S. et al. (2023). Recent advances in titanium carbide MXene-based nanotextures with influential effect of synthesis parameters for solar CO_2 reduction and H_2 production: a critical review. *Desalin. Water Treat.* 76: 295–331.

38 Mirkhani, S.A., Shayesteh Zeraati, A., Aliabadian, E. et al. (2019). High dielectric constant and low dielectric loss via poly(vinyl alcohol)/$Ti_3C_2T_x$ MXene nanocomposites. *ACS Appl. Mater. Interfaces* 11 (20): 18599–18608.

39 Chen, H., Wen, Y., Qi, Y. et al. (2019). Pristine titanium carbide MXene films with environmentally stable conductivity and superior mechanical strength. *Adv. Funct. Mater.* 30 (5).

40 Sang, X., Xie, Y., Lin, M.W. et al. (2016). Atomic defects in monolayer titanium carbide ($Ti_3C_2T_x$) MXene. *ACS Nano* 10 (10): 9193–9200.

41 Zhang, J., Kong, N., Uzun, S. et al. (2020). Scalable manufacturing of free-standing, strong $Ti_3C_2T_x$ MXene films with outstanding conductivity. *Adv. Mater.* 32 (23): 2001093.

42 Lipatov, A., Goad, A., Loes, M.J. et al. (2021). High electrical conductivity and breakdown current density of individual monolayer Ti_3C_2T MXene flakes. *Matter* 4 (4): 1413–1427.

43 Lipatov, A., Alhabeb, M., Lukatskaya, M.R. et al. (2016). Effect of synthesis on quality, electronic properties and environmental stability of individual monolayer Ti_3C_2 MXene flakes. *Adv. Electron. Mater.* 2 (12): 1600255.

44 Zhou, W., Zou, X., Najmaei, S. et al. (2013). Intrinsic structural defects in monolayer molybdenum disulfide. *Nano Lett.* 13 (6): 2615–2622.

45 Yazyev, O.V. and Louie, S.G. (2010). Electronic transport in polycrystalline graphene. *Nat. Mater.* 9 (10): 806–809.

46 Wong, D., Velasco, J. Jr.,, Ju, L. et al. (2015). Characterization and manipulation of individual defects in insulating hexagonal boron nitride using scanning tunnelling microscopy. *Nat. Nanotechnol.* 10 (11): 949–953.

47 Shukla, V. (2020). The tunable electric and magnetic properties of 2D MXenes and their potential applications. *Mater. Adv.* 1 (9): 3104–3121.

48 Wang, H., Wu, Y., Zhang, J. et al. (2015). Enhancement of the electrical properties of MXene Ti_3C_2 nanosheets by post-treatments of alkalization and calcination. *Mater. Lett.* 160: 537–540.

49 Shi, H., Zhang, P., Liu, Z. et al. (2021). Ambient-stable two-dimensional titanium carbide (MXene) enabled by iodine etching. *Angew. Chem. Int. Ed.* 60 (16): 8689–8693.

50 Ghidiu, M., Lukatskaya, M.R., Zhao, M.Q. et al. (2014). Conductive two-dimensional titanium carbide 'clay' with high volumetric capacitance. *Nature* 516 (7529): 78–81.

51 Naguib, M., Kurtoglu, M., Presser, V. et al. (2011). Two-dimensional nanocrystals: two-dimensional nanocrystals produced by exfoliation of Ti_3AlC_2 (Adv. Mater. 37/2011). *Adv. Mater.* 23 (37): 4207–4207.

52 Hope, M.A., Forse, A.C., Griffith, K.J. et al. (2016). NMR reveals the surface functionalisation of Ti_3C_2 MXene. *PCCP* 18 (7): 5099–5102.

53 Han, M., Yin, X., Wu, H. et al. (2016). Ti_3C_2 MXenes with modified surface for high-performance electromagnetic absorption and shielding in the X-band. *ACS Appl. Mater. Interfaces* 8 (32): 21011–21019.

54 Hu, M., Hu, T., Li, Z. et al. (2018). Surface functional groups and interlayer water determine the electrochemical capacitance of $Ti_3C_2T_x$ MXene. *ACS Nano* 12 (4): 3578–3586.

55 Peng, J., Chen, X., Ong, W.-J. et al. (2019). Surface and heterointerface engineering of 2D MXenes and their nanocomposites: insights into electro-and photocatalysis. *Chem* 5 (1): 18–50.

56 Peng, C., Wei, P., Chen, X. et al. (2018). A hydrothermal etching route to synthesis of 2D MXene (Ti_3C_2, Nb_2C): enhanced exfoliation and improved adsorption performance. *Ceram. Int.* 44 (15): 18886–18893.

57 Cheng, L., Chen, Q., Li, J., and Liu, H. (2020). Boosting the photocatalytic activity of $CdLa_2S_4$ for hydrogen production using Ti_3C_2 MXene as a co-catalyst. *Appl. Catal. B-Environ.* 267: 118379.

58 Peng, C., Wang, H., Yu, H., and Peng, F. (2017). (111) TiO_{2-x}/Ti_3C_2: synergy of active facets, interfacial charge transfer and Ti^{3+} doping for enhance photocatalytic activity. *Mater. Res. Bull.* 89: 16–25.

59 Khazaei, M., Arai, M., Sasaki, T. et al. (2015). OH-terminated two-dimensional transition metal carbides and nitrides as ultralow work function materials. *Phys. Rev. B* 92 (7): 075411.

60 Xie, Y., Naguib, M., Mochalin, V.N. et al. (2014). Role of surface structure on Li-ion energy storage capacity of two-dimensional transition-metal carbides. *J. Am. Chem. Soc.* 136 (17): 6385–6394.

61 He, F., Zhu, B., Cheng, B. et al. (2020). 2D/2D/0D $TiO_2/C_3N_4/Ti_3C_2$ MXene composite S-scheme photocatalyst with enhanced CO_2 reduction activity. *Appl. Catal., B* 272: 119006.

62 Li, J., Zhao, L., Wang, S. et al. (2020). In situ fabrication of 2D/3D g-C_3N_4/Ti_3C_2 (MXene) heterojunction for efficient visible-light photocatalytic hydrogen evolution. *Appl. Surf. Sci.* 515: 145922.

63 Cai, T., Wang, L., Liu, Y. et al. (2018). Ag_3PO_4/Ti_3C_2 MXene interface materials as a Schottky catalyst with enhanced photocatalytic activities and anti-photocorrosion performance. *Appl. Catal., B* 239: 545–554.

64 Zhang, Y., Wang, L., Zhang, N., and Zhou, Z. (2018). Adsorptive environmental applications of MXene nanomaterials: a review. *RSC Adv.* 8 (36): 19895–19905.

65 Liu, Q., Tan, X., Wang, S. et al. (2019). MXene as a non-metal charge mediator in 2D layered CdS@ Ti_3C_2@ TiO_2 composites with superior Z-scheme visible light-driven photocatalytic activity. *Environ. Sci. Nano* 6 (10): 3158–3169.

66 Tie, L., Yang, S., Yu, C. et al. (2019). In situ decoration of ZnS nanoparticles with Ti3C2 MXene nanosheets for efficient photocatalytic hydrogen evolution. *J. Colloid Interface Sci.* 545: 63–70.

67 Xu, N., Li, H., Gan, Y. et al. (2020). Zero-dimensional MXene-based optical devices for ultrafast and ultranarrow photonics applications. *Adv. Sci. (Weinh)* 7 (22): 2002209.

68 Zeng, Z., Yan, Y., Chen, J. et al. (2019). Boosting the photocatalytic ability of Cu_2O nanowires for CO_2 conversion by MXene quantum dots. *Adv. Funct. Mater.* 29 (2): 1806500.

69 Yang, X., Jia, Q., Duan, F. et al. (2019). Multiwall carbon nanotubes loaded with MoS_2 quantum dots and MXene quantum dots: non–Pt bifunctional catalyst for the methanol oxidation and oxygen reduction reactions in alkaline solution. *Appl. Surf. Sci.* 464: 78–87.

70 Li, Y., Ding, L., Guo, Y. et al. (2019). Boosting the photocatalytic ability of g-C_3N_4 for hydrogen production by Ti_3C_2 MXene quantum dots. *ACS Appl. Mater. Interfaces* 11 (44): 41440–41447.

71 Wei, H., Jiang, Q., Ampelli, C. et al. (2020). Enhancing N_2 fixation activity by converting Ti_3C_2 MXenes nanosheets to nanoribbons. *ChemSusChem* 13 (21): 5614–5619.

72 Shi, H., Dong, Y., Zheng, S. et al. (2020). Three dimensional Ti_3C_2 MXene nanoribbon frameworks with uniform potassiophilic sites for the dendrite-free potassium metal anodes. *Nanoscale Adv.* 2 (9): 4212–4219.

73 Enyashin, A.N. and Ivanovskii, A.L. (2012). Atomic structure, comparative stability and electronic properties of hydroxylated Ti_2C and Ti_3C_2 nanotubes. *Comput. Theor. Chem.* 989: 27–32.

74 Zuo, G., Wang, Y., Teo, W.L. et al. (2021). Enhanced photocatalytic water oxidation by hierarchical 2D-Bi_2MoO_6@2D-MXene Schottky junction nanohybrid. *Chem. Eng. J.* 403.

75 Sun, L., Fu, Q., and Pan, C. (2021). Hierarchical porous "skin/skeleton"-like MXene/biomass derived carbon fibers heterostructure for self-supporting, flexible all solid-state supercapacitors. *J. Hazard. Mater.* 410: 124565.

76 My Tran, N., Ta, Q.T.H., and Noh, J.-S. (2021). Unusual synthesis of safflower-shaped TiO_2/Ti_3C_2 heterostructures initiated from two-dimensional Ti_3C_2 MXene. *Appl. Surf. Sci.* 538: 148023.

77 Sherryna, A. and Tahir, M. (2022). Role of surface morphology and terminating groups in titanium carbide MXenes ($Ti_3C_2T_x$) cocatalysts with engineering aspects for modulating solar hydrogen production: a critical review. *Chem. Eng. J.* 433: 134573.

78 Tahir, M., Sherryna, A., Mansoor, R. et al. (2022). Titanium carbide MXene nanostructures as catalysts and cocatalysts for photocatalytic fuel production: a review. *ACS Appl. Nano Mater.* 5 (1): 18–54.

79 Huang, K., Li, C., Li, H. et al. (2020). Photocatalytic applications of two-dimensional Ti_3C_2 MXenes: a review. *ACS Appl. Nano Mater.* 3 (10): 9581–9603.

80 Tahir, M., Sherryna, A., and Zakaria, Z.Y. (2021). Facile synthesis of MAX modified graphitic carbon nitride nanocomposite for stimulating hydrogen production through photocatalytic water splitting, chemical. *Eng. Transac.* 89: 571–576.

81 Wang, Y., Wang, X., Li, X. et al. (2019). Engineering 3D ion transport channels for flexible MXene films with superior capacitive performance. *Adv. Funct. Mater.* 29 (14): 1900326.

82 Tahir, M. (2021). Investigating the influential effect of etchant time in constructing 2D/2D HCN/MXene heterojunction with controlled growth of TiO_2 NPs for stimulating photocatalytic H_2 production. *Energy Fuels* 35 (8): 6807–6822.

83 Luo, B., Liu, G., and Wang, L. (2016). Recent advances in 2D materials for photocatalysis. *Nanoscale* 8 (13): 6904–6920.

84 Ma, Y., Xu, D., Chen, W. et al. (2022). Oxygen-vacancy-embedded 2D/2D NiFe-LDH/MXene Schottky heterojunction for boosted photodegradation of norfloxacin. *Appl. Surf. Sci.* 572.

85 Yao, Z., Sun, H., Sui, H., and Liu, X. (2020). 2D/2D heterojunction of R-scheme Ti_3C_2 MXene/MoS_2 nanosheets for enhanced photocatalytic performance. *Nanoscale Res. Lett.* 15 (1): 78.

86 Yang, C., Tan, Q., Li, Q. et al. (2020). 2D/2D Ti_3C_2 MXene/g-C_3N_4 nanosheets heterojunction for high efficient CO_2 reduction photocatalyst: dual effects of urea. *Appl. Catal. B-Environ.* 268: 118738.

87 Du, Z., Cai, H., Guo, Z. et al. (2022). Synergistic photocatalytic of CO_2-to-CO conversion by 2D/1D $Ti_3C_2T_x$/p-BN heterojunction with interfacial chemical bonding. *J. Alloys Compd.* 920.

88 Thirumal, V., Yuvakkumar, R., Kumar, P.S. et al. (2022). Facile single-step synthesis of MXene@CNTs hybrid nanocomposite by CVD method to remove hazardous pollutants. *Chemosphere* 286 (Pt 2): 131733.

89 Xiao, R., Zhao, C., Zou, Z. et al. (2020). In situ fabrication of 1D CdS nanorod/2D Ti_3C_2 MXene nanosheet Schottky heterojunction toward enhanced photocatalytic hydrogen evolution. *Appl. Catal. B-Environ.* 268: 118382.

90 Qin, X., Cao, R., Gong, W. et al. (2022). Hydrothermal growth of ZnCdS/TiO_2 nanoparticles on the surface of the Ti_3C_2 MXene sheet to enhance photocatalytic performance under visible light. *J. Solid State Chem.* 306.

91 Liu, X., Liu, Q., and Chen, C. (2021). Ultrasonic oscillation synthesized ZnS nanoparticles/layered MXene sheet with outstanding photocatalytic activity under visible light. *Vacuum* 183: 109834.

92 Sherryna, A. and Tahir, M. (2021). Recent developments in layered double hydroxide structures with their role in promoting photocatalytic hydrogen production: a comprehensive review. *Int. J. Energy Res.* 46 (3): 2093–2140.

93 Anasori, B. and Gogotsi, Y. (2019). *Metal Carbides and Nitrides (MXenes) Structure, Properties and Applications*. Switzerland AG, Cham, Switzerland: Springer Nature.

94 Kumar, R., Joanni, E., Singh, R.K. et al. (2018). Recent advances in the synthesis and modification of carbon-based 2D materials for application in energy conversion and storage. *Prog. Energy Combust. Sci.* 67: 115–157.

95 Zhao, M.-Q., Torelli, M., Ren, C.E. et al. (2016). 2D titanium carbide and transition metal oxides hybrid electrodes for Li-ion storage. *Nano Energy* 30: 603–613.

3

Synthesis and Characterization of Titanium Carbide (Ti_3C_2) MXenes

Azmat Ali Khan[1], Muhammad Tahir[2], and Nazish Khan[3]

[1]*Baluchistan University of Information Technology, Department of Chemical Engineering, Engineering and Management Sciences (BUITEMS), Airport Road, Beleli, P.O. Box 87100, Quetta, Pakistan*
[2]*UAE University, Chemical and Petroleum Engineering Department, P.O. Box 15551, Al Ain, United Arab Emirates*
[3]*Department of Plant Breeding and Genetics, The University of Agriculture Swat, Takhta Band, Mingora, P.O. Box 19040, Swat, Pakistan*

3.1 Introduction

The use of the MXene material in photocatalysis was initiated in 2011 after the exfoliation of Ti_3C_2 from Ti_3AlC_2 [1]. MXenes are novel two-dimensional layered materials consisting of transition metal carbides, nitrides, and carbonitrides. The MXene material is usually formed from the MAX phase due to unavailability of any natural precursor for this material. $M_{n+1}AX_n$ is the general formula for the MAX phase, where n might be 1, 2, or 3. The X represents the C or N element, along with the early transition metal elements, IIIA or IVA [2]. Layers M–X are more stable than in the MAX structure. In case of the Ti_3AlC_2 layer structure, it is possible to scrape the Al layer in the solutions, with Ti_3C_2 left as a residue. The same procedure can be used to synthesize a large number of MXenes, thanks to the adaptability of the reactant precursors. Theoretical predictions suggested the formation of many more MXenes; however, 30 various types of MXenes have been discovered experimentally so far. Figure 3.1a represents a brief timeline of MXene advancement [5].

Recent studies have shown that using MXenes as support materials enhanced the photothermal effect of metal nanoparticles, and thus the efficacy in photothermal CO_2 catalysis [4], clearly indicating the incredible potential of MXenes for use in CO_2 photocatalysis. MXenes possess the following unique properties: (i) The MXene material's surface-terminating property affects photocatalysis efficiency [6]. (ii) MXenes at elevated temperature are unstable and easily oxidized. (iii) MXene materials can be used as an effective co-catalyst due to its superior electrical conductivity [7]. (iv) The MXene material's interlayer space helps in efficient absorption of gas molecules [8].

The bare MXenes are metallic and lack semiconductor properties, so they cannot be used as a photocatalyst [9]. Although some MXenes alone can be used for

Titanium Carbide MXenes: Synthesis, Characterization, Energy, and Environmental Applications, First Edition. Edited by Muhammad Tahir.

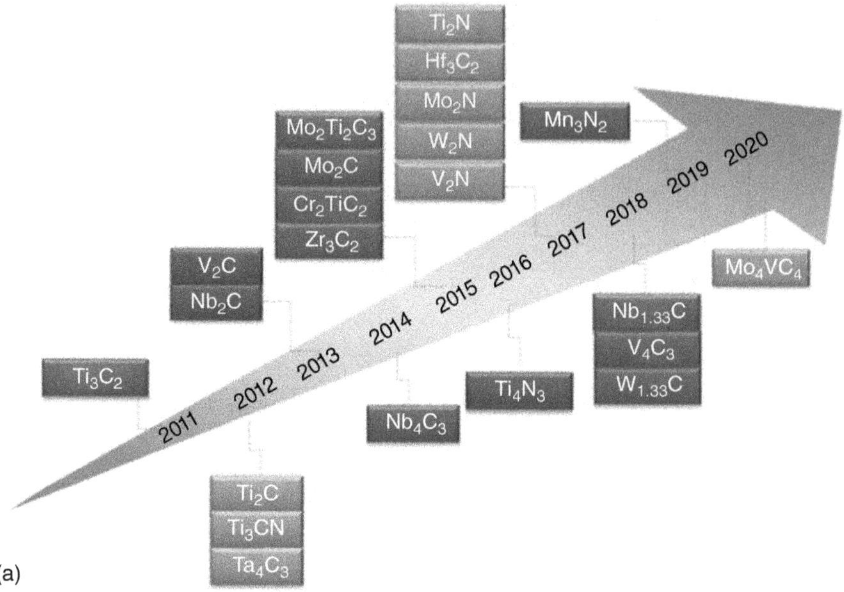

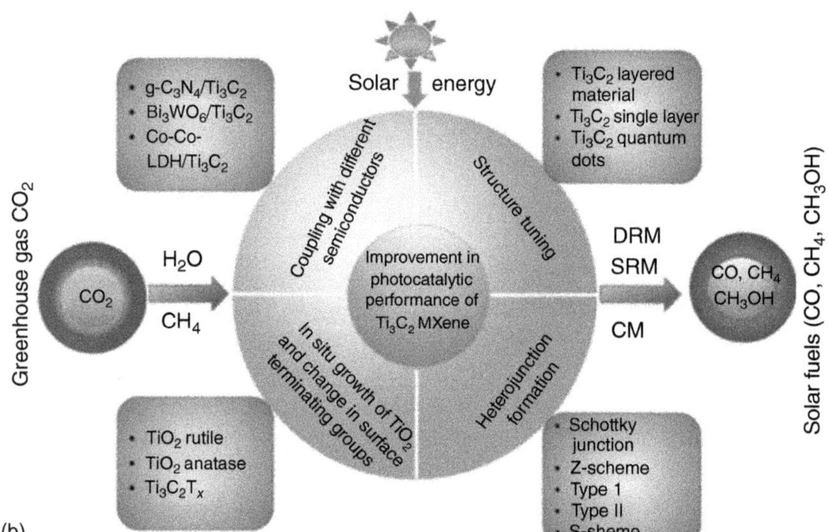

Figure 3.1 (a) Schematic of the working mechanism of MXenes applied in photocatalysis. Source: Adapted from Sun et al. [3]. (b) Enhancement of MXene in the photothermal effect in photothermal CO_2 catalysis. Source: Shen et al. [4]/with permission of Elsevier.

photocatalysis because of the mismatch in band structures, however, some become semiconducting under the effect of surface-terminating groups. Typically, the lower Fermi level of metallic MXenes makes them excellent co-catalysts [10]. Moreover, tremendous interest has been shown in discovering photocatalytic CO_2 reduction using 2D Ti_3C_2 MXene-based semiconductors, which provides a large specific surface area and comparatively high electrical conductivity than graphene and noble metal co-catalysts. The ability of Ti_3C_2 to readily separate photogenerated charge carriers makes it possible to achieve high photoconversion efficiency. This material's two-dimensional layered structure endows it with a large number of active sites and a good interface contact for charge carrier separation, all of which contribute to its overall advantages [11]. Moreover, the Ti_3C_2 layers have various kinds of functional groups that can be activated with MXene substrates, comprising O, OH, C, and F. These useful effects might accelerate photocatalytic electronic transfer from the photocatalyst to Ti_3C_2, consequently reducing charge recombination, increasing charge transfer rate, and boosting photocatalytic conversion of CO_2 to valuable chemicals and fuels [12]. To improve the photocatalytic properties essential for CO_2 conversion to fuels and achieve high efficiency of the Ti_3C_2-based photocatalyst, numerous approaches have been adopted, as demonstrated in Figure 3.1b [11]. However, there is still considerable space for improvement regarding activity, selectivity, and stability. Ti_3C_2 has been the most widely studied MXene material until now, but instead of the overall process, the entire studies mainly focused on the particular role performed by MXenes in that stage. The basic principles for designing MXene-based photocatalysts are still in practice due to insufficient knowledge regarding the surface chemistry and active sites involved in photocatalysis. Hence, there is a need for intense research and further study to advance MXene in terms of the synthesis technique, etching agents, and storage conditions in order to overcome these hurdles.

In this chapter, the details of the synthesis processes employed for the conversion of MAX phase into titanium carbide MXene were explicated. The effects of the synthesis procedure on the properties of MXene and composites formed are also explicated. The characterization techniques employed to study the crystallography, structure, morphology, and optical properties are discussed in detail to study the titanium carbide MXene material. Finally, the conclusion is put forward about the summary of the synthesis and characterizations of the titanium carbide MXene-based materials.

3.2 Different Synthesis Techniques of MXene

In general, there are various methods available to synthesize $Ti_3C_2T_x$ MXenes. The HF etching is the most commonly used technique for synthesizing $Ti_3C_2T_x$ MXenes, which uses hydrofluoric acid as a key etchant. The application of the etchant is important, particularly in the Ti—Al bond breaking in Ti_3AlC_2 MAX, as the Al layer cannot be separated by mechanical separation. Hence, since the pioneer study of $Ti_3C_2T_x$ MXene synthesis through HF etching, numerous techniques have been established to challenge the application of concentrated HF acid, which is non-eco-friendly, particularly in large-scale industrial production. Therefore, safer routes such as acid-containing fluoride etching, alkali etching, halogen etching, electrochemical etching, water-free etching, halogen etching, and molten salt substitution techniques are among the options to synthesize $Ti_3C_2T_x$ MXene. Among them, the in situ HF technique that utilizes the acid containing fluoride

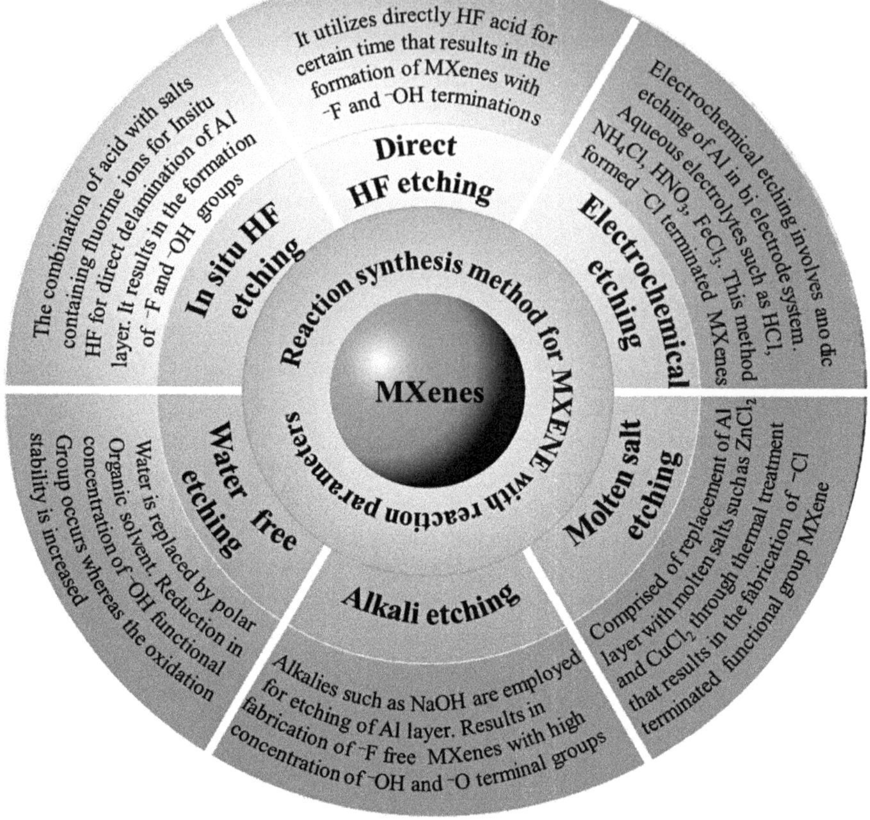

Figure 3.2 Different synthesis techniques of $Ti_3C_2T_x$ MXenes.

etchant has been widely used next to the HF etching technique [13]. Figure 3.2 describes several methods along with their reaction conditions for the preparation of $Ti_3C_2T_x$ MXenes.

The HF, LiF + HCl, and NaOH were used as the etchant's reagent in MAX phase in order to break the weak link between the A sheet and the MX sheet (metallic bond), leading to the formation of MXene [14]. In order to form layered materials with alternating M and X that structurally resembles graphene, the A layers are etched from the MAX phases selectively. MXene multilayer or single-layer nanosheets were also obtained by interpolation and ultrasonic evacuation. Throughout this process, functional groups were forming on the MXene's surface (e.g. –OH, –F, –O, etc.) [15]. An overview of the MXene research and synthesis is displayed in Figure 3.3a [16]. The possible MXene synthesis through two processes, exfoliation and etching, is illustrated in Figure 3.3b.

Finally, the MXenes are ready for use after being separated into ultra-thin nanosheets by sonication [18]. Probably, the basic synthesis parameters can be significantly altered such as reaction temperature, time, and acid concentration

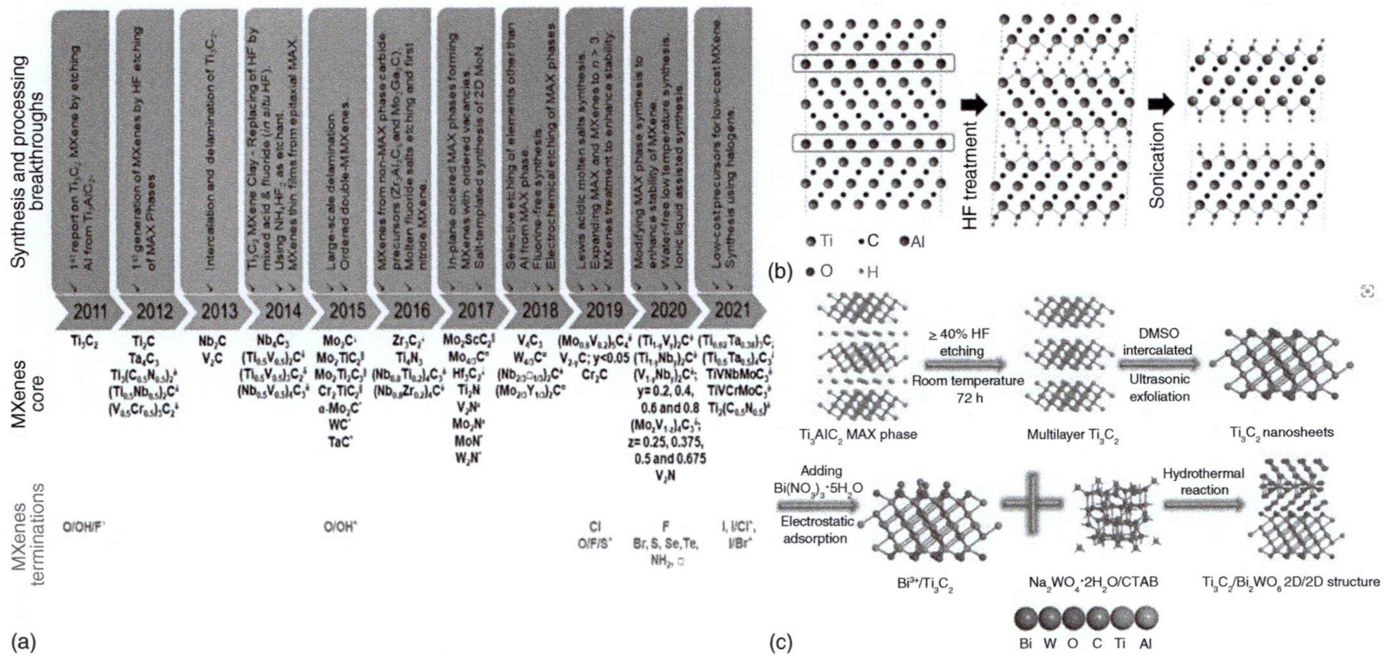

Figure 3.3 (a) Major discoveries in the first 10 years of research by MXenes. Source: Naguib et al. [16]/with permission of John Wiley & Sons. (b) Schematic of the MXene synthesis process. Source: Naguib et al. [1]/John Wiley & Sons. (c) The technique for synthesis of 2D/2D Ti_3C_2/Bi_2WO_6 heterojunction. Source: Cao et al. [17]/John Wiley & Sons.

based on the MAX phase structure of the parent. To prepare Ti_2CT_x from Ti_2AlC, for example, 10% HF was adequate in 10 hours [19]; however, 50% HF was needed for 90 hours at room temperature to synthesize Nb_2CT_x from Nb_2AlC [20]. The bond energies between M and A atoms are supposed to be the source of various essential conditions required to generate MXenes from several MAX precursors [21]. The size reduction of the MAX particles or increase in reaction temperature can considerably reduce the response time needed to eliminate the A layer from the MAX phase [22]. Noticeably, however, the extreme heat may cause the recrystallization of MXenes or the dissolution of carbide [23]. For instance, when Ti_3AlC_2 was etched in HF at 180 °C, the product was not MXenes (Ti_3C_2) but rock-salt structured titanium carbide [22]. Besides HF, numerous etchants have been efficiently used to synthesize MXenes from MAX phase, comprising ammonium fluoride (NH_4HF_2) [24] and HCl^+ LiF solution [25]. The fluorine-free etchant was also identified [14] at elevated response temperatures (270 °C) and high NaOH (27 M) concentrations in a hydrothermal procedure to remove Al and obtain production yield etching. A molten salt (KF, NaF, and LiF) technique has been studied at high temperatures (550 °C) [26] for preserving few-layer MXenes with an O termination. It should be highlighted that exfoliation following etching is essential for attaining single- or multilayer from bulk MXenes. For example, if the etchant is HCl + LiF solution, single- or multilayer MXenes can be readily prepared through sonication or shaking owing to the presence of Li ion [25]. MXenes can only be delaminated by employing an etchant that contains a tiny organic molecule or ion, which is not possible if the etchant is HF alone [27].

Cao et al. currently indicated that the Bi_2WO_6/Ti_3C_2 2D/2D nanocomposite can be created successfully through in situ Bi_2WO_6 sheet formation on the structure of Ti_3C_2 nanosheets [28]. Figure 3.3c illustrates the technique for the 2D/2D Bi_2WO_6/Ti_3C_2 heterostructure photocatalyst formation. The special characteristics of as-prepared 2D/2D Bi_2WO_6/Ti_3C_2 causes efficient photocatalytic CO_2 reduction. To begin, the composite prepared under optimum conditions has greater efficiency for CO_2 adsorption due to high affinity of the CO_2 caused by the large volume and specific surface area of the pores. The reaction's kinetics could be amplified by increasing CO_2 concentration on the catalyst surface. Moreover, the close interaction forms an efficient interspace charge transfer route, prolonging the charge carrier's lifecycle. The unique 2D/2D structure, potential electron attraction ability, and appropriate surface characteristics of Ti MXene are considered to be the key features improving interfacial electron transport and CO_2 reduction [29].

3.2.1 HF Etching

HF etching is regarded as a pioneering and widely used technique to create $Ti_3C_2T_x$ MXenes. HF was primarily considered to be key etching agent owing to its capability to breakdown the large MAX particles into smaller grains and their discrete characteristics to be one of the few chosen etchants that can etch the titanium oxide layer typically presenting as a protective layer of $Ti_3C_2T_x$ MAX [30]. It was proposed that $Ti_3C_2T_x$ MXene might be designed by reacting the Ti_3AlC_2 MAX with 50% of HF acid under continuous stirring at room temperature [1, 19]. The SEM image presented in Figure 3.4a, showed accordion-like structural morphology, proposing the 2D structure of $Ti_3C_2T_x$ MXene formed after 2 hours etching with HF. Research also revealed that several morphological structures such as rolls, nanotubes, and multilayers of $Ti_3C_2T_x$ MXene were also formed after sonication. The effect of etching time could affect the removal of the Al layer of MAX. The peak at 39°, subjected to the MAX phase, was observed to weaken with increasing etching time and completely disappear after some time, as demonstrated in Figure 3.4b. This further indicated that prolonging etching duration could help in elimination of the Al layer and the creation of $Ti_3C_2T_x$ MXene. The additional details on the process of the Ti—Al bond breaking and the development of $Ti_3C_2T_x$ MXene are explained in Eqs. (3.1)–(3.3) [34].

$$M_{n+1}AlX_n + 3HF \rightarrow M_{n+1}X_n + AlF_3 + 1.5H_2 \tag{3.1}$$

$$M_{n+1}X_n + 2H_2O \rightarrow M_{n+1}X_n(OH)_2 + H_2 \tag{3.2}$$

$$M_{n+1}X_n + 2HF \rightarrow M_{n+1}X_nF_2 + H_2 \tag{3.3}$$

Usually, $Ti_3C_2T_x$ MXene is synthesized by eliminating the Al layer sandwiched between the individual layer of $Ti_3C_2T_x$. Equation (3.1) describes the reaction of Ti_3AlC_2 MAX with three HF molecules that results in formation of small white solid particles of AlF_3 with the production of H_2 gas. The loss of the metallic bonding results in the production of layered $Ti_3C_2T_x$ MXene, where the surface of Ti is terminated with functional groups such as –F and –OH, as presented in Eqs. (3.2) and (3.3). The presence of the termination group depends upon the type of etchants used. Hence, the application of HF as a principal etching agent effectively etched away the Al layer of MAX. This procedure is simple and economical, hence preferred. Our research group explored that the etching time will possibly affect the formation of TiO_2 on the surfaces of $Ti_3C_2T_x$ MXene [31]. In this context, the $Ti_3C_2T_x$ MXene was etched by employing 39% and 49% HF concentration, and both were etched for 24 and 96 hours, respectively. It has been proposed that prolonging etching time with HF supported the generation of TiO_2 nanoparticles (NPs) on the surfaces of $Ti_3C_2T_x$ MXene, as shown in Figure 3.4c. Regarding this, the increasing number of active TiO_2 semiconductor creation recommends a strong photocatalytic activity owing to metal–semiconductor synergy. Although the HF etching technique is widely used, the synthesis parameters such as the concentration of HF and the etching time significantly certify the purity of synthesized $Ti_3C_2T_x$ MXene. Moreover, monitoring the synthesis parameters could effectively change the photocatalytic capability of the $Ti_3C_2T_x$ MXene-based materials.

3.2.2 Acid-Containing Fluoride Ions

The acid-containing fluoride ions is another synthesis procedure of $Ti_3C_2T_x$ MXene that has attracted scientific attention. To reduce the utilization of concentrated HF acid, researchers tried to find a new eco-friendly method to form $Ti_3C_2T_x$ MXene. In this regard, acid-containing fluoride ion etchants were employed to etch the Al layer of Ti_3AlC_2 MAX along with the exfoliation of their multilayer structure into

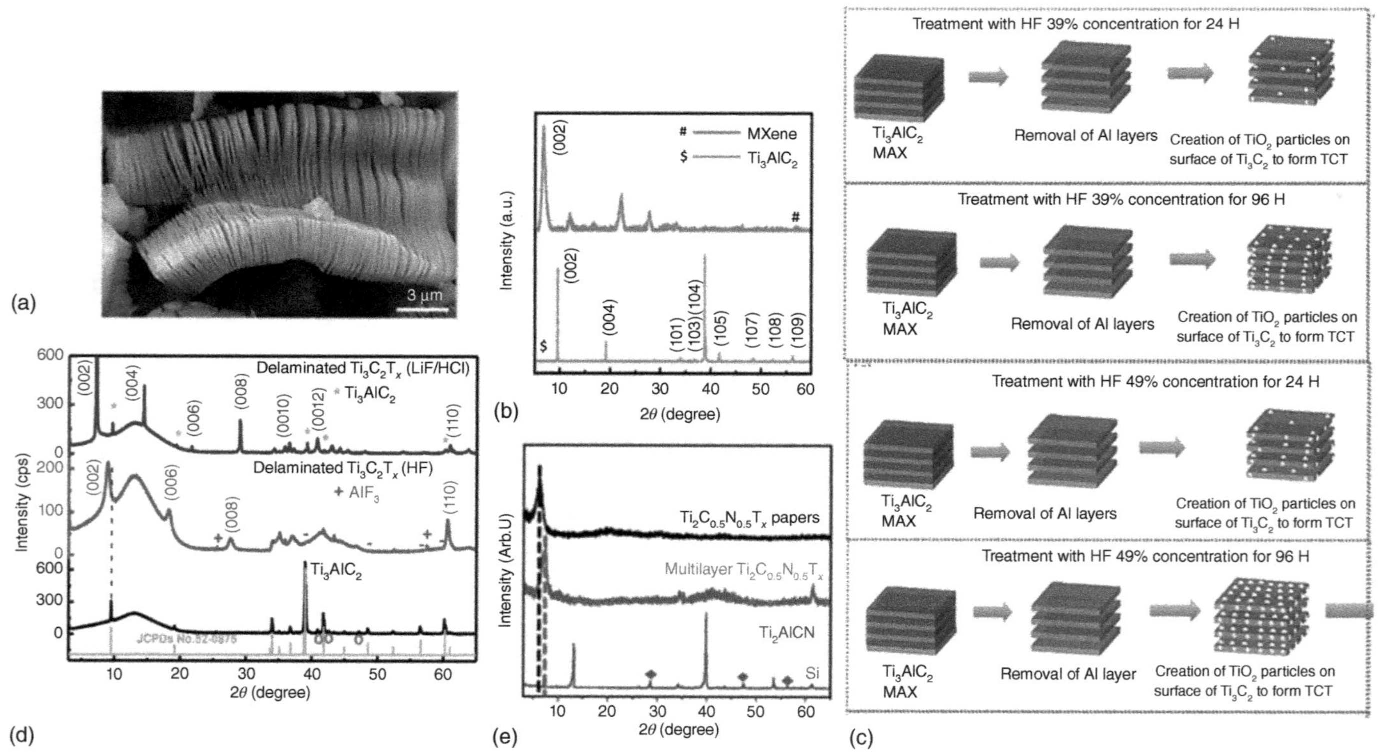

Figure 3.4 (a) SEM of 2D structure of $Ti_3C_2T_x$ MXene after 2 hours etching with HF. Source: Naguib et al. [19]/with permission from American Chemical Society. (b) XRD of $Ti_3C_2T_x$ MXene at various etching times. Source: Mashtalir et al. [22]/with permission of Elsevier. (c) Synthesis of TiO_2 on the $Ti_3C_2T_x$ MXene sheets at different etching times.Source: Khan et al. [31]/with permission of Elsevier. (d) XRD analysis of Ti_3AlC_2, DMSO-delaminated $Ti_3C_2T_x$ (HF), and DMSO-delaminated $Ti_3C_2T_x$ (LiF/HCl). Source: Qian et al. [32]/with permission of John Wiley & Sons. (e) XRD of pristine $Ti_2AlC_{0.5}N_{0.5}$ MAX phase (light grey), multilayer $Ti_2C_{0.5}N_{0.5}T_x$ MXene (dark grey), and freestanding $Ti_2C_{0.5}N_{0.5}T_x$ MXene (black pattern). Source: Liang et al. [33]/with permission of John Wiley & Sons.

the discrete segments [25]. In this respect, the creation and exfoliation of $Ti_3C_2T_x$ MXene could be attained only in one step, irrespective of HF etching that involves intercalating agents to exfoliate the multi-layered structure of $Ti_3C_2T_x$ MXene. The research study reveals that the one-step synthesis technique can be obtained by reacting $Ti_3C_2T_x$ MXene with 6 M HCl with LiF. In this technique, several kinds of acid and fluoride ions could be employed, like using sulfuric acid (H_2SO_4) containing NaF, KF, NH_3F, CsF, and tetrabutylammonium fluoride [35], although every amalgamation may change the characteristics of the resultant $Ti_3C_2T_x$ MXene formed. For example, a research study revealed the increased electrochemical capacity of $Ti_3C_2T_x$ MXenes by applying HCl in comparison to those by HF [36]. Moreover, every fluoride salt used may require different reaction temperatures and etching time. It has been proposed that applying fluoride salts like NaF and KF might complete the etching process under 40 °C for 48 hours, although acid comprising NH_4F complete exfoliation in 24 hours at 30 °C. Comparable morphology can be seen in every synthesized $Ti_3C_2T_x$ MXene irrespective of any type of fluoride ions. In this respect, the accordion-like structure is present, signifying that various acid-containing fluoride ions effectively form the $Ti_3C_2T_x$ MXene. Qian et al. [32] observed from *d*-spacing of 9.29 Å for Ti_3AlC_2, whereas the (002) peak substantially broadened and shifted to *d*-spacing of 9.87 Å ($Ti_3C_2T_x$ (HF)) and 12.25 Å ($Ti_3C_2T_x$ (LiF/HCl)), showing an obviously increased interlayer spacing in surface-functionalized $Ti_3C_2T_x$(LiF/HCl) after dimethylsulfoxide (DMSO) delamination, as demonstrated in Figure 3.4d. Liang et al. [33] used the HCl/KF mixture to etch the MAX phase because KF is water-soluble, and therefore fluoride salt impurities are prevented, contrasting that observed in the case of LiF. As compared to pristine MAX phase, the (002) peak became broad and shifted to a lower angle, showing the removal of Al from layered $Ti_2AlC_{0.5}N_{0.5}$ and full transformation into MXene. The (002) peak shifted from 13.13° to 7.42° after etching, which shows the increase of interlayer distance from 6.8 to 11.9 Å from MAX phase to $Ti_2C_{0.5}N_{0.5}T_x$ MXene, as demonstrated in Figure 3.4e.

3.2.3 Electrochemical Etching

One of the recently established procedures in creating $Ti_3C_2T_x$ MXene is electrochemical etching by which the fluoride-based etchant can be eradicated. This process requires the electrode and electrolytes for the reaction to occur. The Al layer would be etched selectively with the help of electrolytes by anodic etching [37]. Usually, the Ti_3AlC_2 MAX will act as a working electrode where the etching reaction will occur. Mainly, chemical and electrochemical etching is part of the surface reaction comprising electron transfer, signifying the probability of synthesizing $Ti_3C_2T_x$ MXene by electrochemically etching the MAX phase. A study indicated that the Ti_2AlC MAX can be electrochemically etched to form Ti_2CT_x MXene by employing an aqueous HCl (2 M) electrolyte. The unavailability of fluoride ions considerably forms fluoride-free terminated Ti_2CT_x MXene with –Cl, –O, and –OH terminal groups. In this procedure, a three-layered structure of Ti_2CT_x MXene is produced, comprising carbon-derived carbide (CDC), Ti_2CT_x MXene, and unetched MAX.

All parameters that influence the morphology and properties of Ti_2CT_x MXene were studied, like voltage, etching time, and electrolyte concentration as shown in Figure 3.5a–d. Particularly, the HCL concentration and etching time increase, confirming distinct exfoliation of surfaces, although over-etching of Ti_2CT_x MXene may result in the creation of CDC. Although the three-layered structure comprising CDC, MXene, and MAX is synthesized, bath sonication possibly will split the layered structure to get Ti_2CT_x MXene. The synthesis method of Ti_2CT_x MXene through this electrochemical etching is described in Figure 3.5e.

Research study showed that $Ti_3C_2T_x$ MXene was created successfully by using binary aqueous electrolytes in a two-electrode system as presented in Figure 3.5f [37]. In this technique, the electrolyte comprising 1 M NH_4Cl and 0.2 M tetramethylammonium hydroxide (TMAOH) at pH value >9 was applied for 5 hours reaction time to etch the Al layer, leading to intercalation for carbide flake extraction. The Cl ion presence in the electrolyte accelerates the anodic Al etching and break down of the Ti—Al bond. Intercalation by the ammonium hydroxide NH_4OH stimulates the etching lying below the surfaces. This methodology effectively produces 90% of single and double $Ti_3C_2T_x$ MXene layers with larger flakes than the conventional HF technique. Furthermore, the fluoride-free $Ti_3C_2T_x$ MXene proposes good capacitance (220 mF cm^{-2}) than those conventionally prepared. Although this technique proposes $Ti_3C_2T_x$ MXene with Cl termination and successfully removes the –F functional groups, the synthesis of CDC needed much attention. As stated, over-etching the MAX phase in dilute HF, HCl, and NaCl electrolytes might lead to the elimination of both the Al and Ti layer [39]. Moreover, the newly established method needs detailed scientific study to guarantee the successful development of $Ti_3C_2T_x$ MXene with excellent catalytic properties.

3.2.4 Alkali Etching

Apart from using acid, the research also described employing alkali as an etching agent to eliminate the Al layer of Ti_3AlC_2 MAX [14]. An investigation on alkali etching suggested that the hydrothermal-assisted procedure using NaOH as the etching agent efficiently etched the Al layer of Ti_3AlC_2 MAX to create $Ti_3C_2T_x$ MXene. The alkali etching method is a means of forming fluorine-free $Ti_3C_2T_x$ MXene as it does not include any fluoride-containing chemicals. Moreover, this newly designed technique was also discovered to produce highly pure (~92%) $Ti_3C_2T_x$ MXene with a multilayered structure. The intense binding capacity of the NaOH with the Al element promotes their etching ability, thus making it possible. It has been signified that the reaction variables such as temperature and etching concentration are important to certify effective removal of the Al layer. As a result, as explained in Figure 3.6A, the decrease in temperature may restrict the Al abstraction process by the Al (oxide) hydroxides. Further, the incapability to synthesize $Al_4(OH)^-$ due to lattice confinement from the Ti layer, known as the "jamming effect," blocks creation of $Ti_3C_2T_x$ MXene. Moreover, elevated temperature and higher concentration of NaOH cause the oxidation of $Ti_3C_2T_x$ MXene and the creation of NTOs such as $Na_2Ti_3O_7$ and $Na_2Ti_5O_{11}$. Increasing the water amount assisted the

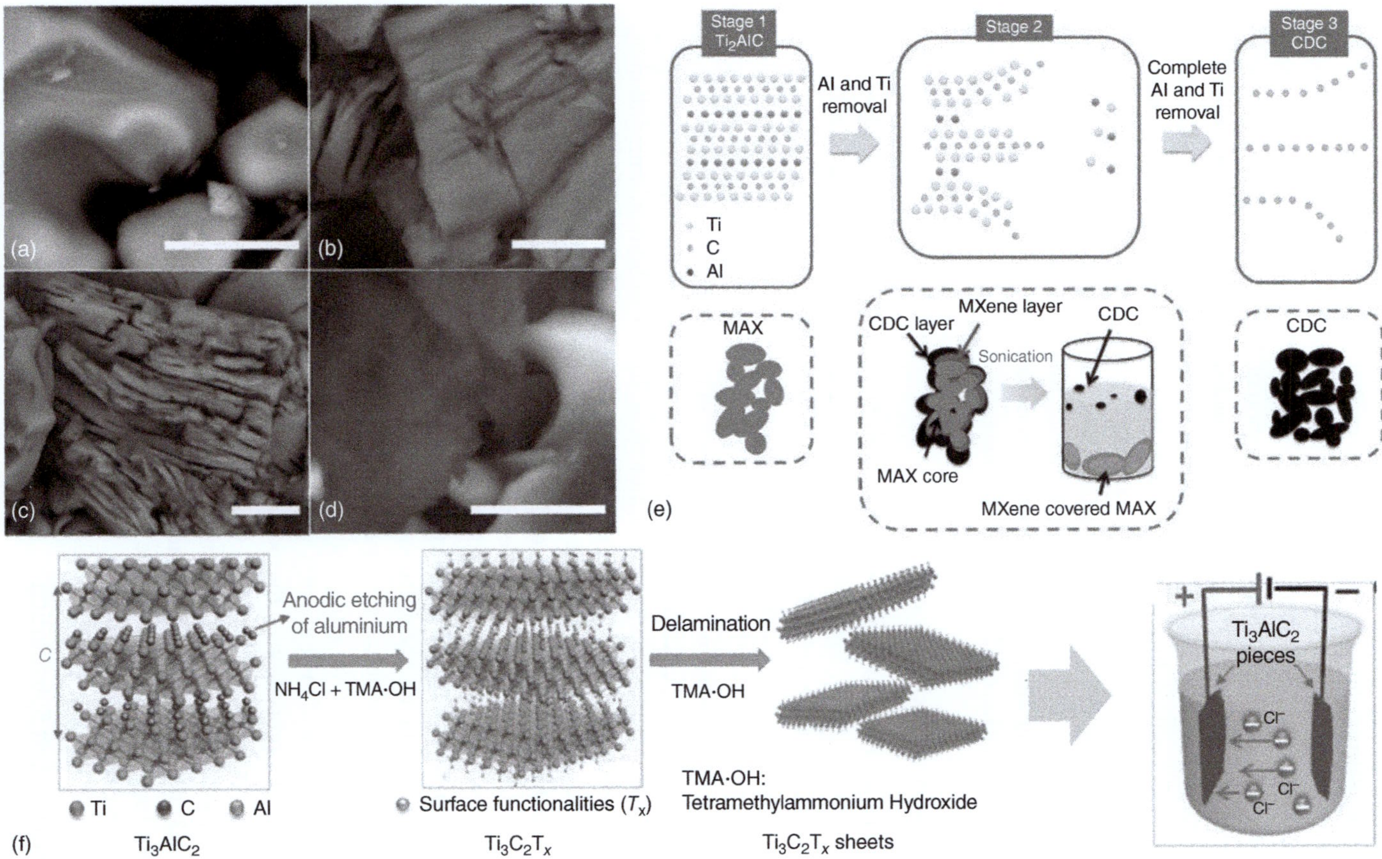

Figure 3.5 SEM images of (a) MAX, (b) Ti_2CT_x MXene treated with 1 M of HCl for 1 day at 0.6 V, (c) 2 M of HCl for 5 days at 0.6 V, (d) 2 M of HCl for 14 days at 0.6 V; (e) Schematic of synthesis of Al removal from MAX phase and fabrication of the three-layered Ti_2CT_x MXene structure. Source: Sun et al. [38]/with permission of Royal Society of Chemistry. (f) Synthesis procedure of electrochemical etching of Ti_2CT_x MXene in two-electrode system. Source: Adapted from Yang et al. [37].

oxidation of Ti to produce NTOs. Hence, an appropriate reaction temperature and alkali concentration are essential to confirm a complete development of $Ti_3C_2T_x$ MXene. The hydrothermal reaction at 270 °C with 27.5 M NaOH concentration successfully formed $Ti_3C_2T_x$ MXene with comparatively greater purity than that of HF-etched MXene. In this hydrothermal NaOH etching technique, monitoring the temperature is crucial to certify a complete creation of $Ti_3C_2T_x$ MXene, though the optimal concentration of $Ti_3C_2T_x$ MXene governed the purity of the synthesized $Ti_3C_2T_x$ MXene. Although alkali-dependent, the hydrothermal technique has been revealed to effectively produce $Ti_3C_2T_x$ MXene, preventing the $Ti_3C_2T_x$ Mxene from over-etching to produce NTOs, which is challenging and needs additional consideration. Also, this technique requires intense research to approve the successful creation of fluoride-free $Ti_3C_2T_x$ MXene. However, this technique can result in –O and –OH terminal groups, and their capacity in enhancing photocatalytic activity and fuel transformation can be further continued.

3.2.5 Water-Free Etching

Water-free etching is an alternative technique to form $Ti_3C_2T_x$ MXene with better oxidation constancy [42]. In this technique, the utilization of water as a principal solvent is changed by the use of the polar organic solvent. The remarkable capability of $Ti_3C_2T_x$ MXene in a broad range of applications signifies them among the most preferred materials. Nevertheless, their hydrophilic nature and propensity to oxidize bound their ability in practices involving an extended storage time. Hence, water-free etching was considered to increase oxidation stability and extended their involvement in any water-sensitive functions. In this technique, ammonium dihydrogen fluoride has been employed as the etching agent because of its ability to separate into NH_4F and HF when dispersed in polar solvents. Three reaction steps are needed: etching, washing, and delamination. Particularly, as described in Figure 3.6B, the etching takes place when Ti_3AlC_2 MAX experienced a reaction with NH_4HF_2 where the Al layer was etched out in the form of AlF_3 and $(NH_4)_3Al_6$. Reaction with propylene carbonate (PC) promotes exfoliation into single-layer $Ti_3C_2T_x$ MXene. It has also been recommended that increase in the interlayer space of $Ti_3C_2T_x$ MXene comparative to water-etched $Ti_3C_2T_x$ MXene is attained [24]. Regulating the termination groups is favorable as it has been shown that by this method, almost 70% of –F and –30% of –O/–OH comprising MXene was obtained [43]. Briefly, this process uses the polar organic solvent PC to substitute water, that decreases the occurrence of the –OH terminating group. The regulation of the terminating group potentially improves the oxidation stability, which increases $Ti_3C_2T_x$ MXene potential in the water-sensitive applications as the finding suggested that increased exposure of the Ti-OH species of $Ti_3C_2T_x$ MXene expedites the oxidation process [44].

3.2.6 Molten-Salt Substitution

Molten salt substitution is another alternate method in creating –Cl-terminated $Ti_3C_2T_x$ MXene [40]. This technique uses molten salts like $ZnCl_2$ and $CuCl_2$ to go

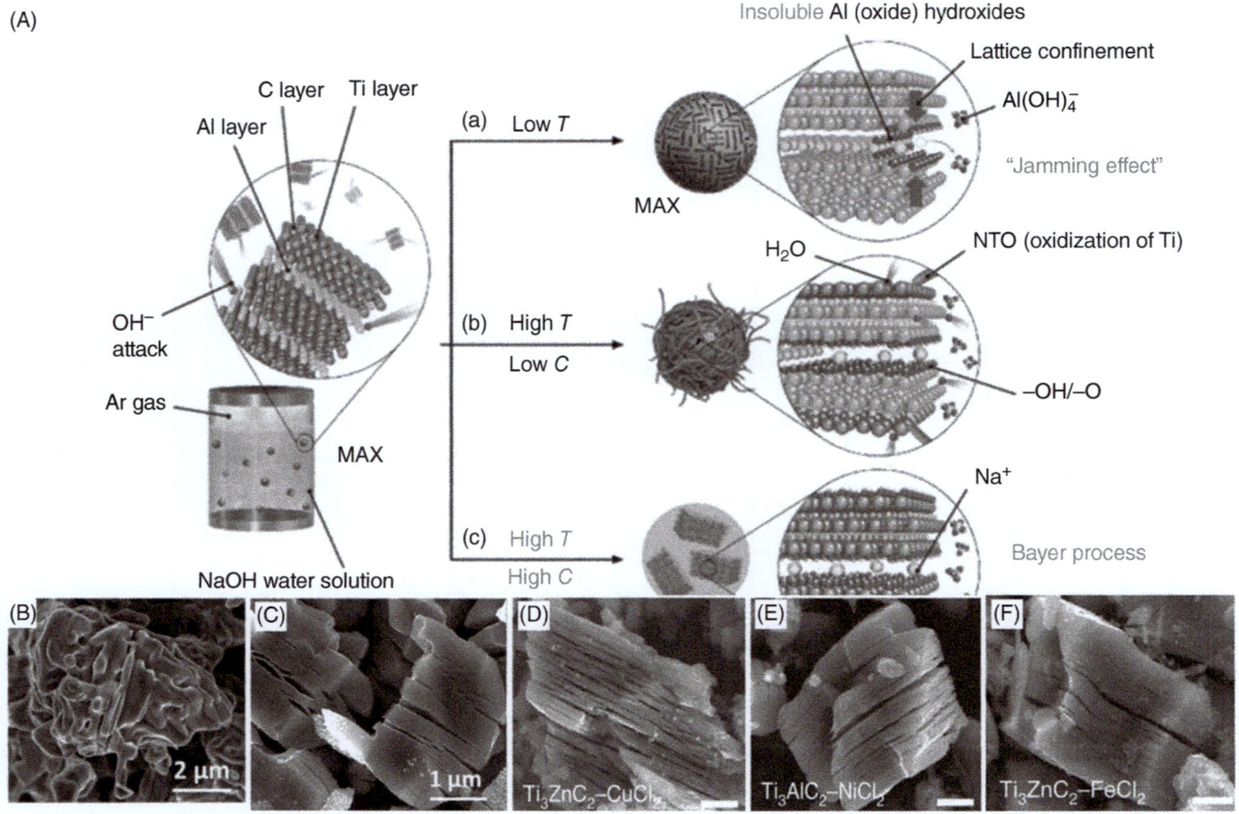

Figure 3.6 (A) Synthesis process of $Ti_3C_2T_x$ MXene through alkali etching. Source: Li et al. [14]/John Wiley & Sons. (B) SEM image of $Ti_3C_2T_x$ MXene prepared through molten salt substitution. Source: Li et al. [40]/American Chemical Society. (C–F) SEM image of $Ti_3C_2T_x$ MXene by utilizing different types of MAX with chloride molten salts. Source: Li et al. [49]/with permission from American Chemical Society.

through a substitution reaction with Ti_3AlC_2 MAX for the elimination of the Al layer. These substitution reactions specifically create much stable nano laminated $Ti_3C_2T_x$ MXene with –Cl terminated than –F terminated $Ti_3C_2T_x$ MXene [45]. The replacement process is described in Eqs. (3.4)–(3.6), that usually includes the substitution of the A element of MAX with the metal salts.

$$Ti_3AlC_2 + 1.5ZnCl_2 \rightarrow Ti_3ZnC_2 + 0.5Zn + AlCl_3 \tag{3.4}$$

$$Ti_3AlC_2 + 1.5ZnCl_2 \rightarrow Ti_3C_2 + 1.5Zn + AlCl_3 \tag{3.5}$$

$$Ti_3C_2 + Zn \rightarrow Ti_3ZnC_2 \tag{3.6}$$

As shown in Eqs. (3.4)–(3.6), the Zn^{2+} from the molten salts go through elemental substitution with the Al^{3+} from the Ti_3AlC_2 MAX. Stronger Lewis acidity by $ZnCl_2$ stimulates the exchange of Zn^{2+} with Al^{3+} making Ti_3ZnC_2 MAX. Moreover, Zn^{2+} work as Lewis acids in the molten salt as they are stronger Cl^- and electron acceptors. Particularly, molten salt of $ZnCl_2$ was ionized into Zn^{2+} and $ZnCl_4{}^{2-}$ in its molten state, even though Al from Ti_3AlC_2 MAX underwent redox conversion into Al^{3+}. The $AlCl_3$ formation took place by the bonding of ionic Al^{3+} with Cl^- where it will be evaporated. Quick evaporation of $AlCl_3$ promotes the outward diffusion of the Al atom for experiencing the replacement reaction. The elemental substitution also occurs when the reduced Zn atoms intercalate into the Al layer in the MAX phase and produce Ti_3ZnC_2 which is then delaminated into Ti_3C_2–Cl_2 MXene. The SEM image in Figure 3.6B indicated the morphology of the Cl-terminated MXene, where it reveals a highly compact assembly with less distinct accordion-sheet in comparison to other traditional synthesis methods. Further research by Li et al. showed that the production of $Ti_3C_2T_x$ MXene from various kinds of MAX phases might be attained by utilizing several chloride molten salts such as $CuCl_2$, $FeCl_2$, $NiCl_2$, and AgCl [41]. Diverse morphology of MXenes etched by different chloride molten salts can be seen in Figure 3.6C–F by which the majority of them displays an eminent accordion-like structure of MXene.

3.3 Characterization of MXenes

3.3.1 X-Ray Diffractions (XRD)

The detailed crystal structure and chemical composition analyses of MXene are conducted by employing XRD analysis, as demonstrated in Figure 3.7. For Ti_3AlC_2, the diffraction peaks at 9.58° (002), 19.17° (004), 33.95° (101), 35.90° (103), 39° (104), 41.75° (105), 48.37° (107), 52.39° (108), 56.41° (109), 60.31° (110), and 70.30° (118) confirmed the Ti_3AlC_2 MAX structure (Card 01-074-8806) [10]. The etching of Ti_3AlC_2 through HF may result in the formation of Ti_3C_2 by removing the Al layer. The XRD peaks at 9.58° (002) and 19.17° (004) shift to lower degrees together with the disappearance of the strongest peak of Ti_3AlC_2 at 39° (104), confirming the removal of Al from MAX phase Ti_3AlC_2. After Ti_3AlC_2 was etched by HF, the diffraction peaks appeared at 9.2° (002), 18.4° (004), 27.7° (006), 34.6°(101), 42.0° (105), and 60.9° (110) [46].

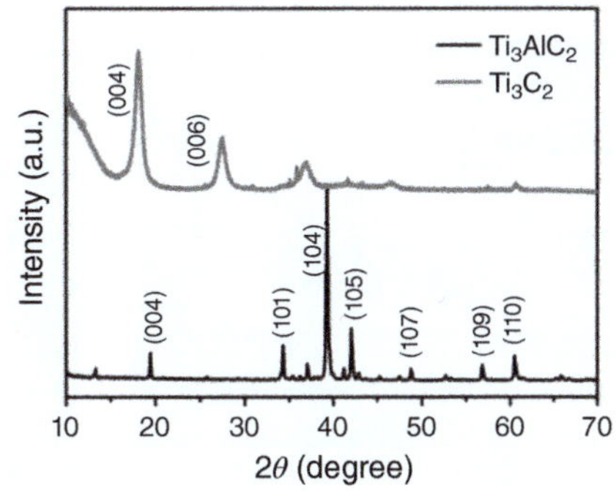

Figure 3.7 XRD of Ti_3AlC_2 MAX and Ti_3C_2 MXene. Source: Wang et al. [10]/with permission of Elsevier.

3.3.2 Field Emission Scanning Electron Microscopy (FESEM) and Transmission Electron Microscopy (TEM)

The structures and morphologies of the samples are characterized by FESEM and TEM analysis. It has been revealed that Ti_3AlC_2 MAX phase possesses block-like structures with a compact layer as demonstrated in Figure 3.8a that is converted to MXene through etching. The TiC MXenes display an accordion-like structure as demonstrated in Figure 3.8b. Wang et al. [53] studied that the etching time can significantly change the crystallinity and morphological properties of the resulting MXenes. With the increase of etching time, the constant removal of Al results in further delamination, which makes the layered sheets thinner. Besides, it is also possible to fabricate MXenes into various other morphological structures with different dimensions such as 0D, 1D, 2D, and 3D during the process of etching. The MXene can be converted into a single layer that is a 2D structure through delamination as demonstrated in Figure 3.8c,d [48]. As revealed in transmission electron microscopy (TEM) images in Figure 3.8e, the monodispersed Ti_3C_2 QDs had an average size of 10.2 ± 0.5 nm [49]. Also, Liu et al. [50] obtained1D TiO_2 nanowires on the 2D Ti_3C_2 nanosheet surface, forming a Ti_3C_2-MXene-derived hierarchical 1D/2D TiO_2/Ti_3C_2 heterostructure as demonstrated in Figure 3.8f. 3D MXene structures can be synthesized by an effective and direct method called template synthesis, in which the resultant 3D materials can be easily customized and adjusted with respect to size and morphology of the templates as demonstrated through the image in Figure 3.8g. Another example of the 3D structure obtained is foam-like structure as demonstrated in the TEM image in Figure 3.8h.

3.3.3 UV–Visible Spectroscopy

UV–Vis DRS is a commonly implemented technique to study optical absorption in MXene-based photocatalysis. It has been observed that Ti_3C_2 like Ti_3AlC_2 does not demonstrate obvious absorption edge owing to the demonstration of outstanding metallic conductivity [54]. However, the oxidation of Ti_3C_2 causes the formation of TiO_2, that leads to the formation of absorption edge in the Ti_3C_2 as observed in Figure 3.9a [55].

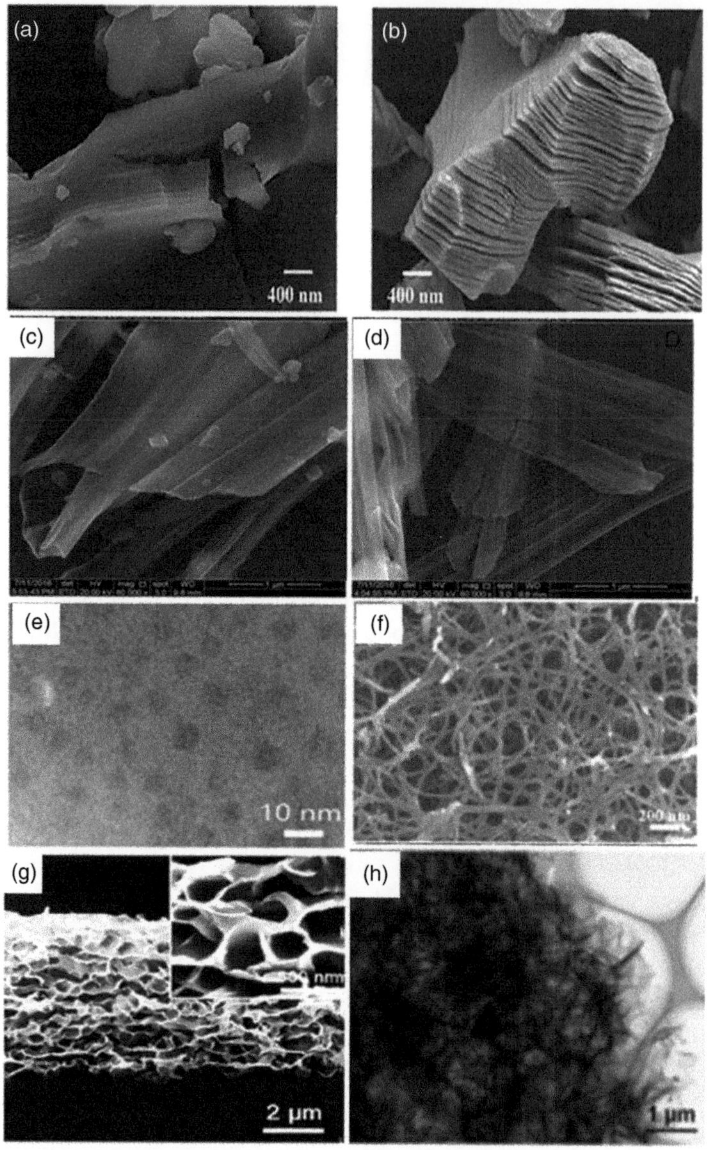

Figure 3.8 (a) SEM image of Ti_3AlC_2 [47]. (b) SEM images of accordion-like structure of Ti_3C_2 [47]. (c, d) 2D single-layer Ti_3C_2 nanosheets. Source: Li et al. [48]/with permission from American Chemical Society. (e) TEM images of 0D quantum dots Ti_3C_2. Source: Chen et al. [49]/with permission from Royal Society of Chemistry. (f) 1D nanowires Ti_3C_2 [50]. (g) SEM image of 3D MXene. Source: Li et al. [51]/John Wiley & Sons. (h) TEM image of foam-like 3D MXene. Source: Zhang et al. [52]/JOHN WILEY & SONS, INC.

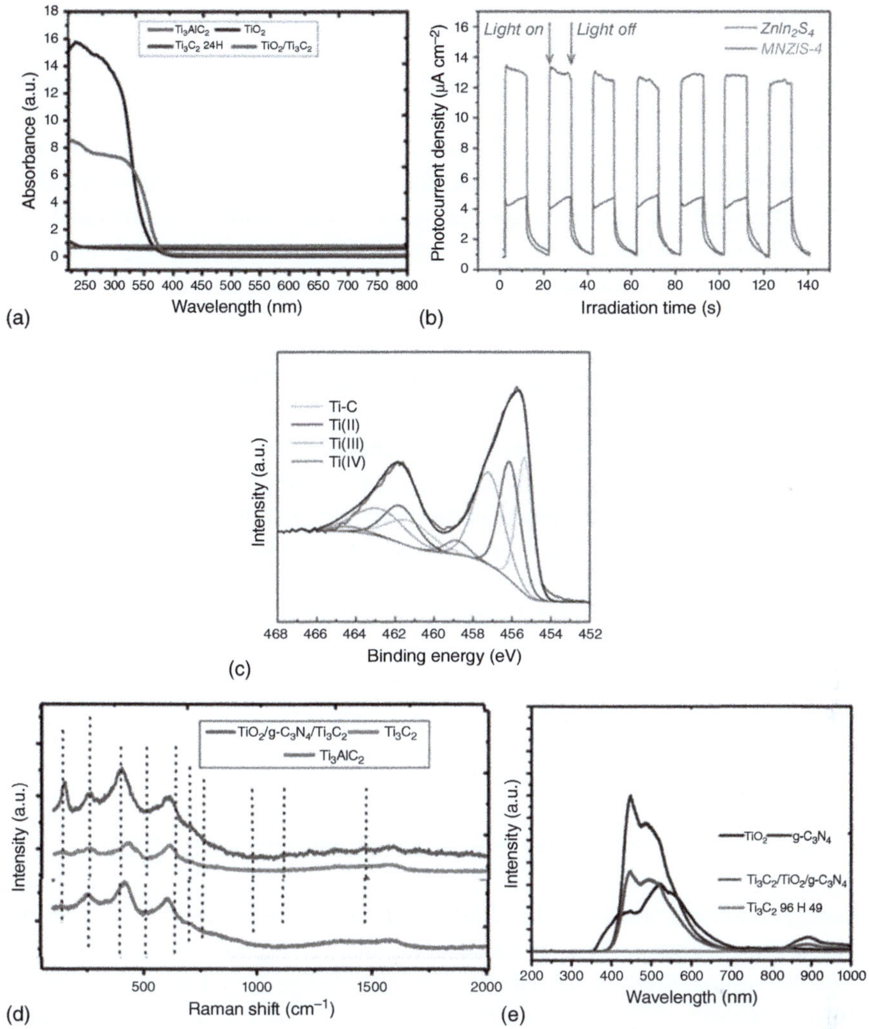

Figure 3.9 (a) UV–vis of Ti_3AlC_2, TiO_2, Ti_3C_2, and TiO_2/Ti_3C_2. Source: Khan et al. [55]/with permission of Elsevier. (b) Photocurrent density of TiO_2/Ti_3C_2 samples. (c) XPS analysis of T2p Ti_3C_2 samples. Source: Zuo et al. [56]/with permission of John Wiley & Sons. (d) RAMAN samples of Ti_3AlC_2 and Ti_3C_2. Source: Khan et al. [57]/with permission of American Chemical Society. (e) PL of TiO_2, $g\text{-}C_3N_4$, and Ti_3C_2 samples. Source: Khan et al. [31]/with permission of Elsevier.

3.3.4 Photocurrent Analysis

The photocurrent density curves can be drawn to examine the photogenerated charge carrier separation and utilization efficiency of the MXene-based photocatalysts. For instance, Low et al. [58] studied the photocurrent response test to evaluate the generation and separation efficiency of the photogenerated charge carrier of the TiO_2/Ti_3C_2 photocatalyst. It was observed that the electrons generated on TiO_2 preferred to migrate to Ti_3C_2 before reaching the cathode owing to excellent electronic conductivity of Ti_3C_2, achieving separation of electron-hole pairs. Similarly, by switching the light off, the behavior of photocurrent decrease was stable, showing the characteristic of the Ti_3C_2 as an electron reservoir. The optimized sample (TT550) exhibited the maximum photocurrent density as compared to all other samples, showing the maximum electron-hole photogeneration efficiency and the lowest electron-hole recombination rate. Similarly, another study demonstrated that the transient photocurrent spectra for $ZnIn_2S_4$ nanosheets anchored on $Ti_3C_2T_x$ MXene had higher photocurrent density than pristine $ZnIn_2S_4$, revealing the improved charge transfer kinetics as demonstrated in Figure 3.9b [56].

3.4 X-Ray Photoelectron Spectroscopy (XPS)

XPS measurements are done to analyze the chemical state and valence information of different species calibrated through C1s at 284.6 eV binding energy. As demonstrated in Figure 3.9c, the XPS spectra in the Ti 2p region of the Ti_3C_2 demonstrated that peaks could be deconvoluted into components attributed to Ti bound to C, Ti(II), Ti(III), and Ti(IV) peaks. The Ti–C $2p_{3/2}$, Ti(II) $2p_{3/2}$, Ti(III) $2p_{3/2}$, and Ti(IV) $2p_{3/2}$ bands were spotted at 455.3, 456.1, 457.2, and 458.8 eV binding energies, respectively. It was observed that the intensity of Ti–C peak was the highest, whereas that of the Ti(IV) was the weakest, combined with Ti–C signals ascending from Ti atoms in the Ti_3C_2 MXene interior nanolayers, which showed that the spectrum was characteristic of Ti_3C_2 [59].

3.5 Raman Spectroscopy and Photoluminescence (PL)

Raman spectroscopy exposes the vibrational modes of molecules to provide structural fingerprints. As Ti_3C_2 is fabricated from Ti_3AlC_2; therefore, the RAMAN analysis of both materials is carried out to see the change in vibrational modes. To identify the phases of Ti_3AlC_2 numerous bands occurred at c. 175, 260, 410, and 605 cm^{-1}, ascribed to the ω_1, ω_2, ω_3, and ω_4 pf the mode-related Raman-active vibration as demonstrated in Figure 3.9d [58, 60]. The Raman spectrum of the Ti_3C_2 material demonstrated a high peak at the location around 154 cm^{-1} accompanied by other modes such as 255, 360, 428, and 620 cm^{-1} that were ascribed to the RAMAN active ω_1, ω_2, ω_3, and ω_4 vibration modes, attributed to the TiC vibration modes as demonstrated in Figure 3.9d [61–63]. Obviously, the Ti_3C_2 ω_3 Raman band modes are blue-shifted as compared to Ti_3AlC_2 that occurred due to the delamination effect of Ti_3AlC_2 due to HF etching [57, 58].

Photoluminescence (PL) spectroscopy is an efficient and commonly employed technique to expose the separation and transfer efficiency of generated charge carriers. Normally, a low steady-state PL indicates an extended lifetime and greater separation efficiency of generated electron-hole pairs, leading to excellent photoactivity. The PL intensity value of all the Ti_3C_2 samples is very less, that is owing the generation of a low number of electron-hole pairs. For instance, Khan et al. [31]

demonstrated that the PL intensity of Ti_3C_2 with the presence of TiO_2 had very low PL intensity due to the presence of Ti_3C_2. It was also observed that coupling Ti_3C_2/TiO_2 with g-C_3N_4 demonstrated the lower PL emission intensity than pristine g-C_3N_4. The comparison of the PL of Ti_3C_2/TiO_2/g-C_3N_4 is demonstrated in Figure 3.9e that showed the lower PL values of the composite as compared to TiO_2 and g-C_3N_4.

3.6 Conclusions

To conclude, $Ti_3C_2T_x$ MXenes can be formed through various techniques by using several etchants and protocols. Since its success in eliminating the A layer of the MAX phase, HF etching was considered among the valuable techniques for producing MXenes. Furthermore, environmental factors drive the progression of advanced production process like mixing acids with fluorine salts, electrochemical etching, molten salt substitution, water-free etching, alkali etching, and halogen etching. These eco-friendly techniques open up innovative ways for synthesizing MXene, specifically bulk production. Besides this, only in situ HF was often used in the production of MXenes among these methods. Photocatalytic hydrogen synthesis by means of Ti_3C_2 MXene photocatalysts offers an exclusive explanation for giving a green, stable, and renewable energy source. Consequently, a considerable amount of consideration has been set into improving these photocatalysts' solar-to-energy transformation. Owing to Ti_3C_2 MXene's distinctive assembly and excellent electrical properties providing it the unique capability to produce the Schottky barrier for confining electrons, the use of Ti_3C_2 MXene as a cocatalyst to give hydrogen has been considered a potential move. It is critical to understand Ti_3C_2 MXene by exploring the basic characteristics of MXene materials, like the adjustment of surface functionalization groups, the stability of MXene in water and air, and the constancy of heat processing. It is also important to be noted that electronic and electrical properties may play a role in improving the functioning of MXene substances. It is assumed that integrating Ti_3C_2 MXene in numerous Ti_3C_2 MXene-based complexes could improve the features of semi-conductor components in various conditions, comprising working as a potential support material, limiting potential barriers, improving carrier dynamics, and acting as electron trappers. The production of Schottky heterojunctions by hybrid Ti_3C_2 and MXene is greatly associated to the metallic features of Ti_3C_2 and MXene. Ti_3C_2 MXene acts as an electron trapper in this setting, prolonging exciton period and enhancing photocatalytic efficiency. Ti_3C_2 MXene, which is accessible and cheap, is the optimal substitute for expensive and non-noble metals. Furthermore, according to some research, integrating Ti_3C_2 MXene can produce a momentous volume of hydrogen that outperforms noble metal incorporation.

The formation of MXenes from MAX phase is a complex process that involves the removal of the Al layer. This removal of the Al layer shows changes in the crystallography, structure, morphology, optical properties, and functional groups on the surface that are characterized through different analysis for the identification of complete conversion. Among these techniques, the most commonly employed are XRD, XPS, Raman spectroscopy, PL, and photocurrent. These techniques were able to identify the crystallographic, structural, morphological conversion of titanium carbide MXenes. Moreover, the formation of composite also resulted in changes in electronic, structural, and optical properties of composites that are explicated.

Acknowledgment

The financial support of this work is provided by United Arab Emirates University (UAEU), UAE under Start up research grant vot no. 12N097.

References

1 Naguib, M., Kurtoglu, M., Presser, V. et al. (2011). Two-dimensional nanocrystals produced by exfoliation of Ti_3AlC_2. *Adv. Mater. Processes* 23 (37): 4248–4253.

2 Lin, Y., Li, Y., Cao, Y., and Wang, X. (2021). Two-dimensional MOFs: design & synthesis and applications. *Chem. Asian J.* 16 (21): 3281–3298.

3 Sun, Y., Meng, X., Dall'Agnese, Y. et al. (2019). 2D MXenes as co-catalysts in photocatalysis: synthetic methods. *Nano-Micro Lett.* 11 (1): 79.

4 Shen, J., Wu, Z., Li, C. et al. (2021). Emerging applications of MXene materials in CO_2 photocatalysis. *FlatChem* 28: 100252.

5 Fan, Y., Li, L., Zhang, Y. et al. (2022). Recent advances in growth of transition metal carbides and nitrides (MXenes) crystals. *Adv. Funct. Mater.* 32: 2111357.

6 Peng, C., Wei, P., Li, X. et al. (2018). High efficiency photocatalytic hydrogen production over ternary $Cu/TiO_2@Ti_3C_2T_x$ enabled by low-work-function 2D titanium carbide. *Nano Energy* 53: 97–107.

7 Su, T., Hood, Z.D., Naguib, M. et al. (2019). 2D/2D heterojunction of $Ti_3C_2/g\text{-}C_3N_4$ nanosheets for enhanced photocatalytic hydrogen evolution. *Nanoscale* 11 (17): 8138–8149.

8 Qin, D., Zhou, Y., Wang, W. et al. (2020). Recent advances in two-dimensional nanomaterials for photocatalytic reduction of CO_2: insights into performance, theories and perspective. *J. Mater. Chem. A* 8 (37): 19156–19195.

9 Peng, J., Chen, X., Ong, W.-J. et al. (2019). Surface and heterointerface engineering of 2D MXenes and their nanocomposites: insights into electro- and photocatalysis. *Chem* 5 (1): 18–50.

10 Wang, B., Wang, M., Liu, F. et al. (2020). Ti_3C_2: an ideal co-catalyst? *Angew. Chem.* 132 (5): 1930–1934.

11 Tahir, M., Ali Khan, A., Tasleem, S. et al. (2021). Titanium carbide (Ti_3C_2) MXene as a promising co-catalyst for photocatalytic CO_2 conversion to energy-efficient fuels: a review. *Energy Fuels* 35 (13): 10374–10404.

12 Cheng, L., Li, X., Zhang, H., and Xiang, Q. (2019). Two-dimensional transition metal MXene-based photocatalysts for solar fuel generation. *J. Phys. Chem. Lett.* 10 (12): 3488–3494.

13 Alhabeb, M., Maleski, K., Anasori, B. et al. (2017). Guidelines for synthesis and processing of two-dimensional titanium carbide ($Ti_3C_2T_x$ MXene). *Chem. Mater.* 29 (18): 7633–7644.

14 Li, T., Yao, L., Liu, Q. et al. (2018). Fluorine-free synthesis of high-purity $Ti_3C_2T_x$ (T = OH, O) via alkali treatment. *Angew. Chem. Int. Ed.* 57 (21): 6115–6119.

15 Zhao, Y., Que, M., Chen, J., and Yang, C. (2020). MXenes as co-catalysts for the solar-driven photocatalytic reduction of CO_2. *J. Mater. Chem. A* 8 (46): 16258–16281.

16 Naguib, M., Barsoum, M.W., and Gogotsi, Y. (2021). Ten years of progress in the synthesis and development of MXenes. *Adv. Mater.* 33 (39): 2103393.

17 Cao, S., Shen, B., Tong, T. et al. (2018). 2D/2D heterojunction of ultrathin MXene/Bi_2WO_6 nanosheets for improved photocatalytic CO_2 reduction. *Adv. Funct. Mater.* 28 (21): 1800136.

18 Verger, L., Xu, C., Natu, V. et al. (2019). Overview of the synthesis of MXenes and other ultrathin 2D transition metal carbides and nitrides. *Curr. Opin. Solid State Mater. Sci.* 23 (3): 149–163.

19 Naguib, M., Mashtalir, O., Carle, J. et al. (2012). Two-dimensional transition metal carbides. *ACS Nano* 6 (2): 1322–1331.

20 Naguib, M., Halim, J., Lu, J. et al. (2013). New two-dimensional niobium and vanadium carbides as promising materials for Li-ion batteries. *J. Am. Chem. Soc.* 135 (43): 15966–15969.

21 Sun, Z., Li, S., Ahuja, R., and Schneider, J.M. (2004). Calculated elastic properties of M_2AlC (M = Ti, V, Cr, Nb and Ta). *Solid State Commun.* 129 (9): 589–592.

22 Mashtalir, O., Naguib, M., Dyatkin, B. et al. (2013). Kinetics of aluminum extraction from Ti_3AlC_2 in hydrofluoric acid. *Mater. Chem. Phys.* 139 (1): 147–152.

23 Dillon, A.D., Ghidiu, M.J., Krick, A.L. et al. (2016). Highly conductive optical quality solution-processed films of 2D titanium carbide. *Adv. Funct. Mater.* 26 (23): 4162–4168.

24 Halim, J., Lukatskaya, M.R., Cook, K.M. et al. (2014). Transparent conductive two-dimensional titanium carbide epitaxial thin films. *Chem. Mater.* 26 (7): 2374–2381.

25 Ghidiu, M., Lukatskaya, M.R., Zhao, M.-Q. et al. (2014). Conductive two-dimensional titanium carbide 'clay' with high volumetric capacitance. *Nature* 516 (7529): 78–81.

26 Urbankowski, P., Anasori, B., Makaryan, T. et al. (2016). Synthesis of two-dimensional titanium nitride Ti_4N_3 (MXene). *Nanoscale* 8 (22): 11385–11391.

27 Verger, L., Natu, V., Carey, M., and Barsoum, M.W. (2019). MXenes: an introduction of their synthesis, select properties, and applications. *Trends Chem.* 1 (7): 656–669.

28 Tang, W., Chen, J., Yin, Z. et al. (2021). Complete removal of phenolic contaminants from bismuth-modified TiO_2 single-crystal photocatalysts. *Chin. J. Catal.* 42 (2): 347–355.

29 Cao, S., Shen, B., Tong, T. et al. (2018). 2D/2D heterojunction of ultrathin MXene/Bi_2WO_6 nanosheets for improved photocatalytic CO_2 reduction. *Adv. Funct. Mater.* 28 (21): 1800136.

30 Anasori, B. and Gogotsi, Y. (2019). *Metal Carbides and Nitrides (MXenes) Structure, Properties and Applications*. Cham, Switzerland: Springer Nature Switzerland AG.

31 Khan, A.A. and Tahir, M. (2021). Well-designed 2D/2D $Ti_3C_2T_{A/R}$ MXene coupled g-C_3N_4 heterojunction with in-situ growth of anatase/rutile TiO_2 nucleates to boost photocatalytic dry-reforming of methane (DRM) for syngas production under visible light. *Appl. Catal., B* 285: 119777.

32 Qian, A., Seo, J.Y., Shi, H. et al. (2018). Surface functional groups and electrochemical behavior in dimethyl sulfoxide-delaminated $Ti_3C_2T_x$ MXene. *ChemSusChem* 11 (21): 3719–3723.

33 Liang, K., Tabassum, A., Majed, A. et al. (2021). Synthesis of new two-dimensional titanium carbonitride $Ti_2C_{0.5}N_{0.5}T_x$ MXene and its performance as an electrode material for sodium-ion battery. *InfoMat* 3 (12): 1422–1430.

34 Feng, A., Yu, Y., Wang, Y. et al. (2017). Two-dimensional MXene Ti_3C_2 produced by exfoliation of Ti_3AlC_2. *Mater. Design* 114: 161–166.

35 Liu, F., Zhou, A., Chen, J. et al. (2017). Preparation of Ti_3C_2 and Ti_2C MXenes by fluoride salts etching and methane adsorptive properties. *Appl. Surf. Sci.* 416: 781–789.

36 Kvashina, T.S., Uvarov, N.F., Korchagin, M.A. et al. (2020). Synthesis of MXene Ti_3C_2 by selective etching of MAX-phase Ti_3AlC_2. *Mater. Today Proc.* 31: 592–594.

37 Yang, S., Zhang, P., Wang, F. et al. (2018). Fluoride-free synthesis of two-dimensional titanium carbide (MXene) using abinary aqueous system. *Angew. Chem. Int. Ed.* 57 (47): 15491–15495.

38 Sun, W., Shah, S.A., Chen, Y. et al. (2017). Electrochemical etching of Ti_2AlC to Ti_2CT_x (MXene) in low-concentration hydrochloric acid solution. *J. Mater. Chem. A* 5 (41): 21663–21668.

39 Lukatskaya, M.R., Halim, J., Dyatkin, B. et al. (2014). Room-temperature carbide-derived carbon synthesis by electrochemical etching of MAX phases. *Angew. Chem. Int. Ed.* 53 (19): 4877–4880.

40 Li, M., Lu, J., Luo, K. et al. (2019). Element replacement approach by reaction with Lewis acidic molten salts to synthesize nanolaminated MAX phases and MXenes. *J. Am. Chem. Soc.* 141 (11): 4730–4737.

41 Li, Y., Shao, H., Lin, Z. et al. (2020). A general Lewis acidic etching route for preparing MXenes with enhanced electrochemical performance in non-aqueous electrolyte. *Nat. Mater.* 19 (8): 894–899.

42 Carey, M., Hinton, Z., Natu, V. et al. (2020). Dispersion and stabilization of alkylated 2D MXene in nonpolar solvents and their pseudocapacitive behavior. *Cell Rep. Phys. Sci.* 1 (4): 100042.

43 Zhao, X., Radovic, M., and Green, M.J. (2020). Synthesizing MXene nanosheets by water-free etching. *Chem* 6 (3): 544–546.

44 Lee, Y., Kim, S.J., Kim, Y.-J. et al. (2020). Oxidation-resistant titanium carbide MXene films. *J. Mater. Chem. A* 8 (2): 573–581.

45 Anasori, B., Lukatskaya, M.R., and Gogotsi, Y. (2017). 2D metal carbides and nitrides (MXenes) for energy storage. *Nat. Rev. Mater.* 2 (2).

46 Wang, H., Zhang, X., and Xie, Y. (2018). Recent progress in ultrathin two-dimensional semiconductors for photocatalysis. *Mater. Sci. Eng. R Rep* 130: 1–39.

47 Liu, J., Jiang, X., Zhang, R. et al. (2019). MXene-enabled electrochemical microfluidic biosensor: applications toward multicomponent continuous monitoring in whole blood. *Adv. Funct. Mater.* 29 (6): 1807326.

48 Li, G., Tan, L., Zhang, Y. et al. (2017). Highly efficiently delaminated single-layered MXene nanosheets with large lateral size. *Langmuir* 33 (36): 9000–9006.

49 Chen, X., Sun, X., Xu, W. et al. (2018). Ratiometric photoluminescence sensing based on Ti_3C_2 MXene quantum dots as an intracellular pH sensor. *Nanoscale* 10 (3): 1111–1118.

50 Liu, N., Yu, L., Liu, B. et al. (2023). Ti_3C_2-MXene partially derived hierarchical 1D/2D TiO_2/Ti_3C_2 heterostructure electrode for high-performance capacitive deionization. *Adv. Sci.* 10 (2): 2204041.

51 Li, K., Wang, X., Li, S. et al. (2020). An ultrafast conducting polymer@MXene positive electrode with high volumetric capacitance for advanced asymmetric supercapacitors. *Small* 16 (4): 1906851.

52 Zhang, P., Peng, Y., Zhu, Q. et al. 3D foam-based MXene architectures: structural and electrolytic engineering for advanced potassium-ion storage. *Energy Environ. Mater.* https://doi.org/10.1002/eem2.12379.

53 Wang, K., Zhou, Y., Xu, W. et al. (2016). Fabrication and thermal stability of two-dimensional carbide Ti_3C_2 nanosheets. *Ceram. Int.* 42 (7): 8419–8424.

54 Ran, J., Gao, G., Li, F.-T. et al. (2017). Ti_3C_2 MXene co-catalyst on metal sulfide photo-absorbers for enhanced visible-light photocatalytic hydrogen production. *Nat. Commun.* 8 (1): 1–10.

55 Khan, A.A., Tahir, M., and Zakaria, Z.Y. (2021). Synergistic effect of anatase/rutile TiO_2 with exfoliated $Ti_3C_2T_R$ MXene multilayers composite for enhanced CO_2 photoreduction via dry and bi-reforming of methane under UV–visible light. *J. Environ. Chem. Eng.* 9 (3): 105244.

56 Zuo, G., Wang, Y., Teo, W.L. et al. (2020). Ultrathin $ZnIn_2S_4$ nanosheets anchored on $Ti_3C_2T_X$ MXene for photocatalytic H_2 evolution. *Angew. Chem.* 132 (28): 11383–11388.

57 Khan, A.A., Tahir, M., and Bafaqeer, A. (2020). Constructing a stable 2D layered Ti_3C_2 MXene cocatalyst-assisted $TiO_2/g\text{-}C_3N_4/Ti_3C_2$ heterojunction for tailoring photocatalytic bireforming of methane under visible light. *Energy Fuels* 34 (8): 9810–9828.

58 Low, J., Zhang, L., Tong, T. et al. (2018). TiO_2/MXene Ti_3C_2 composite with excellent photocatalytic CO_2 reduction activity. *J. Catal.* 361: 255–266.

59 Zhu, J., Tang, Y., Yang, C. et al. (2016). Composites of TiO_2 nanoparticles deposited on Ti_3C_2 MXene nanosheets with enhanced electrochemical performance. *J. Electrochem. Soc.* 163 (5): A785–A791.

60 Hu, T., Wang, J., Zhang, H. et al. (2015). Vibrational properties of Ti_3C_2 and $Ti_3C_2T_2$ (T = O, F, OH) monosheets by first-principles calculations: a comparative study. *Phys. Chem. Chem. Phys.* 17 (15): 9997–10003.

61 Huang, L., Ai, L., Wang, M. et al. (2019). Hierarchical MoS_2 nanosheets integrated Ti_3C_2 MXenes for electrocatalytic hydrogen evolution. *Int. J. Hydrogen Energy* 44 (2): 965–976.

62 Rakhi, R.B., Ahmed, B., Hedhili, M.N. et al. (2015). Effect of postetch annealing gas composition on the structural and electrochemical properties of Ti_2CT_x MXene electrodes for supercapacitor applications. *Chem. Mater.* 27 (15): 5314–5323.

63 Li, X., Yin, X., Han, M. et al. (2017). Ti_3C_2 MXenes modified with in situ grown carbon nanotubes for enhanced electromagnetic wave absorption properties. *J. Mater. Chem. C.* 5 (16): 4068–4074.

4

Synthesis and Characterization of TiC MXene-Based Composites for Energy Storage and Conversion

Azmat A. Khan[1], *Muhammad Tahir*[2], *Areen Sherryna*[3], *Muhammad Madi*[3], *Abdelmoumin Y. Zerga*[3], *Nazish Khan*[4], *and Naveen Kumar*[5]

[1]*Baluchistan University of Information Technology, Engineering and Management Sciences (BUITEMS), Department of Chemical Engineering, Airport Road, Beleli, P.O. Box 87100, Quetta, Pakistan*
[2]*UAE University, Chemical and Petroleum Engineering Department, P.O. Box 15551, Sheik Khalifa Bin Zayed Street, Asharej, Al Ain, United Arab Emirates*
[3]*Universiti Teknologi Malaysia, School of Chemical and Energy Engineering, Faculty of Engineering, Skudai, P. O. BOX 81310, Johor Bahru, Johor, Malaysia*
[4]*The University of Agriculture Swat, Department of Plant Breeding and Genetics, Takhta Band, Mingora, P.O. Box 19040, Swat, Pakistan*
[5]*Maharshi Dayanand University, Department of Chemistry, Delhi road, Near Delhi Bypass, Rohtak 124001, Haryana, India*

4.1 Introduction

Owing to outstanding optoelectronic properties, low cost, and environmental friendliness, 2D ternary transition metal carbide materials, named as "MXenes," have ignited significant research in the fields of energy storage and conversion. Various studies showed that MXenes have the capability to work as competent cocatalysts in photocatalysis for energy storage and conversion and could probably be a feasible option for noble-metal free catalysts [1, 2]. MXenes possess abundant interesting electrical, mechanical, optical, electrochemical, and photochemical properties owing to their distinct chemistry [3]. MXenes shows high flexibility in coordination with their two-dimensional (2D) shape and layered structure that allows it to form composites with other materials, extending a combination of the outstanding qualities of the conjugate materials. Therefore, MXenes and MXene-based composites have ignited enormous research concern, as they present excellent capability for a range of applications, comprising energy storage and conversion [4, 5].

In the modern world, constructing composites is an appealing approach to build stable and useful materials. Owing to the 2D layered structures, MXenes are deemed as an exceptional material for the fabrication of multipurpose composites, which has prompted a rise in the research of MXene-based composites. The most extensively utilized techniques for the fabrication of MXene-based materials include

Titanium Carbide MXenes: Synthesis, Characterization, Energy, and Environmental Applications, First Edition. Edited by Muhammad Tahir.

mechanical/ultrasonic mixing, hydrothermal/solvothermal treatment, electrostatic self-assembly, and calcination processes. TiC MXenes have been combined with other materials to form MXene-based hybrids that achieved unique enhancement of material properties employed for energy storage and conversion.

The TiC MXene material possesses high electronic conductivity, outstanding mechanical flexibility, and adaptable surface chemistry and therefore is considered perfect 2D substrates for energy storage and conversion applications. TiC MXene as a matrix facilitates fast electron and ion transport and provides sufficient active sites for low-dimensional nanomaterials loading/bonding and inhibits aggregation during the electrochemical or photochemical reaction process. In recent years, several studies have been conducted to combine the MXene substrates with different materials such as TiO_2, g-C_3N_4 layered double hydroxides (LDHs), transition metal dichalcogenides (TMDs), metal-organic frameworks (MOFs), and graphitic carbon nitrides.

Herein, synthesis procedures to fabricate TiC MXene-based composites are summarized. The enhancement in properties and morphological improvement to obtain the hybrid materials of TiC MXene-based hybrids is also discussed. The characterization techniques involved to study the properties of the TiC-based composite materials are discussed. The enhancement obtained is revealed through these characterization techniques and is also discussed. Finally, the conclusion of the chapter is put forward.

4.2 Synthesis of TiC-Based Composites

TiC MXenes have been coupled with other materials for the construction of hybrids that have been used for energy storage and conversion. Outstandingly, the key role of 2D TiC MXenes has been of an excellent material that provides a platform for the homogeneous growth of conjugate materials, excellent dispersion and preventing agglomeration, holds extraordinary chemical and thermal stability, plentiful surface-terminating groups (–O, –OH, and –F), and having a large specific surface area. The appropriate role of TiC MXene as a viable cocatalyst is due to its unique 2D layered structure, exceptional electrical conductivity, superior carrier mobility, and the presence of several surface terminal sites with superb hydrophilicity. Nevertheless, sometimes, TiC MXene acts both as a support material and cocatalyst by offering huge exposed active sites and a strong interfacial contact with conjugate materials, resulting in short charge transfer distance effective for the photoinduced charge carrier separation for improved photocatalytic activity [6].

4.2.1 Synthesis and Processing of TiO_2/Ti_3C_2 MXene Composite

The synthesis of TiO_2/Ti_3C_2 MXene composite can be broadly divided into two approaches. First, through the mixing of TiO_2 with TiC MXene and second through in situ conversion of TiC MXene to TiO_2 through oxidation.

The first approach can be obtained through simple mixing or electrostatic interaction. For instance, Ye et al. [7] used Ti_3C_2 synthesized P25/Ti_3C_2 by simply mixing P25 with Ti_3C_2 hybrids, resulting in exceptional electrical conductivity, charge carrier separation, high CO_2 adsorption, and activation sites of surface-alkalinized Ti_3C_2 MXene.

Zhuang et al. [8] employed the electrostatic self-assembly method for the fabrication of 1D/2DTiO_2/Ti_3C_2 composite. Owing to the effective electrostatic action due to surface charges present on the surface the interfaces were increased with more load. The fabrication of the composite was done using the electrostatic self-assembly method, charge separation, and abundant active sites. In the electrostatic method, a quantity of TiO_2 nanofibers was added to distilled water and sonicated. Subsequently, $Ti_3C_2T_x$ MXene was added to distilled water and sonicated. Later, the TiO_2 solution was added to the $Ti_3C_2T_x$ solution and stirred. Finally, the obtained TiO_2/Ti_3C_2 powder was washed with distilled water three times and dried. The complete fabrication process of TiO_2/Ti_3C_2 is demonstrated in Figure 4.1a.

The second approach of in situ growth of TiO_2 through treatment of TiC MXene has been widely studied. Different techniques have been employed to achieve the TiO_2/TiC composites through in situ growth. Contrary to mechanical mixing and self-assembly, the in situ growth method comprised fabrication of a material directly over the surface of the MXenes, resulting in in situ fabrication of materials over the surface of MXenes, leading to a chemically bonded structure that has a significant advantage over other methods. By employing experimental techniques, the in situ growth of various partially oxidized MXenes at the interface of TiC (TiO_2/TiC), with superior charge carrier generation, transportation, and separation was carried out. In this way, numerous TiO_2 nanostructures including, nanosheets, nanoparticles, nanoflowers, nanowires, nanotubes, nanofibers, nanospindles, nanocrystals, nanodots, and truncated octahedral bipyramidals were derived on TiC MXenes [9]. The synthesis approaches that have been employed to synthesize in situ grown TiO_2/TiC from TiC MXene include hydrothermal treatment, solvothermal treatment, calcination, annealing, microwave-assisted treatment, and controlled etching.

In the hydrothermal treatment method, heating is done usually in the presence of water in a Teflon-lined autoclave as demonstrated in Figure 4.1b [10]. In the solvothermal technique, the heating is done in the presence of a solution such as dimethylformamide (DMF) [11]. In the annealing process, a certain amount of the as-prepared TiC MXene was added to a quartz boat and annealed in a tube furnace under a certain gas environment such as air and N_2, as demonstrated in Figure 4.1c [12]. In the calcination process, the as-prepared sample of TiC MXene is calcined under an atmosphere of air or any gas at a heating rate with a certain flow rate and then naturally cooled to get the fabricated TiO_2/TiC MXene [13]. A typical schematic is presented to show the steps in the calcination process in Figure 4.1d. In microwave-assisted synthesis, the Ti_3C_2/TiO_2 heterostructure was synthesized by a one-step microwave oxidation process in a microwave at certain power. The microwave-assisted synthesis may enhance the synergistic effect of the Ti_3C_2/TiO_2 nanocomposite structure by providing a large specific surface area and a good TiO_2 absorption mechanism [14].

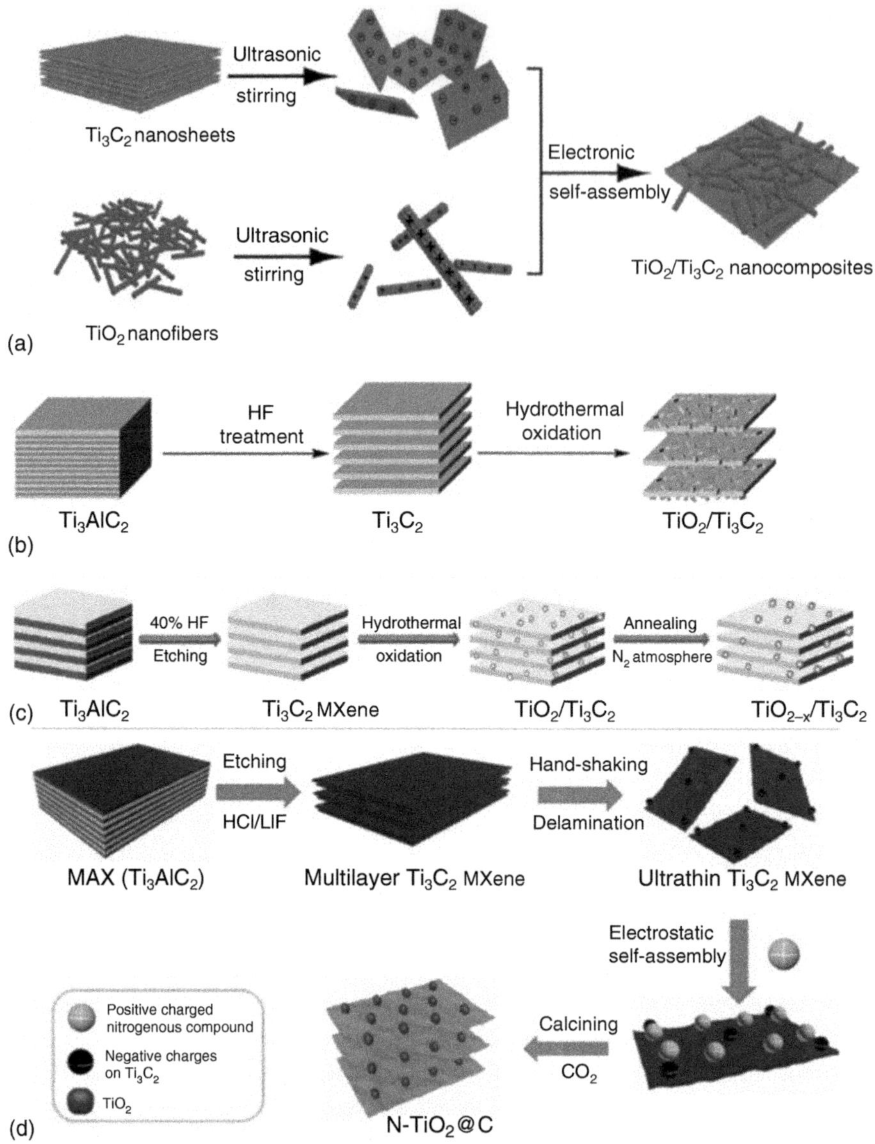

Figure 4.1 Synthesis of TiO_2/Ti_3C_2 composites through (a) electrostatic assembly, (b) hydrothermal treatment, (c) annealing, and (d) calcination.

4.2.2 Synthesis of TiC/g-C_3N_4 Composite

Many synthesis approaches have been proposed recently for gC_3N_4 or MXene with different morphologies. With assembly strategies for gC_3N_4 and TiC MXene or MXene-derived nanostructures, the most important synthesis approaches are electrostatic self-assembly, in situ thermopolymerization/calcination, solution mixing, and chemical vapor deposition [6, 15]. The g-C_3N_4 and MXene materials possess

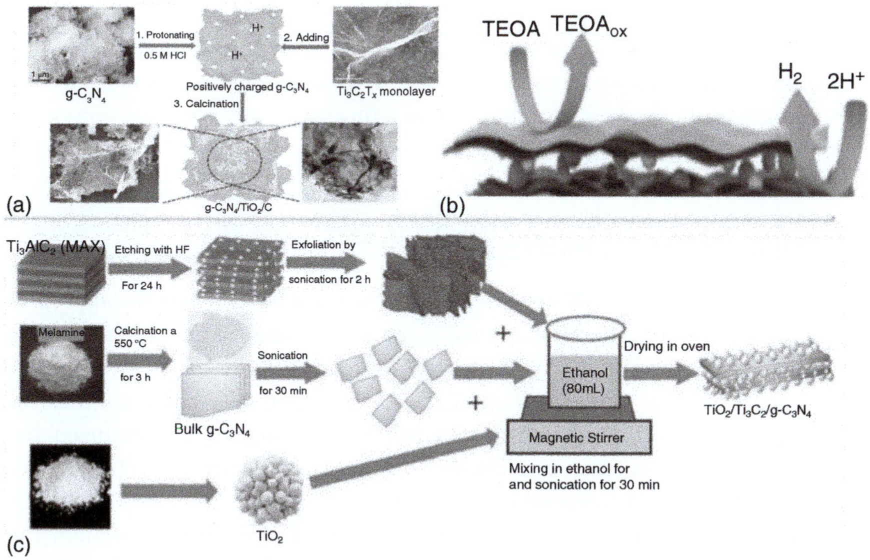

Figure 4.2 Synthesis of $g\text{-}C_3N_4/Ti_3C_2$ by (a) electrostatic assembly. Source: Yang et al. [17]/John Wiley & Sons, Inc. (b) In situ thermopolymerization/calcination. Source: Zhang et al. [19]. (c) Simple mixing method. Source: Khan et al. [20]/with permission of American Chemical Society.

opposite charges and therefore have the possibility of electrostatic attraction. The hybrid is obtained by the electrostatic interaction between opposite charges present on TiC MXene and $g\text{-}C_3N_4$ in an aqueous solution [16]. The schematic illustration of electrostatic assembly is demonstrated in Figure 4.2a [17].

In the method of in situ thermopolymerization/calcination, TiC MXene is pre-synthesized by an etching process, whereas urea or melamine molecules are added to TiC MXene. Since the surface of TiC MXene has the presence of terminal groups (i.e. –F, –O, and –OH), their robust interaction allows urea or melamine molecules to be connected strongly to the surface of TiC MXene. The inclusion of $g\text{-}C_3N_4$ with TiC MXene or TiC MXene-derived support is achieved by successive calcination under air or atmosphere [18]. The schematic illustration is demonstrated in Figure 4.2b [19]. TiC MXenes are very sensitive to high temperature and are easily oxidized, resulting in the formation of TiO_2, reducing the excellent conductivity. Therefore, the selection of calcination temperature is a crucial parameter to determine the stability of TiC MXenes. Therefore, the protection of mono- and few-layered MXenes from oxidation is important by employing mild conditions.

Solution mixing is one of the most employed methods to assemble TiC MXene with $g\text{-}C_3N_4$ possessing morphological features. In this approach, $g\text{-}C_3N_4$ and TiC MXene powders are usually pre-synthesized. The obtained materials of the $g\text{-}C_3N_4$ and TiC MXenes are then mixed in water or an organic solvent solution. The blending of the $g\text{-}C_3N_4$/TiC MXenes composites is obtained by evaporation of water or the organic solvent solution, as demonstrated in Figure 4.2c [20].

In the method of chemical vapor deposition (CVD), a substrate is exposed to volatile precursors that react and/or decompose on the surface of the substrate to fabricate the anticipated film. For instance, Diao et al. [21] fabricated a ternary g-C_3N_4/Ti_3C_2/TiO_2 through the CVD method by employing a quartz boat comprising the mixture that was annealed at 550 °C for three hours in a tube furnace under the atmosphere of N_2 to improve the adhesion of the obtained composite material.

4.2.3 General Synthesis and Processing of LDH/TiC MXene Composites

The fabrication of LDH/TiC MXene nanocomposite can be achieved through three promising techniques. In the first approach, the TiC MXene is first synthesized and LDH is dispersed on its surface. In the second approach, LDH is in situ grown in the presence of TiC MXene. In the third approach, LDH and TiC MXene are synthesized and spread like in the first approach; however, the TiC MXene with reduced size is then spread on the surface of LDH. For example, to acquire LDH/TiC MXene composites with LDH distributed on the surface of TiC MXene, Chen et al. [22] fabricated CoCo-LDH/$Ti_3C_2T_x$ by sonicating $Ti_3C_2T_x$ (TNS) sonicated in methanol and subsequent addition of $Co(NO_3)_2{\cdot}6H_2O$ and stirring. Then, 2-methylimidazole (2-MIM) was added to methanol, stirred, and sonicated to obtain $Co(NO_3)_2$/TNS. Subsequently, the solution was added to a Teflon-lined autoclave and heated at the temperature of 120 °C for one hour through hydrothermal treatment. Finally, centrifugation of the solution was done to obtain a dark green precipitate that was washed with water and methanol several times and vacuum-dried. In the in situ approach, Zhao et al. [23] obtained NiAl-LDH/$Ti_3C_2T_x$ nanosheet (NiAl-LDH/TNS) obtained through the hydrothermal technique as shown in Figure 4.3a. First, a specific quantity of $Ti_3C_2T_x$ MXene nanosheets was dispersed through ultrasonication in deionized (DI) water. Subsequently, specific quantities of $Ni(NO_3)_2{\cdot}6H_2O$, $Al(NO_3)_3{\cdot}9H_2O$, urea, and NH_4F were added to the solution and dissolved through stirring, followed by hydrothermal treatment in a Teflon-lined autoclave. The precipitate obtained after hydrothermal treatment was cleansed with DI water and ethanol several times to normalize the solution. Similarly, Jo et al. [24] fabricated nitrogen-doped carbon dot/CoAl-LDH/graphitic carbon nitride (NCD/LDH/CN) hybrid heterojunctions formed through a hydrothermal treatment. Certain quantities of the as-prepared CN and NCD powders were dispersed through the ultrasonication solution containing $Co(NO_3)_2{\cdot}6H_2O$ and $Al(NO_3)_3{\cdot}9H_2O$. Then, specific quantities of NH_4F and urea were added to the solution and stirred. The mixture was added to a Teflon-lined-stainless steel autoclave for thermal treatment. Finally, the obtained precipitate was washed with water and ethanol and dried.

Furthermore, Khan et al. [25] obtained $Ti_3C_2T_{A/R}$ through etching with HF that was then normalized by treating with DI water and methanol. The obtained $Ti_3C_2T_{A/R}$ MXene was stirred in a methanol solution. Concurrently, CoAlLa-LDH was obtained through coprecipitation and hydrothermal treatment method. The obtained precipitate after normalization was added to the methanol solution. The solution comprising CoAlLa-LDH was added to the solution of $Ti_3C_2T_{A/R}$, which was additionally stirred for one hour to achieve appropriate blending through the self-assembly technique. Finally, the achieved solution was dried in an oven to

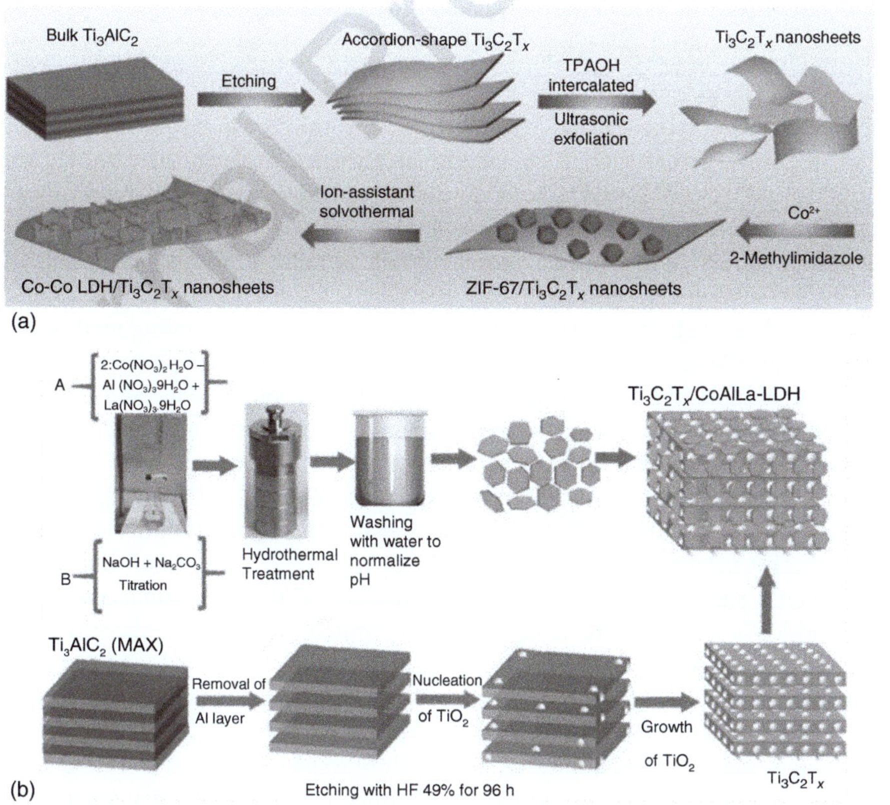

Figure 4.3 (a) Synthesis of CoCo-LDH/TNS. Source: Zhao et al. [23]/with permission of American Chemical Society. (b) Synthesis mechanism of $Ti_3C_2T_{A/R}$/CoAlLa-LDH heterostructure. Source: Ali Khan et al. [25]/permission from American Chemical Society.

achieve the $Ti_3C_2T_{A/R}$/CoAlLa-LDH heterostructure. The detailed mechanism of preparation is demonstrated in Figure 4.3b.

We attained the LDH/TiC MXene composites through the combination of the second and third techniques while the TiC MXene is loaded on the surface of LDH. For instance, Shi et al. [26] in situ loaded Ti_3C_2 MXene on NiAl-LDH to get Ti_3C_2/NiAl-LDH nanocomposite. In particular, the attained Ti_3C_2 MXene was distributed in dimethyl sulfoxide (DMSO) and stirred and ultrasonicated in DI water. The suspension was then centrifuged for the collection of supernatants. Successively, the dispersed Ti_3C_2 MXene was mixed with $Ni(NO_3)_2{\cdot}6H_2O$, $Al(NO_3)_3{\cdot}9H_2O$, urea, and NH_4F. The obtained mixture was then ultrasonicated and strongly stirred to obtain distribution. The obtained solution was then added to a Teflon-lined autoclave for hydrothermal treatment. During crystallization, the Ti_3C_2 was placed on the scaffold of the NiAl-LDH material. The achieved photocatalyst sample was cleaned with DI water and methanol for pH normalization. In this manner, NiAl-LDH/Ti_3C_2 nanohybrids with different quantity of Ti_3C_2 MXene dispersions were achieved.

4.2.4 Synthesis of Ti_3C_2 MXene-Based Perovskite Composites

The coupling of perovskite with TiC MXene has found widespread application in the field of energy storage and conversion. A well-defined perovskite with TiC MXene heterostructure has improved electron transfer more pervasive than an unordered structure such as scattered nanoparticles. Majorly, the coupling of perovskite with TiC MXene has been done through the in situ thermal treatment approach.

For instance, 2D Ti_3C_2 MXene nanosheets with metallic features synthesized via etching and subsequent exfoliation processes were anchored on $MAPbI_3$ perovskite in aqueous hydroiodic acid (HI) solution based on the in situ coupling method to fabricate stable MXene/$MAPbI_3$-HI. The obtained composite rapid photogenerated carrier separation/transport occurred at the interface of $MAPbI_3$ and MXene nanosheets and higher hydrogen evolution capacity in the two-layered MXene nanosheets with abundant active sites [27].

Similarly, formamidinium lead bromide quantum dots on Ti_3C_2 heterojunction nanosheets (0D/2D $FAPbBr_3/Ti_3C_2$) were synthesized through the in situ technique. To synthesize the $FAPbBr_3$ quantum dot/Ti_3C_2 nanosheet composites, the Ti_3C_2 monolayered nanosheets were added to the mixture of $PbBr_2$, 1-octadecene, and oleic acid (OA). The $FAPbBr_3/Ti_3C_2$ were then purified to remove the excessive ligands and dispersed into the hexane solvent [28].

Furthermore, $Ti_3C_2@SrTiO_3$ composites were prepared by in situ hydrothermal treatment. Ti_3C_2 nanosheets and $Sr(OH)_2$ were dissolved in DI water in a Teflon-lined stainless steel autoclave. The mixture was hydrotreated to 160 °C and kept for 24 hours in an oven. The products obtained were filtered and washed with DI and ethanol several times and freeze-dried for 24 hours [29].

Additionally, the MXene/$MAPbI_3$ hybrids were also formed through a straightforward HI aqueous solution immersion through the in situ coupling method. Initially, the 2D layered MXene nanosheets were added to the $MAPbI_3$-saturated aqueous HI solution to achieve the MXene–HI dispersion. The MXene–HI dispersion and $MAPbI_3$ were added into the $MAPbI_3$-saturated aqueous HI solution, respectively. After that, the solutions of MXene/$MAPbI_3$ were degassed with Ar to remove air, thermally treated to completely dissolve the $MAPbI_3$, and re-precipitated upon cooling. During the precipitation–crystallization cycle of $MAPbI_3$ in the HI solution, the MXene nanosheets were in situ anchored on the surface of $MAPbI_3$ to form the MXene/$MAPbI_3$-HI system. The obtained precipitate was washed and centrifugated with ethyl acetate and diethyl ether three times respectively, followed by drying.

Similarly, $CsPbBr_3$/MXene nanohybrids were synthesized by growing $CsPbBr_3$ nanocrystals on MXene nanosheets by employing a modified in situ growth method. It was hypothesized that the surface terminal groups (i.e. –F, –O, and –OH) on the surface of MXene interact with oleic acid (OA), and oleylamine (OLA), by forming kinetically favorable nucleation points for capturing $PbBr_2$ precursors. The expected nanocrystals showed growth at the nucleation sites while consuming the cesium and $PbBr_2$ precursors in the solution [30].

4.2.5 Synthesis of MXene-Based MOF Composite

The TiC MXene-based MOF composite can be prepared through two major approaches, either through physical mixing of MOF and TiC MXene or through growth of the MOF particles on the TiC MXene surface. Also, MOFs can be used as sacrificial templates to synthesize different materials such as carbon materials and LDHs, which are then distributed on the TiC MXene surface. Various synthesis approaches have been employed that are discussed, whereas the overview of the approaches is summarized as shown in Figure 4.4.

One of the more common approaches to prepare the Ti_3C_2 MXene-based MOF composite is physical mixing of TiC MXene and MOF. The MOFs and TiC MXenes are bound and form contact with each other. One of the promising approaches is to obtain physical mixing through ultrasonication to successfully fabricate MXene/MOF composites. As in ultrasonication, the TiC MXene nanosheets are exfoliated from multi-layer to single-layer materials that are then spread on MOF materials [31]. For example, Qin et al. obtained TiC MXene-based MOF composite (Ti_3C_2-QD/Ni-MOF) by employing a facile ultrasonic approach as illustrated in Figure 4.5a. First, Ti_3C_2 quantum dots (QDs) were synthesized through etching of MAX phase to Ti_3C_2 multilayers, followed by hydrothermal treatment to obtain the Ti_3C_2 QDs. Then, the as-prepared Ni-based MOF was amalgamated with Ti_3C_2 QDs under continuous ultrasonication for four hours. Uniform dispersion of QDs on the surface of 2D Ni-MOF was achieved as demonstrated in the HRTEM images in Figure 4.5b,c. The defects may also be noted on the Ti_3C_2 surface that

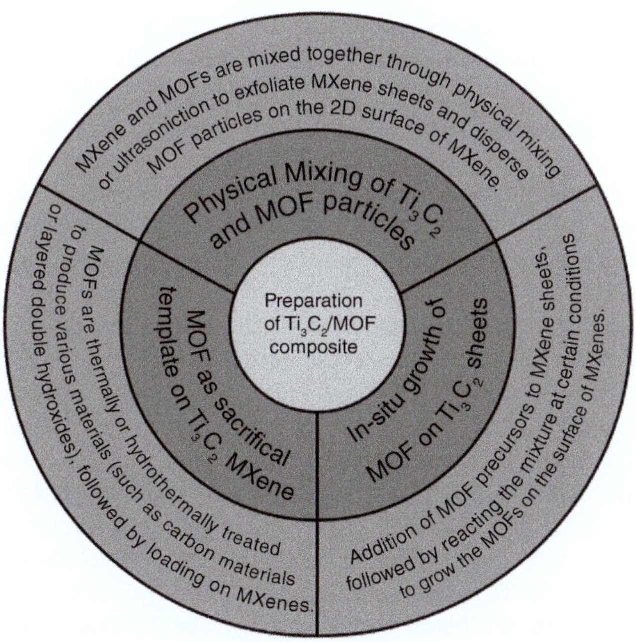

Figure 4.4 Overview of approaches to prepare Ti_3C_2 MXene-based MOF composites.

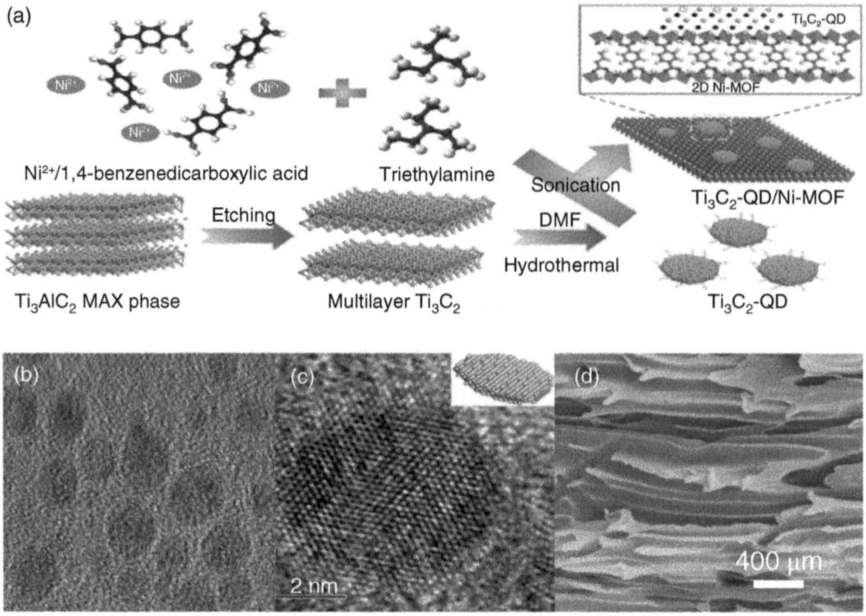

Figure 4.5 (a) Illustration of Ti_3C_2-QD/Ni-MOF synthesis via the ultrasonication approach, (b, c) HRTEM images of Ti_3C_2-QD/Ni-MOF. Source: Qin et al. [32]/with permission of American Chemical Society, and (d) FE-SEM image of layered Ti_3C_2/NiCo-MOF. Source: Liu et al. [33]/MDPI/CC BY 4.0.

provide unsaturated coordination sites for catalytic reactions [32]. Also, Liu et al. implemented the sonication approach to fabricate the MXene/MOF composite. First, NiCo-MOF nanosheets were prepared through the precipitation approach to obtain the MOF. The MOF powders attained were then distributed in a Ti_3C_2 aqueous solution under sonication. The ultrasonication encourages the intercalation of 2D structure MOF and MXene due to their 2D structures as demonstrated in the FESEM image in Figure 4.5d. The formation of interconnected porous networks between the MXene/MOF composite impedes the self-restacking of the and Ti_3C_2 sheets and NiCo MOF [33].

Another approach to synthesize the MOF-based TiC MXene composite is through in situ growth of MOF on the surface of MXenes. Growing MOF particles on the TiC MXenes results in more intimate contact between the two materials. For instance, Zheng et al. anchored a 3D pillared Ni-MOF 2D layered Ti_3C_2 MXene surface. The thiophene-2,5-dicarboxylate (Tdc) and 4,4′-bipyridine (Bpy) linkers were first incorporated into the Ti_3C_2 sheets, followed by the addition of $NiCl_2{\cdot}6H_2O$. The role of the linker's addition to the MXene was for binding organic ligands to the MXene interlayer functional groups. Finally, the adding of Ni^{II} coordinated ligands was conducted over the Ti_3C_2 surface to fabricate the Ti_3C_2@Ni-MOF composite. The effective growth of Ni-MOF was allowed on the 2D Ti_3C_2 sheet surface. The growth of Ni-MOF on MXenes prevented the oxidation of MXenes due to the interlayer functional groups that reduced the exposure of surface atoms on

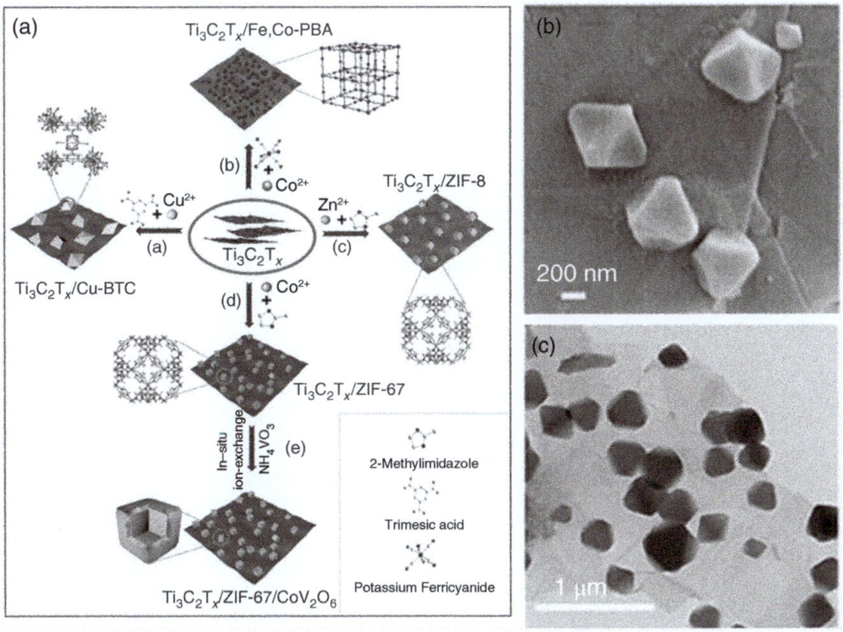

Figure 4.6 (a) Schematic of the synthesis of $Ti_3C_2T_x$/MOF. (b) SEM image of $Ti_3C_2T_x$/MOF. (c) TEM image of $Ti_3C_2T_x$/MOF. Source: Liu et al. [35]/John Wiley & Sons, Inc.

MXenes [34]. Similarly, Liu et al. [35] carried out in situ growth of MOF particles on $Ti_3C_2T_x$ nanosheets, as demonstrated in Figure 4.6a. Through SEM analysis and figure. The 3D MOFs played an important role in avoiding the accumulation of MXene nanosheets and promoted the transport of electrolytes. The SEM image in Figure 4.6b and TEM image in Figure 4.6c demonstrated that MXene/MOF comprised MXene nanosheets and MOF particles. Another example of the in situ growth is through dual heterojunction NH_2-MIL-125 modulation with the Ti_3C_2 hybrid obtained by employing the in situ construction approach. The Ti_3C_2 nanosheets were added dropwise into the NH_2-MIL-125 precursor solution, trailed by following hydrothermal treatment. The TiO_2 is formed in the process due to the hydrolysis of unreacted tetrabutyltitanate, The type II heterojunction was fabricated between NH_2-MIL-125 and TiO_2 forming the Schottky junction with Ti_3C_2 acting as the cocatalyst to trap photocatalytic generated electrons [36].

The MXene-based MOF composites can also be synthesized by employing MOFs as sacrificial agents. For instance, MOFs can be employed to derive various carbon materials that are employed for electrochemical applications for energy storage and conversion [37]. N-doped porous carbon (N-PC) was synthesized through carbonization of Zn-based MOF-5-NH_2 in an Ar atmosphere. Then, the N-PC material was added to the Ti_3C_2 MXene through amalgamation and ultrasonic treatment, resulting in the formation of MOF-derived (Ti_3C_2/N-PC) composite. From the images of the composite, it is noted that the Ti_3C_2 MXene interlamellar spacing was enlarged, owing to the effective incorporation of N-PC between the

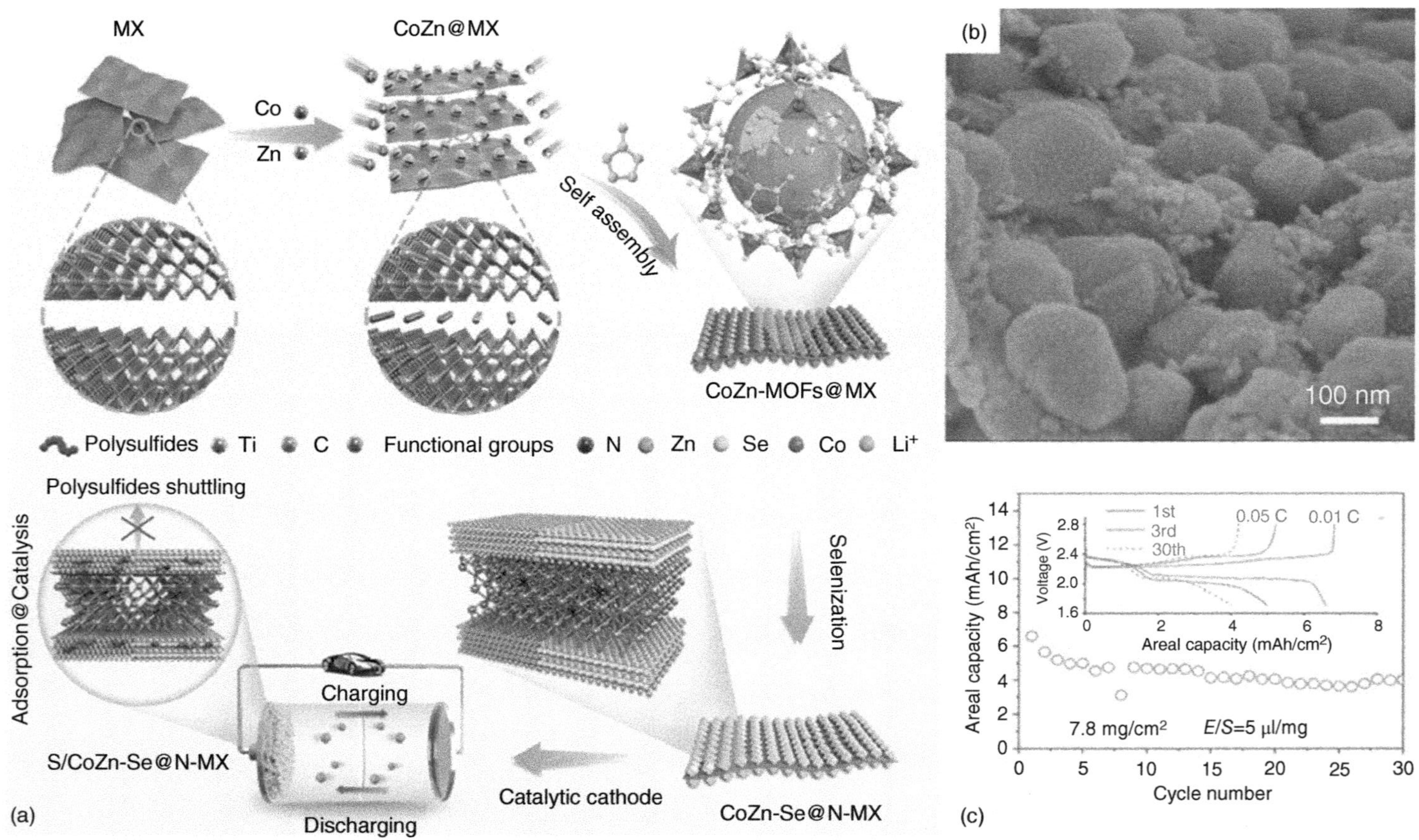

Figure 4.7 (a) Schematic illustration of the synthesis of CoZn-Se@N-MXene, and Li–S battery configuration applying the catalytic cathode. (b) SEM image of CoZn-Se@N-MXene. (c) Cyclic activities of the S/CoZn-Se@N-MXene cathode at a high sulfur loading and less electrolyte process (inset: corresponding galvanostatic discharge–charge profiles). Source: Ye et al. [39]/with permission of John Wiley & Sons.

gaps of the sheets [22, 38]. Similarly, 0D bimetallic selenides on 2D nitrogen-doped MXene (CoZn-Se@N-MX) were obtained MOFs derived using the in situ selenization technique as demonstrated in Figure 4.7a. The MXene nanosheets served as the catalysis substrate and enabled rapid electron and ion transport and prevented the aggregation of the MOF-derived selenides during the LiPS conversion reaction. The MOF-derived selenides acted as a spacer for the isolation of the MX nanosheets and prevented restacking, that preserved the active areas employed for polysulfide electrocatalysis. Significantly, the MOF-derived selenides effectively accelerated polysulfide redox reaction and enhanced sulfur utilization. The obtained CoZn-Se@N-MXene heterostructure possessing hierarchical porous architecture and greater active area as demonstrated in Figure 4.7b enabled fast Li^+ diffusion and lowered the energy barriers for Li_2S precipitation and dissolution, leading to improved LiPS conversion kinetics as demonstrated in Figure 4.7c [39].

4.3 Characterization of Ti_3C_2-Based Composites

4.3.1 Characterization of TiO_2/TiC MXene

From the XRD patterns, the conversion of Ti_3AlC_2 into Ti_3C_2 MXene nanosheets can be verified. The strongest diffraction peak of Ti_3AlC_2 at 39 disappears the (002) at 9.6° and (004) at 19.2° diffraction peaks shifts to lower degrees that shows the transformation from Ti_3AlC_2 to Ti_3C_2. In addition, the conversion of Ti_3C_2 to Ti_3C_2/TiO_2 nanosheets is demonstrated in Figure 4.8a. It is inferred that by augmenting the HF concentration, increase of TiO_2 (anatase) was observed, whereas the extension etching time caused the increase in the quantity of TiO_2 (rutile). It is obvious that with the increase in etching time and concentration, increase in the intensity of anatase and rutile phases of TiO_2 was observed, demonstrating the oxidation degree of the Ti_3C_2 and the formation of TiO_2 [40]. Figure 4.8b demonstrates the RAMAN spectra of Ti_3AlC_2 and obtained $TiO_2@Ti_3C_2$ that have been etched for 24 and 96 hours with HF having a concentration of 39% and 49%. The Ti_3AlC_2 demonstrated bands 148.5 cm^{-1} (E_g), 267.5 cm^{-1} (B_{1g}), 414 cm^{-1}, and 614.5 cm^{-1} (E_g) ascribed to the Raman-active phonon vibration modes of ω_1, ω_2, ω_3, and ω_4, respectively. For Ti_3C_2 24H@39 sample etched the Ti_3AlC_2 for 24 hours with 39% HF, the bands were located at 140, 211, 431, and 613 cm^{-1} ascribed to the ω_1, ω_2, ω_3, and ω_4, respectively. The ω_3 Raman band modes of Ti_3C_2 are blue-shifted due to the delamination of Ti_3AlC_2. Also, the occurrence of TiO_2 rutile is detected in the Ti_3C_2 24 H 39 with peaks at 443 cm^{-1}, which demonstrate the occurrence of TiO_2 rutile phase. In Ti_3C_2 24 H 49, an intense peak at 156 cm^{-1} occurred showing the TiO_2 anatase phase, the rutile TiO_2 occurred at 443 cm^{-1}, showing the creation of anatase and rutile phase TiO_2. In a similar way, the blue shift was observed for 423 cm^{-1}, demonstrating the layered structure of Ti_3C_2. For Ti_3C_2 96 H 39, five peaks occurred at 154, 256, 440, 467, and 613 cm^{-1}. The mild peak at 154 cm^{-1}, confirms the presence of the TiO_2 anatase phase and the peak at 440 cm^{-1} demonstrated rutile, whereas the other three peaks showed the occurrence of

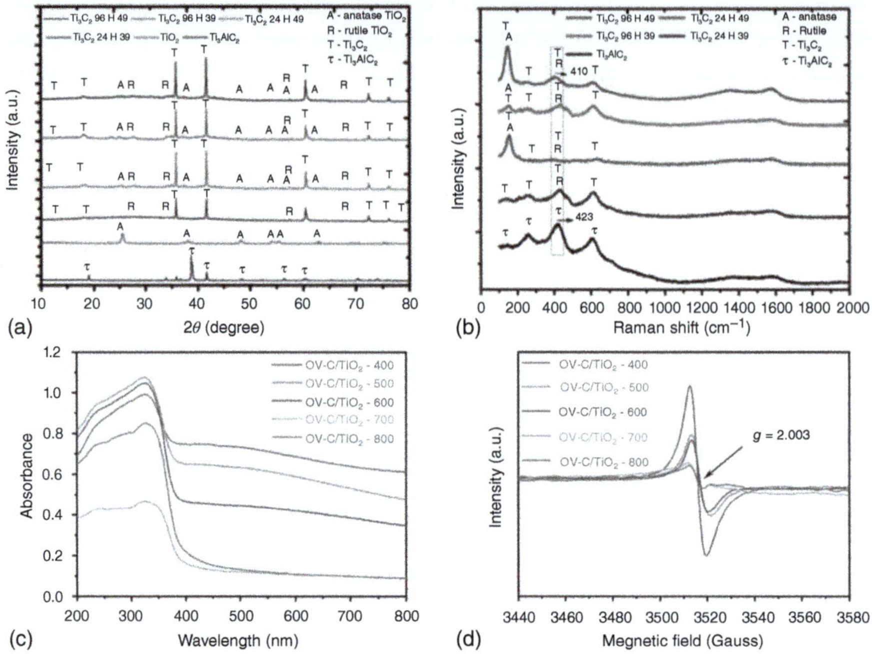

Figure 4.8 (a) XRD spectra of (a) TiO_2, Ti_3AlC_2, and TiO_2@Ti_3C_2 samples. (b) RAMAN spectra of TiO_2, Ti_3AlC_2, and TiO_2@Ti_3C_2 samples. Source: (a, b) Khan et al. [40]/with permission of Elsevier. (c) UV–vis DRS spectra of Ti_3C_2 samples annealed at 400–800 °C. (d) ESR spectra of OV-C/TiO_2 derived by Ti_3C_2 MXenes annealed at 400–800 °C. Source: Qian et al. [42]/with permission of John Wiley & Sons.

Ti_3C_2. For Ti_3C_2 96 H 49, the increase in the quantity of both anatase and rutile phases confirmed by high intensities of anatase and rutile phases was observed. The Ti_3C_2 96 H 49 had five peaks at 148, 203, 251, 410, 513, and 615 cm^{-1}. The peak at 148 cm^{-1} was due anatase TiO_2, the peak at 410 cm^{-1} was due to rutile, whereas the other peaks were related to Ti_3C_2. The shift in Raman frequency was due to the synergistic effect of rutile, anatase, and Ti_3C_2 [40].

The electronic band structure and optical properties of Ti_3C_2 and Ti_3C_2/TiO_2 are carried out through UV–vis diffuse reflectance spectra (UV–vis DRS). For instance, Li et al. [41] conducted UV–vis DRS spectra. The difference in optical properties of Ti_3C_2 MXene, Ti_3C_2–TiO_2 nanoflowers, and TiO_2 nanobelts can be ascertained easily. For pure TiO_2 nanobelts, a robust absorption edge shorter than 400 nm is attributed to the bandgap absorption of 3.2 eV, whereas the Ti_3C_2 MXene demonstrated no absorption of the UV region, but there is an enhanced visible absorption due to the dark color. The Ti_3C_2–TiO_2 nanoflowers exhibited a substantial enhancement in the absorption in the visible light region ranging from 400 to 800 nm. Besides, the light absorption of Ti_3C_2–TiO_2 nanoflowers increases with the increase of synthesizing temperature, namely, the decreased loading of Ti_3C_2. Qian et al. [42] obtained oxygen vacancy-rich C/TiO_2 (OV-C/TiO_2) synthesized through one-step calcination technique employing Ti_3C_2 MXene as a precursor and employed for the N_2 photoreduction. The prepared Ti_3C_2 MXene was annealed at different temperatures ranging from 400, 500, 600, 700, and 800 °C in air. The obtained samples were termed as OV-C/TiO_{2-X}, where X represented the annealing temperature as 400, 500, 600, 700, and 800. From UV–vis DRS as demonstrated in Figure 4.8c, all the catalysts exhibited extended absorption of light n in the area of 400–700 nm. The absorption threshold at about 400 nm was due to the intrinsic bandgap absorption of TiO_2,whereas the absorption in the range of 400–700 nm is ascribed to the presence of oxygen vacancies. Its decrease indicates reduction of the oxygen vacancy with the increase of annealing temperature. Also, the electron spin resonance (ESR) was employed to characterize the occurrence of oxygen vacancies in the TiO_2. The distinct responses near the g value of 2.003 in Figure 4.8d was due to the unpaired electrons of oxygen vacancies, demonstrating a decreasing trend with the increase of annealing temperature.

4.3.2 Characterization of g-C_3N_4/TiC MXene

One of the techniques employed to illustrate the structure of materials is X-ray diffraction that characterizes the crystal forms of as-prepared materials. The XRD patterns of g-C_3N_4, Ti_3C_2, and mixtures. For the g-C_3N_4 sample, the peaks at 13.1° and 27.5° are correlated to the (100) and (002) crystal planes [43]. The in-plane shape of g-C_3N_4 is for the mild peak at 13.1°, while the peak at 27.5° is attributed to the aromatic stacking [44]. The hybrids' of Ti_3C_2 and g-C_3N_4 have showed distinctive peaks showing the presence of both Ti_3C_2 and g-C_3N_4 in hybrids. The intensity of peaks related to g-C_3N_4 show increase in intensity with increasing g-C_3N_4 quantity. Moreover, the peak at 25° related to Ti_3C_2 is masked by the g-C_3N_4 (002) peak [45]. Notably, the (002) peaks of all CNT_x samples are reduced as compared to pure Ti_3C_2, signifying that the layer separation of Ti_3C_2 has showed enhancement owing to the addition of g-C_3N_4 [46]. Similarly, another research as demonstrated in Figure 4.9a demonstrates the XRD of prepared Ti_3C_2, g-C_3N_4 powder and g-C_3N_4/Ti_3C_2. It is observed that the pristine Ti_3C_2 film had a peak at 6.7°, typical for Ti_3C_2 structure having a 1.32 nm layer spacing. The peak for g-C_3N_4 was observed at 27.8° due to the g-C_3N_4 aromatic system interlayer stacking, attributed to (002) crystal plane. Theg-C_3N_4/Ti_3C_2had peaks of both g-$C_3N_4$and Ti_3C_2, Additionally, the (002) peak shifted from 6.7° for pure Ti_3C_2 to 5.9° for g-C_3N_4/Ti_3C_2 hybrid with the increase of g-C_3N_4 content from 0 to 10 wt%. This was due to increase in interlayer spacing from 1.31 to 1.52 nm, demonstrating that g-C_3N_4 nanosheets interleave MXene nanosheets layers [47]. The Fourier transform infrared spectroscopy (FTIR) has been employed to show the spectra of Ti_3C_2, g-C_3N_4, and CNT_x samples. In FTIR analysis, the peaks at 1625 and 3434 cm^{-1} are owing to the H_2O or –OH termination on the surface of Ti_3C_2 MXene [44, 48]. The peak at the 804 cm^{-1} sample is ascribed to the symmetric stretching of tris-triazine units in g-C_3N_4. Additionally, the peaks at 1241–1580 cm^{-1} are ascribed to the stretching vibration of C—N and C=N heterocycles. The band in the area of 3000–3500 cm^{-1} range is due to N–H stretching (–NH_2 or = NH) and surface-absorbed water or –OH termination [49, 50]. It is crucial to notice that the FTIR peaks of hybrids are identical to those of g-C_3N_4,

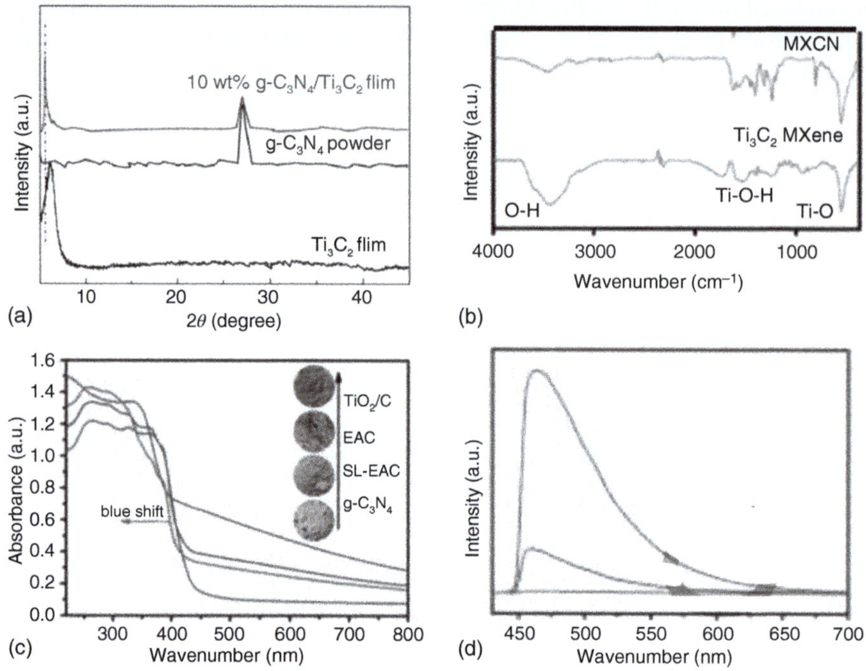

Figure 4.9 (a) XRD scans of Ti_3C_2, $g\text{-}C_3N_4$, and $g\text{-}C_3N_4/Ti_3C_2$. Source: Xu et al. [47]/with permission of John Wiley & Sons. (b) FTIR spectra of Ti_3C_2 and $g\text{-}C_3N_4/Ti_3C_2$ (MXCN). Source: Zhang et al. [51]/with permission of John Wiley & Sons. (c) UV–vis DRS of $g\text{-}C_3N_4$, TiO_2/C, $g\text{-}C_3N_4/TiO_2/C$ (EAC), and ultrathin $g\text{-}C_3N_4/TiO_2/C$ (SL-EAC). (d) PL spectra of Ti_3C_2, $g\text{-}C_3N_4$, and $g\text{-}C_3N_4/Ti_3C_2$. Source: (c, d) Yang et al. [17]/with permission of John Wiley & Sons.

whereas the peaks of CNT_x are roughly similar to those of $g\text{-}C_3N_4$. Furthermore, when the melamine quantity is increased, the increase in intensities is noticed. Similarly, in another work, FTIR analysis was done for the evaluation of composition and bonding of $g\text{-}C_3N_4$, $Ti_3C_2T_x$, and. $g\text{-}C_3N_4/Ti_3C_2T_x$ samples as demonstrated in Figure 4.9b. The Ti–O and Ti–O–H bending vibrations of $Ti_3C_2T_x$ are located at 555 and 1523 cm^{-1}, respectively. The $g\text{-}C_3N_4/Ti_3C_2T_x$ spectra show peak at 811 cm^{-1} ascribed to the triazine ring bending mode in the carbon nitride system. The thick bands near 960–1800 cm^{-1} are attributed to the C—N and C=N bond stretching vibrations. The wide adsorption band located at the 2896–3374 cm^{-1} area is credited to the N—H bonds stretching modes of primary and secondary amines in $g\text{-}C_3N_4$ [51].

From the UV–vis DRS, the absorption edge of $g\text{-}C_3N_4/Ti_3C_2$ is 515 nm, showing a clear shift to the visible area as compared to $g\text{-}C_3N_4$ nanosheets that showed the absorption edge at 450 nm. In addition, it is also observed that $g\text{-}C_3N_4/Ti_3C_2$ showed a high absorbance for visible light as compared to $g\text{-}C_3N_4$, enhancing the photothermal effect and the red shift of absorption band and increase of light absorbability resulting in promotion of visible-light photocatalytic performance. UV–vis spectra of $g\text{-}C_3N_4$, TiO_2/C, monolayer $g\text{-}C_3N_4/TiO_2/C$ (EAC), and ultrathin

$g\text{-}C_3N_4/TiO_2/C$ (SL-EAC) of another work are demonstrated in Figure 4.9c. Comparing the absorption edge of EAC and $g\text{-}C_3N_4$ at 468 nm and SL-EAC, a shift was observed for SL-EAC to 441 nm. The SL-EAC blue shift was due to the quantum confinement of ascribed to thickness decrease of $g\text{-}C_3N_4$, that confirmed the stacked $g\text{-}C_3N_4$ layer separation into ultrathin nanosheets. Moreover, the addition of TiO_2/C caused higher absorption of EAC and SL-EAC in visible-light area [17].

The recombination rate of excited electron–hole pairs is a key factor in evaluating the photocatalytic performance. A strong PL emission peak for $g\text{-}C_3N_4$ demonstrates a high recombination rate of the charge carriers. However, the PL peak is substantially weaker for $g\text{-}C_3N_4/Ti_3C_2$, signifying the formation of a heterojunction between Ti_3C_2 and $g\text{-}C_3N_4$, that efficiently suppressed the charge recombination [52]. In Figure 4.9d, it is demonstrated that in SL-EAC, the electron transfer is stalled by the built-in electric field of EAC and SL-EAC from $g\text{-}C_3N_4$ to TiO_2/C. The PL signal of SL-EAC is observed to be weaker than that of EAC, which indicates that the ultrathin $g\text{-}C_3N_4$ nanosheets in SL-EAC accelerated the transfer of electrons to its surface, thus enhancing its carrier separation.

The characterization of the surface chemical states and the resulting bands are conducted through an X-ray photoelectron spectrometer (XPS) shown in

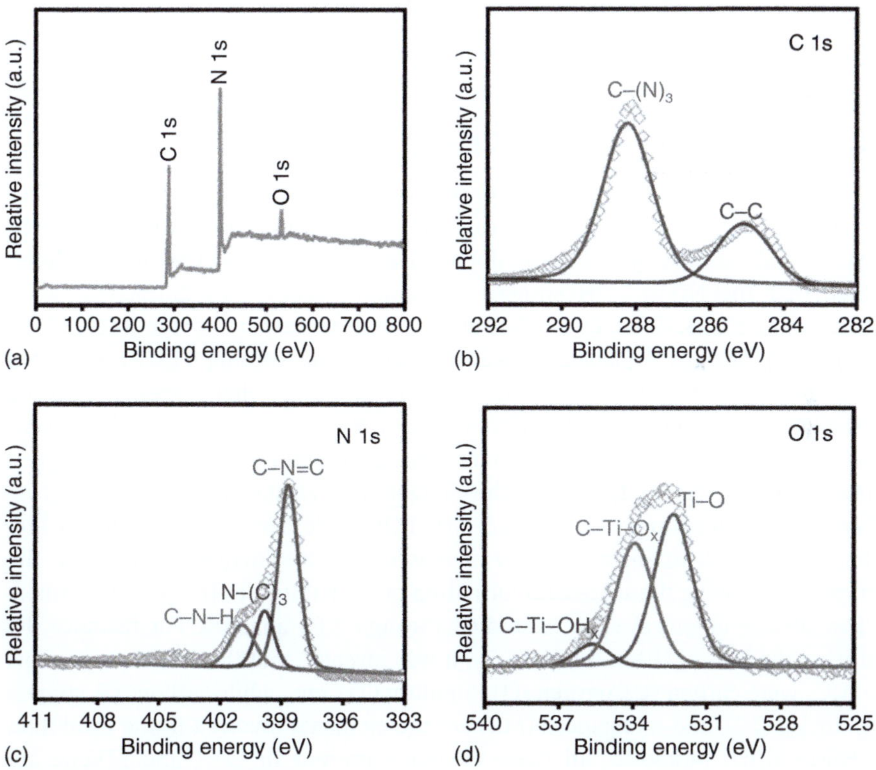

Figure 4.10 (a) XPS survey spectroscopy of $g\text{-}C_3N_4/Ti_3C_2$, (b) C1s spectrum of $g\text{-}C_3N_4/Ti_3C_2$, (c) N 1s spectrum of $g\text{-}C_3N_4/Ti_3C_2$, (d) Ti 2p spectrum of $g\text{-}C_3N_4/Ti_3C_2$. Source: Li et al. [56]/with permission of Elsevier.

Figure 4.10a–d. The XPS survey spectrum for the g-C_3N_4/$Ti_3C_2T_x$ sample is shown in Figure 4.10a. Due to the low quantity of $Ti_3C_2T_x$ MXene, the sample demonstrated only three peaks at peaks of 285 eV binding energies 399 eV, and 532 eV for C1s, N1s, and O1s, respectively. Figure 4.10b demonstrates C 1s XPS spectra of the g-C_3N_4/Ti_3C_2 sample. The peak located at 285.2 eV is allotted to C–C band. The major peak at 288.2 eV, attributed to N—C=N_2 coordination, is ascribed to C atoms with one double and two single bonds with three N neighbors in g-C_3N_4. The N 1s XPS spectra in Figure 4.10b demonstrated three peaks for g-C_3N_4, comprising the detected C–N=C for 398.5 eV, N–$(C)_3$ for 399.9 eV, and C–N–H for 401.6 eV. Figure 4.10d demonstrates the O 1s XPS spectra of the g-C_3N_4/$Ti_3C_2T_x$ [47].

Field-emission scanning electron microscopy (FESEM) and transmission electron microscopy (TEM) are employed to study the morphology and structure of materials. The morphology of as-prepared $Ti_3C_2T_x$ MXene, g-C_3N_4, and g-C_3N_4/$Ti_3C_2T_x$ photocatalyst is exhibited in Figure 4.11. In Figure 4.11a, the $Ti_3C_2T_x$ MXene demonstrates an accordion-like structure. Figure 4.11b demonstrates the TEM image of the g-C_3N_4 sample. The TEM images of the g-C_3N_4/$Ti_3C_2T_x$ photocatalyst are demonstrated in Figure 4.11c,d. Notably, the photocatalyst possesses a layered structure. The layered $Ti_3C_2T_x$ MXene is connected to g-C_3N_4, fabricating a sandwich-like structure. The specific surface area and gap between layers of g-C_3N_4 can be increased due to the insertion of $Ti_3C_2T_x$ MXene beneficial for enhancing photocatalytic activity [47].

4.3.3 Characterization of LDH/TiC MXene Composites

The crystal structure of materials is studied by employing XRD analysis. For instance, Zhang et al. [53] conducted the XRD analysis of NiMn-LDH, Ti_3C_2 MXene, and NiMn-LDH/Ti_3C_2 as demonstrated in Figure 4.12a. It was determined that the NiMn LDH showed hexagonal lattices with diffraction peaks located at 11.5°, 23.3°, 34.3°, 39.0°, and 60.2° that correspond to the (003), (006), (012), (015), and (013) planes, respectively. The peaks of bulk Ti_3C_2 are located at 6.2°, 19.3°, 34.1°, 36.3°, 38.6°, 42.2°, 45.0°, and 56.9° that correspond to the (002), (004), (101), (103), (104), (105), (107), (108) and (110) planes, respectively. For NiMn–LDH/Ti_3C_2 composites, characteristic diffraction peaks of both Ti_3C_2 and NiMn–LDH are present. Especially, the characteristic crystal planes (002) and (105) of are clearer after compositing with NiMn–LDH as demonstrated in Figure 4.12b, whereas the peak intensity of (105) increases with the increase of Ti_3C_2 in the composite, showing the successful formation of the NiMn–LDH/Ti_3C_2 composite.

The surface chemical state is studied through XPS analysis. For instance, by studying the XPS of NiMn–LDH/Ti_3C_2, it was revealed that the main components of Ti_3C_2 were carbon (C), oxygen (O), titanium (Ti), etc., while carbon (C), oxygen (O), nickel (Ni), and manganese (Mn) are the elements involved in the formation of NiMn–LDH. Moreover, all these elements present in LDH and MXene are also present in NiMn–LDH/Ti_3C_2 composites as demonstrated in Figure 4.12c. The Ti $2p_{3/2}$ spectra of Ti_3C_2 are divided into distinct four peaks that comprise Ti–C, Ti^{2+}, Ti^{3+}, and Ti–O as demonstrated in Figure 4.12d. The Ti $2p_{3/2}$ spectra of NiMn–LDH/Ti_3C_2 are divided into three peaks of Ti–C, Ti^{2+}, and Ti–O. The peak for Ti^{3+} fades and the ratio of Ti–C and Ti^{2+} is reduced, whereas the Ti–O intensity is enhanced, showing the replacement of terminated groups (Ti–C and T_x) by Ti–O. Similarly for Ti $2p_{1/2}$ spectra, the NiMn-LDH/Ti_3C_2 demonstrate a high quantity of TiO. Moreover, the shift in the locations of Ti $2p_{3/2}$ and Ti $2p_{1/2}$ peaks in NiMn-LDH/Ti_3C_2 to the lower binding energy is observed possibly due to some chemical reaction occurring to Ti in NiMn-LDH/Ti_3C_2. For the C 1s demonstrated in Figure 4.12e, the peak for C–Ti at 281.41 eV vanishes for Ti_3C_2 and NiMn-LDH combined, consistent with Ti 2p spectra. The quantity of C–OH, C=O, and COOH in NiMn-LDH/Ti_3C_2 is larger than that of Ti_3C_2, and these oxygen-containing functional groups are due to the chemical bonds of Ti_3C_2 and NiMn-LDH. Another important O 1s spectra of is demonstrated in Figure 4.12f, with the presence of three peaks of metal-O, CO_3^{2-}, and C–O. It is observed that the ratio of M–O increased in NiMn LDH/Ti_3C_2, as compared to NiMn-LDH that demonstrates that apart from the bonding of Mn atoms and Ni atoms with O atoms, the bonding with Ti atoms of Ti_3C_2 is also formed [53].

The material composition and morphology of LDH/TiC MXene composites may be analyzed using SEM, TEM, and energy-dispersive EDS analysis. Usually, a hierarchal structure is formed through the coupling of LDH/TiC MXene, whereas the TiC acts as a substrate and LDH is deposited on the surface of TiC. For instance,

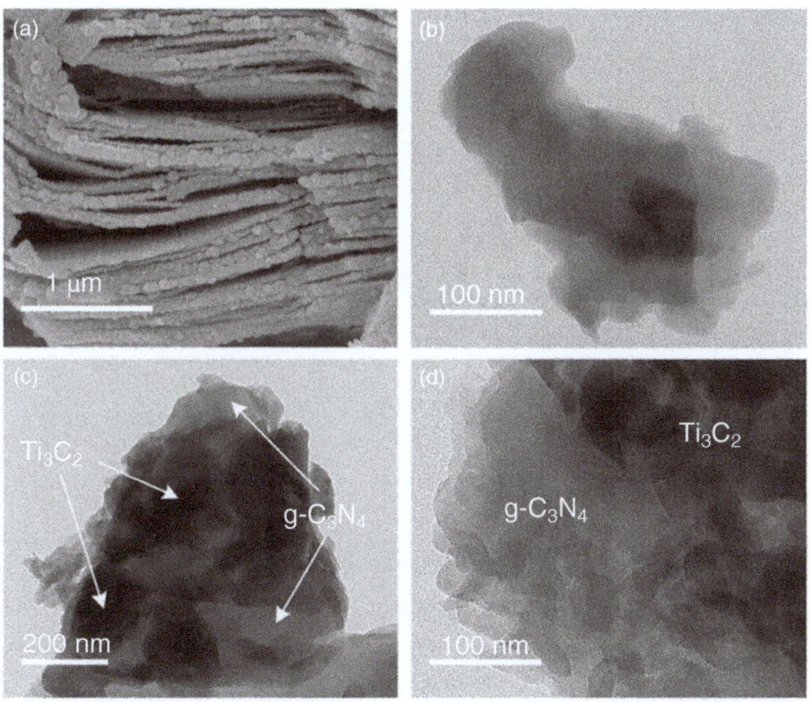

Figure 4.11 (a) FESEM analysis of Ti_3C_2, (b) TEM image of g-C_3N_4, and (c, d) TEM images of g-C_3N_4/Ti_3C_2. Source: Xu et al. [47]/John Wiley & Sons, Inc.

Zhao et al. [23] fabricated a core/shell 2D/2D nanocomposite of $Ti_3C_2T_x$ and NiAl-LDH. From the images of SEM as demonstrated in Figure 4.13a,b, the NiAl–LDH nanosheets are evenly dispersed on the multilayered $Ti_3C_2T_x$ surface, implying the strong contact between $Ti_3C_2T_x$ and NiAl-LDH. It is significant that an extremely thin NiAl-LDH shell is in situ grown on the TiC surface to fabricate a core/shell 2D/2D structure that is attributed to the restriction effect acquired from the ultrathin MXene with negative charges.

For example, Khan et al. [25] fabricated $Ti_3C_2T_{A/R}$/CoAlLa-LDH nanocomposite with CoAlLa-LDH nanosheets scattered on the surface and in the $Ti_3C_2T_{A/R}$ layers,

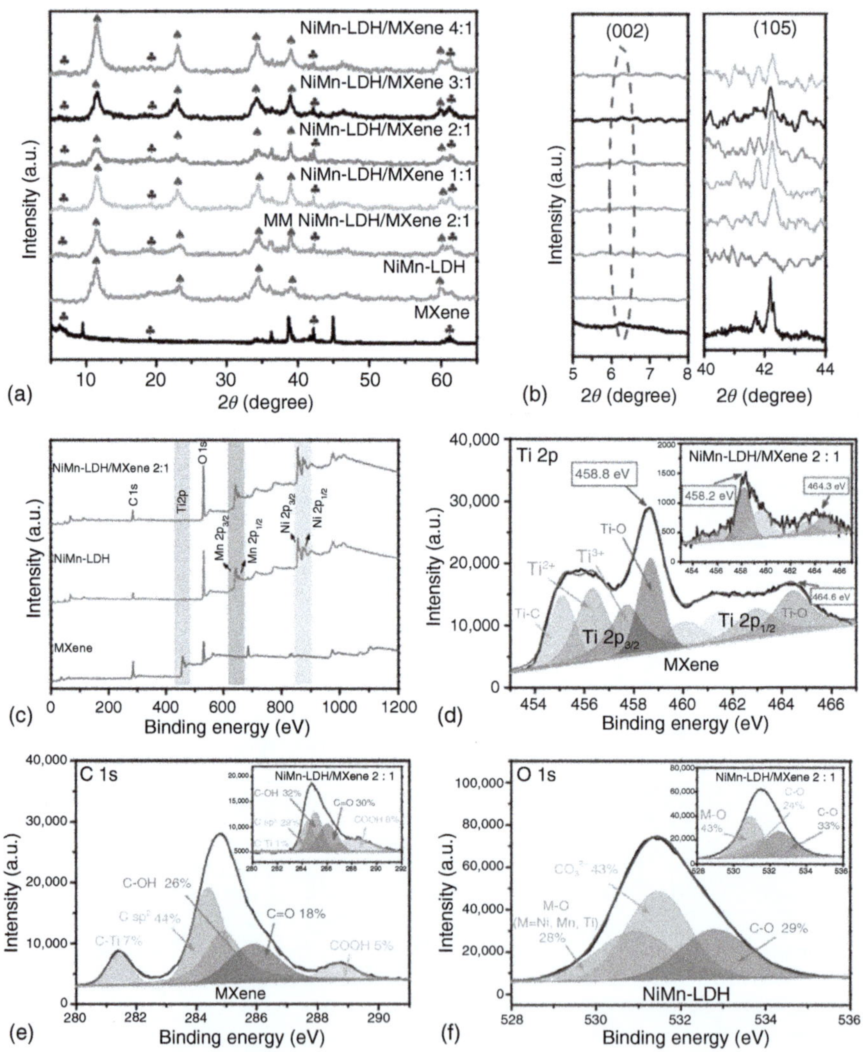

Figure 4.12 (a, b) XRD analysis of NiMn-LDH, Ti_3C_2 MXene and NiMn-LDH/Ti_3C_2; (c) overall XPS survey of NiMn-LDH/Ti_3C_2; High-resolution XPS of (d)Ti 2p, (e) C 1s, (f) O 1s.

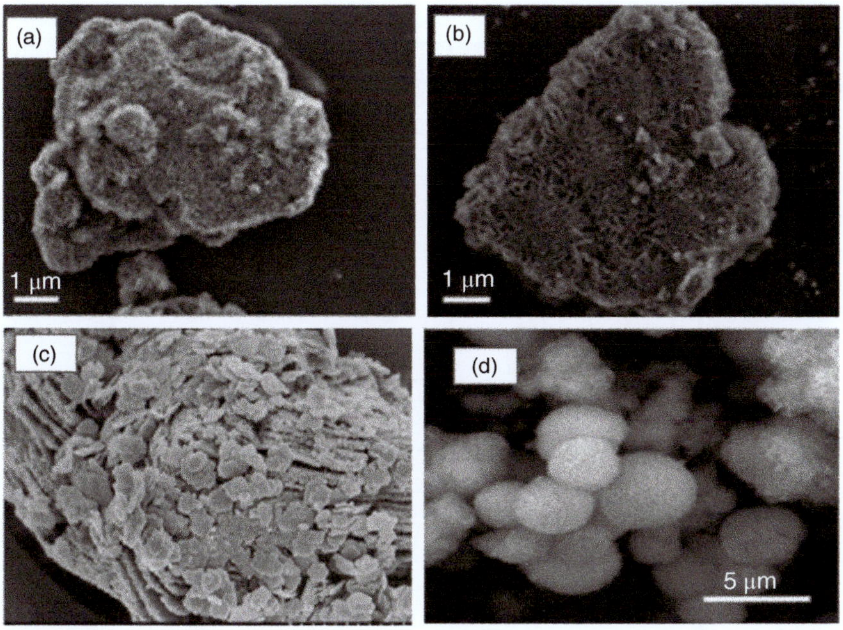

Figure 4.13 (a, b) Core/shell 2D/2D heterostructure of $Ti_3C_2T_x$/NiAl-LDH, (c) CoAlLa-LDH dispersed on $Ti_3C_2T_{A/R}$. (d) Flower-like structure of TiC/NiAl-LDH.

as indicated in Figure 4.13c. It is revealed that the distribution was achieved through effortless mixing in methanol solution. Still, Shi et al. [26] constructed a flower-like hierarchical structure of TiC/NiAl-LDH with NiAl-LDH showing flower-like morphology constructed from a huge number of nanoplates. After the inclusion of Ti_3C_2, the spaces in the NiAl-LDH nanoplates were filled with the Ti_3C_2 sheets. The 3D hierarchical LDH/TiC hybrid photocatalyst with core/shell structure was constructed through in situ loading of Ti_3C_2 to the NiAl-LDH as demonstrated in Figure 4.13d.

4.3.4 Characterization of Ti_3C_2 MXene-Based Perovskite Composites

The crystalline structure of Ti_3C_2 MXene-based perovskite composites can be validated by employing XRD analysis. The coexistence of two different phases perovskite and MXene may confirm the formation of the composite. Usually, the composite exhibits both characteristic diffraction peaks of MXene and perovskite. For instance, the MXene/$MAPbI_3$ composite exhibited both the characteristic diffraction peaks of Ti_3C_2 MXene and $MAPbI_3$ [27]. However, in some cases, due the subtle amount, one component may not be detected [28]. Another work demonstrates the XRD analysis of the $MAPbBr_3/Ti_3C_2T_x$ heterostructures as shown in Figure 4.14a. The peaks of pristine $MAPbBr_3$ NCs were detected at the angles of 14.9°, 21.1°, 30.0°, and 33.6°, which are in correspondence with the cubic-phased $MAPbBr_3$ with 5.96 Å. The $MAPbBr_3/Ti_3C_2T_x$ hybrid demonstrate the peaks of $MAPbBr_3$ NCs without impurity phases. A distinct peak at around 6.3° was observed

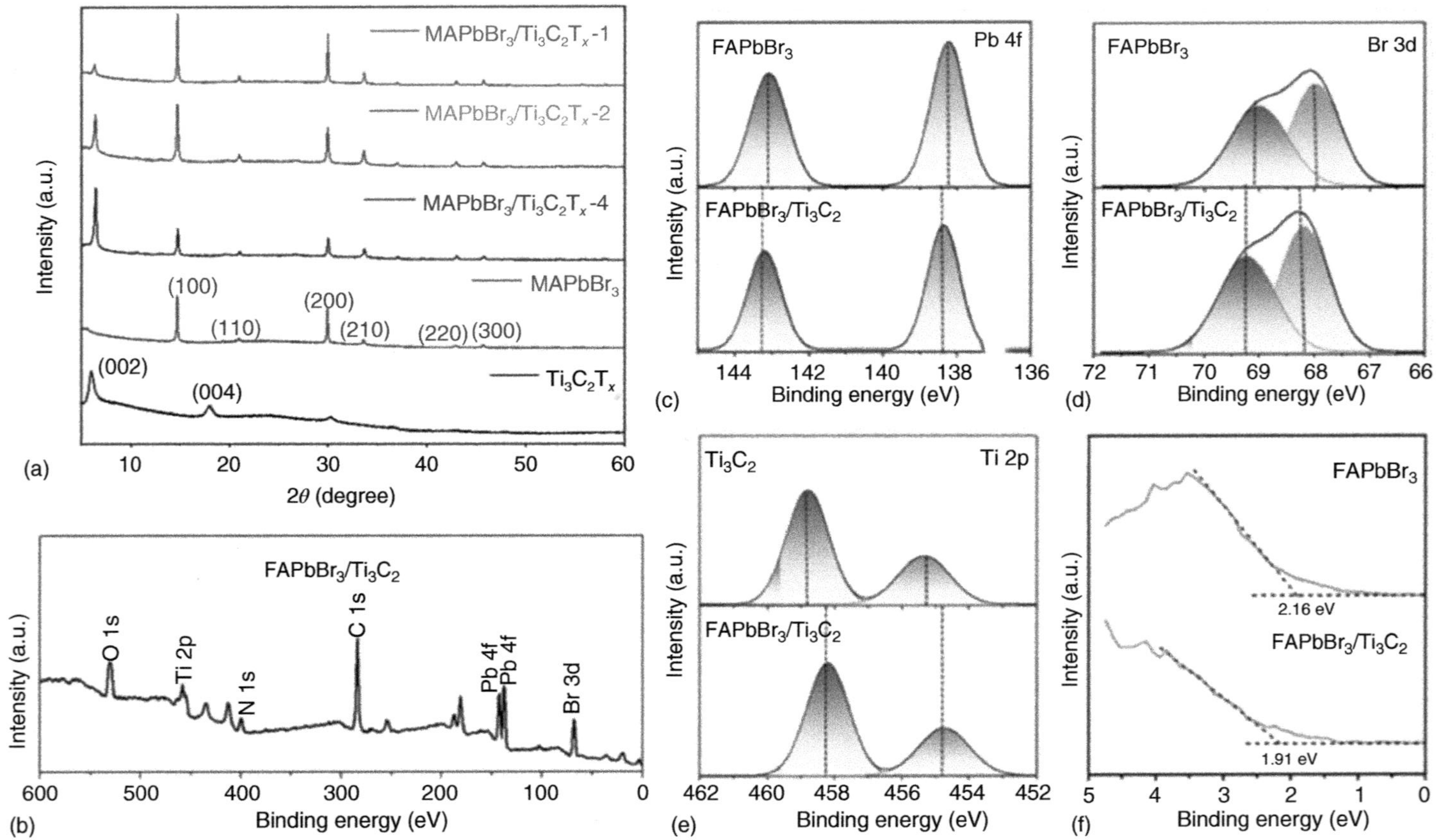

Figure 4.14 (a) XRD analysis of the perovskite on SnO_2 with and without the presence of MXene. Source: [54]/with permission of John Wiley & Sons; (b) XPS survey of $FAPbBr_3/Ti_3C_2$; High-resolution XPS spectra of (c) Pb 4f, (d) Br 3d, (e) Ti 2p, (f) Valence spectra of $FAPbBr_3/Ti_3C_2$ and $FAPbBr_3$. Source: Que et al. [28]/with permission of American Chemical Society.

in the $MAPbBr_3/Ti_3C_2T_x$ hybrid assigned to the (002) planes of $Ti_3C_2T_x$ as the intensity increased with the enhanced quantity of $MAPbBr_3/Ti_3C_2T_x$ nanosheets. The presence of characteristic peaks of both $MAPbBr_3$ and $Ti_3C_2T_x$ confirms the successful fabrication of the $MAPbBr_3/Ti_3C_2T_x$ hybrid [54].

The detailed chemical bonding behavior and the elemental composition of samples may be investigated by conducting XPS analysis. To discover the interfacial interaction between $FAPbBr_3$ and Ti_3C_2, the XPS spectra of Pb 4f, Br 3d, and Ti 2p in the $FAPbBr_3/Ti_3C_2$heterojunction were studied. The XPS survey spectrum of the $FAPbBr_3/Ti_3C_2$ sample showed the presence all the elements of both components as demonstrated in Figure 4.14b. As compared to $FAPbBr_3$, the binding energies of Pb 4f and Br 3d for the $FAPbBr_3/Ti_3C_2$ sample were observed to be positively shifted by 0.18 and 0.69 eV as demonstrated in Figure 4.14c,d, respectively. These shifts indicated that $FAPbBr_3$ works as the electron donor after coupling with Ti_3C_2 in the $FAPbBr_3/Ti_3C_2$ heterojunction. For Ti 2p, the spectrum of Ti_3C_2 showed the presence of two peaks located at the positions of 455.35 and 458.80 eV. The Ti 2p showed a negative shift (0.61 eV) as compared to Ti_3C_2 shown in Figure 4.14e,f. The energy gap (ΔE) of the valence band (VB) between the Fermi energy (E_f) of $FAPbBr_3$ and $FAPbBr_3/Ti_3C_2$ was also measured with valence XPS. The ΔE values of $FAPbBr_3$ and $FAPbBr_3/Ti_3C_2$ were found to be 2.16 and 1.91 eV, respectively, resulting in reduction of bandgap of $FAPbBr_3/Ti_3C_2$ [28].

4.3.5 Characterization of MOF-Based TiC MXene Nanocomposite

The XRD analysis is used to determine the crystalline phase of MOF/TiC MXene, for instance, the XRD patterns of pure UiO-66-NH_2, Ti_3C_2/TiO_2/UiO-66-NH_2, and $Ti_3C_2T_x$/UiO-66-NH_2. It can be observed that the addition of Ti_3C_2/TiO_2 has no obvious effect on the crystal structure of UiO-66-NH_2. Moreover, no Ti_3C_2 or TiO_2 diffraction peak was realized in Ti_3C_2/TiO_2/UiO-66-NH_2, due to the low Ti_3C_2/TiO_2 loading.

In another research work, XRD tests were employed to confirm the successful preparation of the UiO-66-NH_2(Zr/Ti) (UZT), carboxyl-functionalized MXene

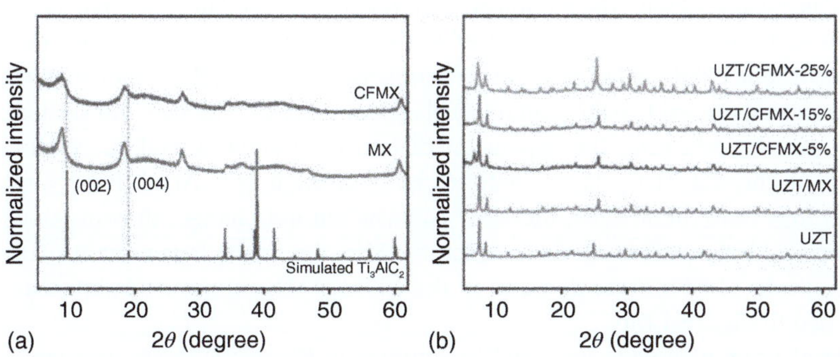

Figure 4.15 (a) XRD patterns of simulated Ti_3AlC_2, MXene(MX)and carboxyl-functionalized MXene (CFMX). (b) UiO-66-NH_2(Zr/Ti) (UZT), UZT/MX, and UZT CFMX samples. Source: Tian et al. [57]/with permission of Elsevier.

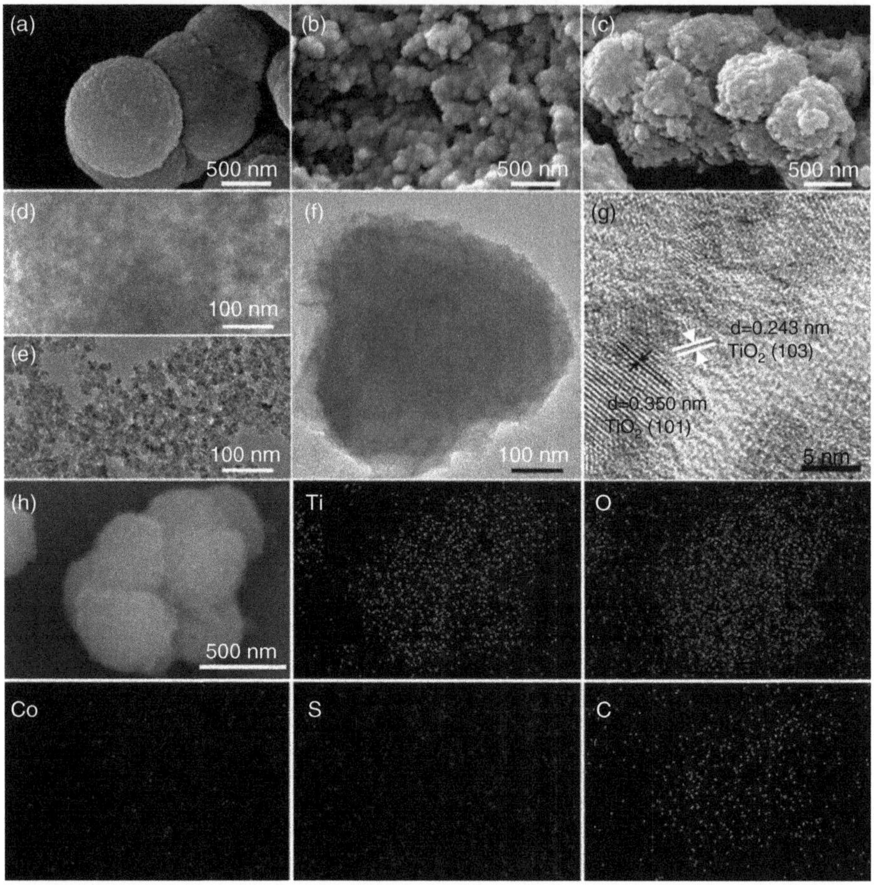

Figure 4.16 The SEM and TEM of (a, d) TiO_2, (b, e) TiO_2/CoS_x, (c, f) TiO_2-Ti_3C_2-CoS_x samples, (g) HRTEM, and (h) elemental mapping of TiO_2-Ti_3C_2-CoS_x.

(CFMX), and their composites as demonstrated in Figure 4.15a,b. The XRD analysis of the UZT is consistent with the simulated patterns, which validates their purity as demonstrated in Figure 4.15a. As compared to the Ti_3AlC_2 simulated patterns, the peak of the MXene exfoliated samples at 39° vanished, indicating the removal of Al layers. The peaks, related to (002) and (004) planes of Ti_3AlC_2 widened and shifted to lower degrees, indicating the expansion of lattice parameters and successful fabrication of the 2D Ti_3C_2 layers with etching through HF. After dispersing carboxyl groups onto the MXene, the XRD patterns did not change, showing that the benzoic acid has no effect on their structure. Also, due to the lower quantity of MXene, no peaks of CFMX are detected in the UZT/CFMX composite samples as demonstrated in Figure 4.15b.

The optical properties were analyzed by employing UV–vis DRS spectroscopy. Both $Ti_3C_2T_x$ and annealed $Ti_3C_2T_x$ showed no absorption edge due to the metallic nature. However, the weakened light absorption of annealed $Ti_3C_2T_x$ as compared to $Ti_3C_2T_x$ was observed ascribed to the decreased amount of $Ti_3C_2T_x$ under annealing treatment. The absorption intensity of $Ti_3C_2/TiO_2/UiO\text{-}66\text{-}NH_2$ was enhanced in the visible region due to the loading of black $Ti_3C_2T_x$. Additionally, no obvious shift of the absorption edge is observed. The bandgap energy of UiO-66-NH_2, Ti_3C_2/UiO-66-NH_2, and Ti_3C_2/TiO_2/UiO-66-NH_2 was found to be 2.75 eV, showing no difference in bandgap energy.

The SEM analysis was used to study the morphologies of as-synthesized samples of pristine UiO-66-NH_2, $Ti_3C_2T_x$, and annealed $Ti_3C_2T_x$ Ti_3C_2/TiO_2/UiO-66-NH_2. The UiO-66-NH_2 exhibited a cubical structure. The $Ti_3C_2T_x$ exhibited accordion-like structures, whereas the annealed $Ti_3C_2T_x$ displayed a structure similar to that of $Ti_3C_2T_x$, as no TiO_2 can be observed due to lower concentration. UiO-66-NH_2 was covered around $Ti_3C_2T_x$ due to the very hydrophilic surfaces, causing high agglomeration of UiO-66-NH_2. Contrarily, Ti_3C_2/TiO_2/UiO-66-NH_2 showed homogeneous attaching of UiO-66-NH_2 to the annealed $Ti_3C_2T_x$ surface with more available reactive sites. The elemental distribution of Ti_3C_2/TiO_2/UiO-66-NH_2 was revealed using the corresponding EDS spectrum. The elements C, N, O, F, Ti, and Zr with even distribution in Ti_3C_2/TiO_2/UiO-66-NH_2, indicating that UiO-66-NH_2 was appropriately distributed on the annealed $Ti_3C_2T_x$ surface.

To study the morphology and structure, TEM analysis was used. TiO_2crystals anchored on the surface of layered annealed $Ti_3C_2T_x$ were formed with intimate contact. The contact between UiO-66-NH_2, TiO_2, and Ti_3C_2 with each other was to form interfacial contact of Ti_3C_2/TiO_2/UiO-66-NH_2, Ti_3C_2/TiO_2, and Ti_3C_2/UiO-66-NH_2. The HRTEM image exhibited a lattice spacing of 1.01 nm that corresponds to (101) plane of Ti_3C_2. The co-existence of Ti_3C_2/TiO_2/UiO-66-$NH_2$and intimate contact with each other determined that the enlarged could cause better charge transfer and improve the reactive centers for better performance.

The images of elemental mapping demonstrated that Ti, Zr, C, and O are detected and are densely dispersed in the composite, indicating the successful formation of UZT/CFMX. The distribution of the F element is not obvious, demonstrating that the HF etchant has been cleaned up in the composite (Figure **1i–n**). Moreover, the content of the Ti ions in UZT has also been demonstrated. According to the inductively coupled plasma and X-ray photoelectron spectroscopy spectrum (XPS) results, the proportion of Zr and Ti can be calculated to be about 1 : 2, which demonstrates that the Ti ions have been successfully doped into the UiO-66-NH_2 (Table **S1**, Supporting Information).

Similarly, Zhao et al. [55] synthesized ternary TiO_2-Ti_3C_2-CoS_x hybrid by employing a solvothermal approach. The SEM and TEM were performed for morphology and structure study of the produced samples. The pure TiO_2 showed smooth round block shape having a radius of around of 450 nm as demonstrated in Figure 4.16a. Contrary to the pure TiO_2, the TiO_2-Ti_3C_2-CoS_x sample showed properly distributed nanoparticles due to the presence of porous ZIF-derived CoS_x demonstrated in Figure 4.16b. The TiO_2-CoS_x particles were incorporated inside Ti_3C_2 nanosheets, resulting in the formation of a rough block structure formed of smaller size nanoparticles demonstrated in Figure 4.16c. Similarly, the TEM approach was employed to study the structures of as-synthesized TiO_2-Ti_3C_2-CoS_x samples.

The agglomeration of TiO_2 particles is observed in Figure 4.16d; however, when CoS_x was added, the TiO_2 particles were uniformly scattered when CoS_x is introduced as demonstrated in Figure 4.16e. The Ti_3C_2 nanosheets worked as a scaffold for the formation of the TiO_2–CoS_x heterostructures that were hybridized TiO_2-Ti_3C_2-CoS_x samples as demonstrated in TEM analysis in Figure 4.16f. Furthermore, it benefited charge carrier transportation. Additionally, the HRTEM method was used to study the phase structure of the heterostructure as demonstrated in Figure 4.16g. The lattice gap of the TiO_2 nanoparticle was observed to be 0.35 nm, corresponding to the (101) crystal plane, while a unique 0.243 nm fringe space was ascribed to the (103) crystal plane related to TiO_2. The component distributions of the TiO_2-Ti_3C_2-CoS_x are demonstrated in Figure 4.16h, with homogeneous distribution of Ti, O, C, Co, and S elements being found.

4.4 Conclusion

In this chapter, the development on the synthesis of TiC MXene-based composites has been explained. It has been revealed that different strategies have been employed for synthesis of TiC MXene-based composites such as simple mixing, electrostatic assembly, hydrothermal treatment, solvothermal treatment, calcination, annealing, and in situ growth. Through these treatment methods, TiC MXene has been hybridized with materials such as TiO_2, g-C_3N_4, LDH, perovskite, and MOF materials to form composite materials for energy storage and conversion. The morphological properties of composites are strongly dependent upon the route of synthesis. The characterizations of the composites of the composites have been studied in detail, and it was found that TiC MXene can form a composite with intimate interaction with conjugate materials. From the crystal structural study, it was observed that TiC MXene has the capability to form very strong composites with high stability. The morphological analysis conducted through SEM and TEM showed that TiC MXene can inhibit the agglomeration of conjugate materials, resulting in formation of highly efficient hybrid materials. From the optical study conducted through UV–vis DRS study and PL spectroscopy, it was determined that addition of TiC MXene materials to the conjugate material results in better charge carrier transfer. The efficient interaction results in efficient charge transfer, important for energy storage and conversion materials.

References

1 Qin, T., Wang, Z., Wang, Y. et al. (2021). Recent progress in emerging two-dimensional transition metal carbides. *Nano-Micro Lett.* 13 (1): 183.
2 Sharma, S.K., Kumar, A., Sharma, G. et al. (2022). MXenes based nano-heterojunctions and composites for advanced photocatalytic environmental detoxification and energy conversion: a review. *Chemosphere* 291: 132923.
3 Han, F., Luo, S., Xie, L. et al. (2019). Boosting the yield of MXene 2D sheets via a facile hydrothermal-assisted intercalation. *ACS Appl. Mater. Interfaces* 11 (8): 8443–8452.

4 Han, X., An, L., Hu, Y. et al. (2020). Ti_3C_2 MXene-derived carbon-doped TiO_2 coupled with g-C_3N_4 as the visible-light photocatalysts for photocatalytic H_2 generation. *Appl. Catal., B* 265: 118539.

5 Gao, Y., Wang, L., Zhou, A. et al. (2015). Hydrothermal synthesis of TiO_2/Ti_3C_2 nanocomposites with enhanced photocatalytic activity. *Mater. Lett.* 150: 62–64.

6 Biswal, L., Nayak, S., and Parida, K. (2021). Recent progress on strategies for the preparation of 2D/2D MXene/g-C_3N_4 nanocomposites for photocatalytic energy and environmental applications. *Catal. Sci. Technol.* 11 (4): 1222–1248.

7 Ye, M., Wang, X., Liu, E. et al. (2018). Boosting the photocatalytic activity of P25 for carbon dioxide reduction by using a surface-alkalinized titanium carbide MXene as cocatalyst. *ChemSusChem* 11 (10): 1606–1611.

8 Zhuang, Y., Liu, Y., and Meng, X. (2019). Fabrication of TiO_2 nanofibers/MXene Ti_3C_2 nanocomposites for photocatalytic H_2 evolution by electrostatic self-assembly. *Appl. Surf. Sci.* 496: 143647.

9 Sreedhar, A. and Noh, J.-S. (2021). Recent advances in partially and completely derived 2D Ti_3C_2 MXene based TiO_2 nanocomposites towards photocatalytic applications: a review. *Sol. Energy* 222: 48–73.

10 Zhang, X., Liu, Y., Dong, S. et al. (2017). One-step hydrothermal synthesis of a TiO_2-$Ti_3C_2T_x$ nanocomposite with small sized TiO_2 nanoparticles. *Ceram. Int.* 43 (14): 11065–11070.

11 Feng, Y., Zhou, F., Deng, Q., and Peng, C. (2020). Solvothermal synthesis of in situ nitrogen-doped Ti_3C_2 MXene fluorescent quantum dots for selective Cu^{2+} detection. *Ceram. Int.* 46 (6): 8320–8327.

12 Miao, Z., Wang, G., Zhang, X., and Dong, X. (2020). Oxygen vacancies modified TiO_2/Ti_3C_2 derived from MXenes for enhanced photocatalytic degradation of organic pollutants: the crucial role of oxygen vacancy to Schottky junction. *Appl. Surf. Sci.* 528: 146929.

13 Huang, H., Song, Y., Li, N. et al. (2019). One-step in-situ preparation of N-doped TiO_2@C derived from Ti_3C_2 MXene for enhanced visible-light driven photodegradation. *Appl. Catal., B* 251: 154–161.

14 Li, Z., Liu, Y., Guo, D. et al. (2019). Synthesis of Ti_3C_2/TiO_2 heterostructure by microwave heating with high electrochemical performance. *Mater. Res. Express* 6 (6): 065056.

15 Hou, H., Shao, G., and Yang, W. (2021). Recent advances in g-C_3N_4-based photocatalysts incorporated by MXenes and their derivatives. *J. Mater. Chem. A* 9 (24): 13722–13745.

16 Yang, Z.-z., Wei, J.-j., Zeng, G.-m. et al. (2019). A review on strategies to LDH-based materials to improve adsorption capacity and photoreduction efficiency for CO_2. *Coord. Chem. Rev.* 386: 154–182.

17 Yang, Y., Zhang, D., Fan, J. et al. (2021). Construction of an ultrathin S-scheme heterojunction based on few-layer g-C_3N_4 and monolayer $Ti_3C_2T_x$ MXene for photocatalytic CO_2reduction. *Solar RRL* 5 (2): 2000351.

18 Yang, C., Tan, Q., Li, Q. et al. (2020). 2D/2D Ti_3C_2 MXene/g-C_3N_4 nanosheets heterojunction for high efficient CO_2 reduction photocatalyst: dual effects of urea. *Appl. Catal., B* 268: 118738.

19 Zhang, M., Qin, J., Rajendran, S. et al. (2018). Heterostructured d-Ti_3C_2/TiO_2/g-C_3N_4nanocomposites with enhanced visible-light photocatalytic hydrogen production activity. *ChemSusChem* 11 (24): 4226–4236.

20 Khan, A.A., Tahir, M., and Bafaqeer, A. (2020). Constructing a stable 2D layered Ti_3C_2 MXene cocatalyst-assisted TiO_2/g-C_3N_4/Ti_3C_2heterojunction for tailoring photocatalytic bireforming of methane under visible light. *Energy Fuels* 34 (8): 9810–9828.

21 Biswal, L., Nayak, S., and Parida, K. (2021). Recent progress on strategies for the preparation of 2D/2D MXene/gC_3N_4 nanocomposites for photocatalytic energy and environmental applications. *Catal. Sci. Technol.* 11 (4): 1222–1248.

22 Chen, W., Han, B., Xie, Y. et al. (2020). Ultrathin Co-Co LDHs nanosheets assembled vertically on MXene: 3D nanoarrays for boosted visible-light-driven CO_2 reduction. *Chem. Eng. J.* 391: 123519.

23 Zhao, S., Pan, D., Liang, Q. et al. (2021). Ultrathin NiAl-layered double hydroxides grown on 2D $Ti_3C_2T_x$ MXene to construct core–shell heterostructures for enhanced photocatalytic CO_2reduction. *J. Phys. Chem. C* 125 (19): 10207–10218.

24 Jo, W.-K., Kumar, S., and Tonda, S. (2019). N-doped C dot/CoAl-layered double hydroxide/g-C_3N_4 hybrid composites for efficient and selective solar-driven conversion of CO_2 into CH_4. *Composites Part B* 176: 107212.

25 Ali Khan, A. and Tahir, M. (2021). Construction of an S-scheme heterojunction with oxygen-vacancy-rich trimetallic CoAlLa-LDH anchored on titania-sandwiched Ti_3C_2multilayers for boosting photocatalytic CO_2reduction under visible light. *Ind. Eng. Chem. Res.* 60 (45): 16201–16223.

26 Shi, Q., Zhang, X., Yang, Y. et al. (2021). 3D hierarchical architecture collaborating with 2D/2D interface interaction in NiAl-LDH/Ti_3C_2 nanocomposite for efficient and selective photoconversion of CO_2. *J. Energy Chem.* 59: 9–18.

27 Li, H., Lv, X., Li, R. et al. (2022). Stable and efficient Ti_3C_2 MXene/$MAPbI_3$-HI system for visible-light-driven photocatalytic HI splitting. *J. Power Sources* 522: 231006.

28 Que, M., Zhao, Y., Yang, Y. et al. (2021). Anchoring of formamidinium lead bromide quantum dots on Ti_3C_2nanosheets for efficient photocatalytic reduction of CO_2. *ACS Appl. Mater. Interfaces* 13 (5): 6180–6187.

29 Hui, X., Zhang, P., Wang, Z. et al. (2022). Vacancy defect-rich perovskite $SrTiO_3$/Ti_3C_2 heterostructures in situ derived from Ti_3C_2 MXenes with exceptional oxygen catalytic activity for advanced Zn–air batteries. *ACS Appl. Energy Mater.* 5 (5): 6100–6109.

30 Pan, A., Ma, X., Huang, S. et al. (2019). $CsPbBr_3$perovskite nanocrystal grown on MXene nanosheets for enhanced photoelectric detection and photocatalytic CO_2reduction. *J. Phys. Chem. Lett.* 10 (21): 6590–6597.

31 Tahir, M. and Tahir, B. (2020). 2D/2D/2D O-C_3N_4/Bt/$Ti_3C_2T_x$ heterojunction with novel MXene/clay multi-electron mediator for stimulating photo-induced CO_2 reforming to CO and CH_4. *Chem. Eng. J.* 400: 125868.

32 Qin, J., Liu, B., Lam, K.-H. et al. (2020). 0D/2D MXene quantum dot/Ni-MOF ultrathin nanosheets for enhanced N_2photoreduction. *ACS Sustainable Chem. Eng.* 8 (48): 17791–17799.

33 Liu, Y., He, Y., Vargun, E. et al. (2020). 3D porous Ti_3C_2 MXene/NiCo-MOF composites for enhanced lithium storage. *Nanomaterials* 10 (4).

34 Zheng, S., Zhou, H., Xue, H. et al. (2022). Pillared-layer Ni-MOF nanosheets anchored on Ti_3C_2 MXene for enhanced electrochemical energy storage. *J. Colloid Interface Sci.* 614: 130–137.

35 Liu, C., Bai, Y., Li, W. et al. (2022). In situ growth of three-dimensional MXene/metal–organic framework composites for high-performance supercapacitors. *Angew. Chem. Int. Ed.* 61 (11): e202116282.

36 Wu, Y., Li, X., Yang, Q. et al. (2020). Mxene-modulated dual-heterojunction generation on a metal-organic framework (MOF) via surface constitution reconstruction for enhanced photocatalytic activity. *Chem. Eng. J.* 390: 124519.

37 Luo, X., Abazari, R., Tahir, M. et al. (2022). Trimetallic metal–organic frameworks and derived materials for environmental remediation and electrochemical energy storage and conversion. *Coord. Chem. Rev.* 461: 214505.

38 Huang, R., Liao, D., Chen, S. et al. (2020). A strategy for effective electrochemical detection of hydroquinone and catechol: decoration of alkalization-intercalated Ti_3C_2 with MOF-derived N-doped porous carbon. *Sens. Actuators, B* 320: 128386.

39 Ye, Z., Jiang, Y., Li, L. et al. (2021). Self-assembly of 0D–2D heterostructure electrocatalyst from MOF and MXene for boosted lithium polysulfide conversion reaction. *Adv. Mater.* 33 (33): 2101204.

40 Khan, A.A. and Tahir, M. (2021). Well-designed 2D/2D $Ti_3C_2T_{A/R}$ MXene coupled g-C_3N_4 heterojunction with in-situ growth of anatase/rutile TiO_2 nucleates to boost photocatalytic dry-reforming of methane (DRM) for syngas production under visible light. *Appl. Catal., B* 285: 119777.

41 Li, Y., Deng, X., Tian, J. et al. (2018). Ti_3C_2 MXene-derived Ti_3C_2/TiO_2 nanoflowers for noble-metal-free photocatalytic overall water splitting. *Appl. Mater. Today* 13: 217–227.

42 Qian, J., Zhao, S., Dang, W. et al. (2021). Photocatalytic nitrogen reduction by Ti_3C_2 MXene derived oxygen vacancy-rich C/TiO_2. *Adv. Sustainable Syst.* 5 (4): 2000282.

43 Su, Y., Chen, P., Wang, F. et al. (2017). Decoration of TiO_2/g-C_3N_4 Z-scheme by carbon dots as a novel photocatalyst with improved visible-light photocatalytic performance for the degradation of enrofloxacin. *RSC Adv.* 7 (54): 34096–34103.

44 Wang, J., Wang, G., Wang, X. et al. (2019). 3D/2D direct Z-scheme heterojunctions of hierarchical TiO_2 microflowers/g-C_3N_4 nanosheets with enhanced charge carrier separation for photocatalytic H_2 evolution. *Carbon* 149: 618–626.

45 Ding, X., Li, Y., Li, C. et al. (2019). 2D visible-light-driven TiO_2@Ti_3C_2/g-C_3N_4 ternary heterostructure for high photocatalytic activity. *J. Mater. Sci.* 54 (13): 9385–9396.

46 Zheng, M., Guo, R., Liu, Z. et al. (2018). MoS_2 intercalated p-Ti_3C_2 anode materials with sandwich-like three dimensional conductive networks for lithium-ion batteries. *J. Alloys Compd.* 735: 1262–1270.

47 Xu, G., Gong, D., Yang, L. et al. (2022). Synthesis of free-standing flexible g-C_3N_4/MXene film as electrode materials for supercapacitors. *Int. J. Energy Res.* 46 (10): 13308–13315.

48 Peng, Q., Guo, J., Zhang, Q. et al. (2014). Unique lead adsorption behavior of activated hydroxyl group in two-dimensional titanium carbide. *J. Am. Chem. Soc.* 136 (11): 4113–4116.

49 Junwei, F., Xu, Q., Low, J. et al. (2019). Ultrathin 2D/2D WO_3/g-C_3N_4 step-scheme H_2-production photocatalyst. *Appl. Catal., B* 243: 556–565.

50 Sun, Y., Jin, D., Sun, Y. et al. (2018). G-C_3N_4/$Ti_3C_2T_x$ (MXenes) composite with oxidized surface groups for efficient photocatalytic hydrogen evolution. *J. Mater. Chem. A* 6 (19): 9124–9131.

51 Zhang, S., Huang, Y., Han, X. et al. (2023). $Ti_3C_2T_x$/g-C_3N_4/CNTs ternary hybrid film for all-solid flexible supercapacitors and superb bandwidth electromagnetic wave absorber. *ChemNanoMat* 9 (4): e202200511.

52 Liu, N., Lu, N., Su, Y. et al. (2019). Fabrication of g-C_3N_4/Ti_3C_2 composite and its visible-light photocatalytic capability for ciprofloxacin degradation. *Sep. Purif. Technol.* 211: 782–789.

53 Zhang, D., Cao, J., Zhang, X. et al. (2020). NiMn layered double hydroxide nanosheets in-situ anchored on Ti_3C_2 MXene via chemical bonds for superior supercapacitors. *ACS Appl. Energy Mater.* 3 (6): 5949–5964.

54 Zhang, Z., Li, Y., Liang, C. et al. (2020). In situ growth of $MAPbBr_3$nanocrystals on few-layer MXene nanosheets with efficient nergy transfer. *Small* 16 (17): 1905896.

55 Zhao, J., Liu, L.-W., Liu, F. et al. (2019). Conductive Ti_3C_2 and MOF-derived CoS_x boosting the photocatalytic hydrogen production activity of TiO_2. *CrystEngComm* 21: 2416–2421.

56 Li, J., Zhao, L., Wang, S. et al. (2020). In situ fabrication of 2D/3D g-C_3N_4/Ti_3C_2 (MXene) heterojunction for efficient visible-light photocatalytic hydrogen evolution. *Appl. Surf. Sci.* 515: 145922.

57 Tian, P., He, X., Zhao, L. et al. (2018). Enhanced charge transfer for efficient photocatalytic H_2 evolution over UiO-66-NH_2 with annealed $Ti_3C_2T_x$ MXenes. *Int. J. Hydrogen Energy* 44: 788–800.

5

Titanium Carbide (TiC) MXene-Based Titanium Dioxide Composites for Energy and Environment Applications

Riyadh R. Ikreedeegh[1] *and Muhammad Tahir*[2]

[1] *Arabian Gulf Oil Company, Sarir Oil Refinery, Department of Analysis and Quality Control, P.O. Box 263, El Kish, Benghazi, Libya*
[2] *UAE University, Chemical and Petroleum Engineering Department, P.O. BOX 15551, Sheik Khalifa Bin Zayed Street, Asharej, Al Ain, United Arab Emirates*

5.1 Introduction

The excessive consumption of fossil fuels has led to a serious depletion in the global energy resources [1, 2]. Burning these fuels has also led to a detrimental effect on the environment due to the massive atmospheric emissions of generated greenhouse gases and other pollutants [3–5]. The conversion of these toxic pollutants into fuels and valuable chemicals is considered one of the most promising approaches for environmental remediation and energy production. This could be achieved through a solar light-assisted process known as photocatalysis, making this strategy as a sustainable, environmentallyfriendly, and free-cost route toward solving the environmental and energy problems. The photocatalysis process is conducted over a surface of a semiconductor material named as photocatalyst, and when light having an equal or greater bandgap energy of that material is subjected to its surface, it generates a huge population of electron–hole pairs (e^-/h^+). These photogenerated charge carriers then undergo a series of reduction and oxidation (redox) reactions to convert these toxic pollutants and produce the final solar products [6–8].

Titanium dioxide material is considered the most extensively investigated photocatalysis material owing to its photocatalytic activity, good chemical stability, abundancy, and low cost. However, because of its wide bandgap energy (3.2 eV), the TiO_2 photocatalyst is UV-active only and functions poorly under visible light, thus lowering its photoactivity [9]. Moreover, it suffers from a high recombination rate of photogenerated charge electron–hole pairs (e^-/h^+), resulting in limited photocatalytic efficiency [9]. Several strategies have been employed to solve these issues and enhance the various aspects of photocatalytic TiO_2-based systems such as visible light response and efficient charge separation. These strategies include, but are not limited to, doping with noble-metals [10, 11], surface photosensitization [12], and coupling with other metals [13] and non-metals [14]. However, many of these used

Titanium Carbide MXenes: Synthesis, Characterization, Energy, and Environmental Applications, First Edition. Edited by Muhammad Tahir.

materials are expensive and less efficient, making these methods uneconomical. Recently, two-dimensional, low cost, layered nanomaterials known as titanium carbides (MXenes) have gained a great concern in improving the photocatalytic activity of other semiconductors, in particular TiO_2, due to their exceptional chemical and physical properties. MXenes are also known to have excellent electrical and photothermal conductivity, making them efficient materials in converting light energy to heat energy, and thus strong potential photocatalysts for energy and environmental applications.

Herein, the properties and characteristics of the benchmark TiO_2 photocatalysts and the highly conductive 2D MXene (Ti_3C_2) materials are discussed. The main synthesis methods of MXene with their reaction conditions are also discussed in detail in a separate section. The role of MXene in enhancing the photocatalytic performance of TiO_2 photocatalysts is summarized in three main sections; the formation of Schottky heterojunctions, enhancing light harvesting, and enhancing reactant adsorption. Finally, the recent developments of MXene/TiO_2 nanocomposites for environmental remediation and energy production applications such as photocatalytic CO_2 reduction, photocatalytic H_2 production, and pollutant photodegradation are summarized and deeply discussed.

5.2 Recent Developments in TiC-Based TiO_2Composites

5.2.1 General Overview

Among all the semiconductor materials, titanium dioxide (TiO_2) has attracted much attention due to its availability, low cost, non-toxicity, chemical stability, corrosion resistance, and powerful oxidation properties [6, 15]. Moreover, TiO_2 can be synthesized in different morphologies and nanostructures of microspheres [16], nanosheets [17], nanoparticles [18], nanotubes [19], nanowires [20], and nanorods [21] as shown in Figure 5.1. Owing to all these unique characteristics, TiO_2 has been widely applied as a photocatalyst for different applications, including CO_2 reduction [9, 22, 23], hydrogen production [24–26], and pollutant degradation [27–29]. The first attempt of using TiO_2 as a photocatalyst was reported by Fujishima and Honda in 1972 [30] which highly promoted the field of solar energy applications, and many approaches were then followed by other researchers by using various semiconductor materials. Although the TiO_2 material is considered a benchmark photocatalyst, it still has some significant issues.

In the context, the conductive MXene material, a family of 2D transition metal carbides and/or nitrides, has gained wide interest after its discovery by Gogosti's group in 2011 [31]. The MXene nanomaterials have been introduced for the modification of TiO_2; this includes MXene-derived TiO_2 (in situ growth of TiO_2 on MXene) and the TiO_2/MXene-based composites (by external addition of TiO_2 to MXene).

The structural composition of MXenes could be signified by the general chemical formula of $M_{n+1}X_nT_x$, where M represents the early transition metals such as titanium (Ti), vanadium (V), chromium (Cr), zirconium (Zr), niobium (Nb), molybdenum (Mo), hafnium (Hf), tantalum (Ta), and tungsten (W); n is equal to the

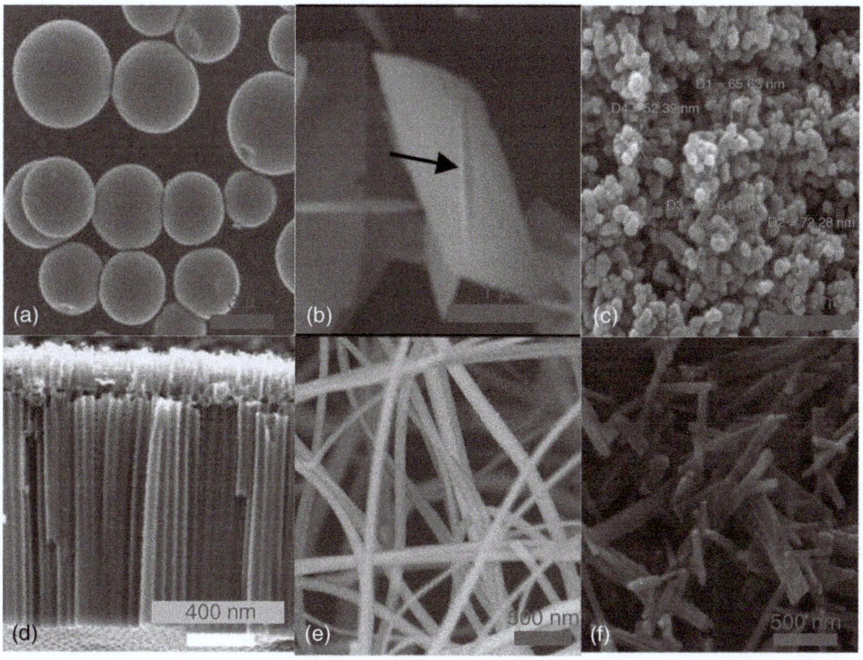

Figure 5.1 Different morphologies of the TiO_2 photocatalyst: (a) microsphere. Source: Reproduced with permission from Zheng et al. [16]. © 2009, John Wiley & Sons, Inc. (b) nanosheets. Source: Reproduced with permission from Fang et al. [17]. © 2011, John Wiley & Sons, Inc. (c) nanoparticles. Source: Reproduced with permission from Hajizadeh et al. [18]. © 2020, John Wiley & Sons, Inc. (d) nanotubes. Source: Reproduced with permission from Xu et al. [19]. © 2007, John Wiley & Sons, Inc. (e) nanowires. Source: Reproduced with permission from Diao et al. [20]. © 2020, John Wiley & Sons, Inc. (f) nanorods. Source: Reproduced with permission from Sokhandani et al. [21]. © 2018, John Wiley & Sons, Inc.

number 1, 2, or 3; X represents either carbon (C) or nitrogen (N); T_x represents the surface terminations in which x can be oxygen (O_2), hydroxyl group (OH^-), or a halogen group element like fluorine (F) and chlorine (Cl), bromine (Br), iodine (I), and astatine (As) [32]. Figure 5.2 illustrates the general formulae with the main crystal structures and compositions of MXenes.

5.2.2 Synthesis of MXenes

Generally, MXenes are synthesized through selective acid etching of certain layers from their layered precursors like MAX phases which are considered the most commonly employed precursors for MXene production. MAX represents a huge family of ternary carbides and nitrides with a general formula of $M_{n+1}AX_n$, where M represents the transition metals; n is equal to 1, 2, or 3; A is for group 13 and 14 elements in the periodic table; X is either carbon or nitrogen. However, the hydrofluoric acid and fluoride-containing acids are the most used solutions for obtaining the MXenes through the etching of MAX layers. Figure 5.3 shows the main synthesis approaches for the preparation of MXene with their reaction conditions.

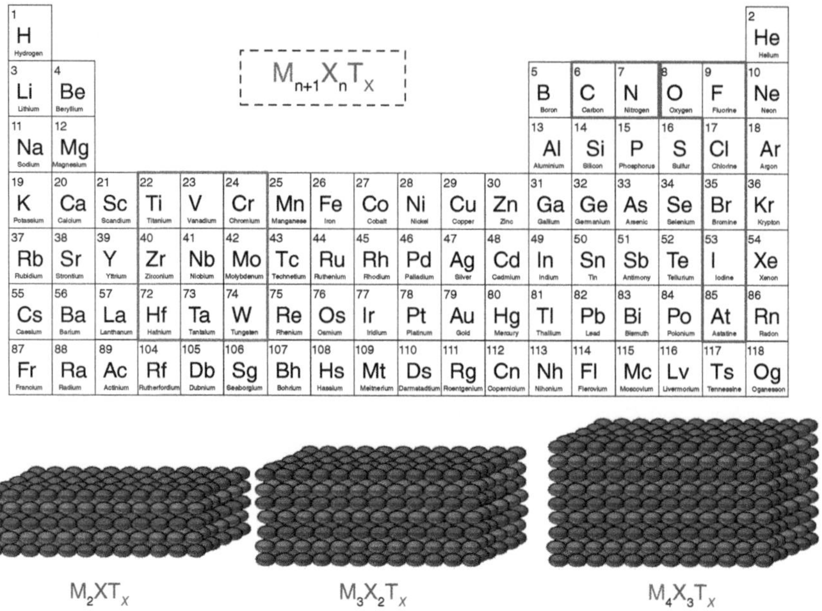

Figure 5.2 A schematic illustration for the compositional structures with the general formulae of MXene.

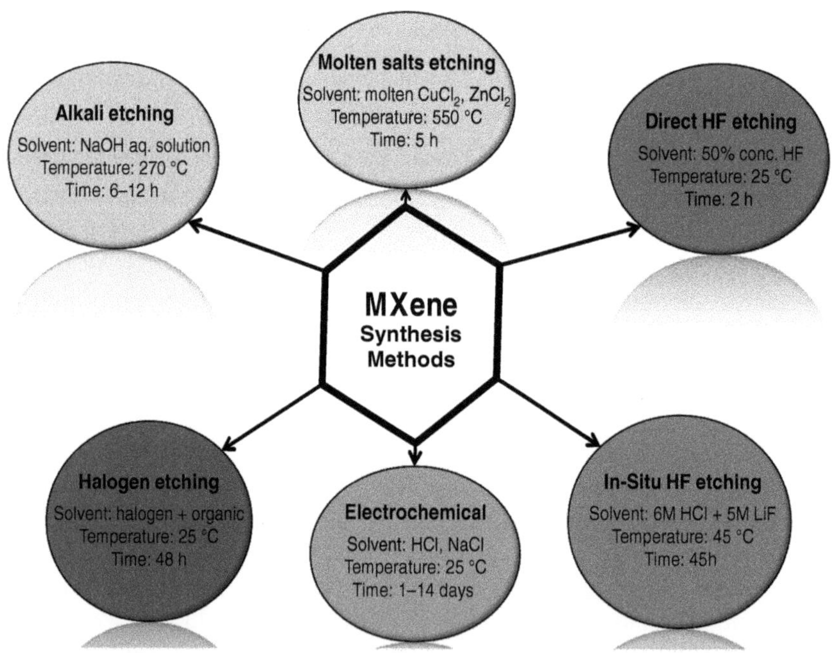

Figure 5.3 The main synthesis methods of MXene with their reaction conditions.

5.2.2.1 Direct HF Etching

The M-A bond in MAX phases is chemically higher active thanthe M-X bond, which makes it possible for selective removal of A atoms from the MAX precursors. Using hydrofluoric acid as an etchant is considered the most powerful technique for producing MXenes. In brief, a specific amount of Ti_3AlC_2 is immersed in a 50% concentrated hydrofluoric (HF) solution for twohours at a temperature of 20–25 °C. After the HF treatment, the Al atoms are replaced with F or OH^- functional groups [33].

5.2.2.2 In Situ HF Etching

In direct HF etching, the highly concentrated HF acid is regarded as an extremely toxic and corrosive chemical which can harmfully penetrate the skin, tissues, muscles, and bones, making its handling and disposal extremely dangerous. Therefore, milder concentration conditions should be applied. This could be achieved with an alternative etchant made by using a mixed solution of HCl acid and solid fluoride salts such as LiF, NH_4F, and KF. As reported, the Mxene is produced more favorably after eliminating the Al layer from the MAX phase by using a solution of 6 M HCl with LiF at a temperature of 45 °C for 45 hours [34]. Compared to direct HF etching, this modified insitu HF etching technique can also provide new surface terminations of the–Cl group.

5.2.2.3 Alkali Etching

Using alkali solution as an etchant is considered a new and effective method to reduce the excessive use of HF acid in conventional MXene synthesis methods. Highly pure (92%) MXenes can be obtained through this route by utilizing NaOH aqueous solution with a hydrothermal-assisted reaction for the removal of the Al layer in the MAX phase to be replaced with -OH and -O surface termination groups [35].

5.2.2.4 Halogen Etching

The halogen etching is considered a new and environmental-friendly synthesis route which can be employed for improving the electrical and mechanical properties of the produced MXene. This method is conducted by selectively etching the A-layer from the ternary layered MAX phase through stirring in a halogen-based etching agent such as I_2, Br, IBr, and ICl at room temperature [36]. It was also revealed that extra stripping of the A layer in the MAX phase could be obtained with the use of organic solvents like acetonitrile and tetrahydrofuran [37].

5.2.2.5 Molten SaltEtching

The aqueous etching methods fail with the non-Al MAX and nitride MAX phases since most of aqueous systems are only suitable for the Al-containing MAX which is used to produce hydrophilic MXenes under low operating temperatures [38]. Therefore, molten salts such as $CuCl_2$ and $ZnCl_2$ can be used to obtain MXenes from all MAX phase with -Cl surface termination groups which exhibited higher stability than other terminal groups [39].

5.2.2.6 Electrochemical Etching

The electrochemical etching route is regarded as one of the best technique for minimizing the hazardous content of fluoride-containing solutions during the MAX etching process. This approach involves the selective removal of the Al layer by employing the MAX phase as a working electrode with the aid of HCl solution acting as an electrolyte [40]. Different electrolytic systems such as NaCl and NH_4Cl aqueous solutions can also be used. It is worth mentioning that the MAX etching process is greatly affected by some factors including the etching time, the applied voltage, and the electrolyte concentration.

5.2.3 Role of MXenes in Enhancing TiO_2 Photocatalytic Performance

The Ti atoms on the MXene (Ti_3C_2) surface are comparatively active and behave as nucleation spots, which facilitate the formation of interfacial heterojunction between the MXene and TiO_2 [41]. However, the role of the MXene material as a cocatalyst for enhancing the TiO_2 photocatalytic performance can be summarized in the formation of Schottky heterojunctions, enhancing the light harvesting and enhancing reactant absorption as explained in the following sections:

5.2.3.1 Formation of Schottky Heterojunctions

During the photocatalyst excitation, electrons (e^-) are generated and transferred from the valence band to the conduction band, leaving holes (h^+) in the valence band. During this electron journey, a huge population of these electrons recombines with the holes in an unfavorable process called charge recombination. However, the formation of Schottky heterojunction limits the backward return of the photo-induced electrons to their semiconductor valence bands and indirectly prolonging the charge lifetime. The formation of Schottky heterojunctions is usually related with the incorporation of noble metals such as Pt, Au, Ag, and Pd that have shown good results in improving the photocatalytic performance of other semiconductors. However, the extremely high cost of these metals limits their application as a practical and efficient material. Therefore, tremendous efforts have been devoted to find an efficient alternative and MXene was one of the best candidates owing to its ability to form a Schottky junction and its role as a solid electron mediator.

As mentioned previously, the TiO_2 photocatalyst greatly suffers from the problem of charge recombination and thus lowers its photocatalytic performance. However, this issue can be resolved through the introduction of the conductive MXene material which can promote the transfer of charges carriers on the photocatalyst surface. Moreover, the redox reactions could be enhanced by the increased number of exposed active sites obtained by the unique MXene morphological nanostructure.

TiO_2–MXene composites can be formed by using the thermal oxidation of MXene in which TiO_2 is formed as a result of the oxidized titanium (Ti) atoms present in MXene. The (001)TiO_2/MXene heterostructure was hydrothermally synthesized by Zhang et al. [42]. The $Ti_3C_2T_x$ MXene was obtained through HF-etching of MAX Ti_3AlC_2 phase for twohours at room temperature. The obtained MXene was then stirred in $HCl/NaBF_4$ solution for controlling the crystal surface,

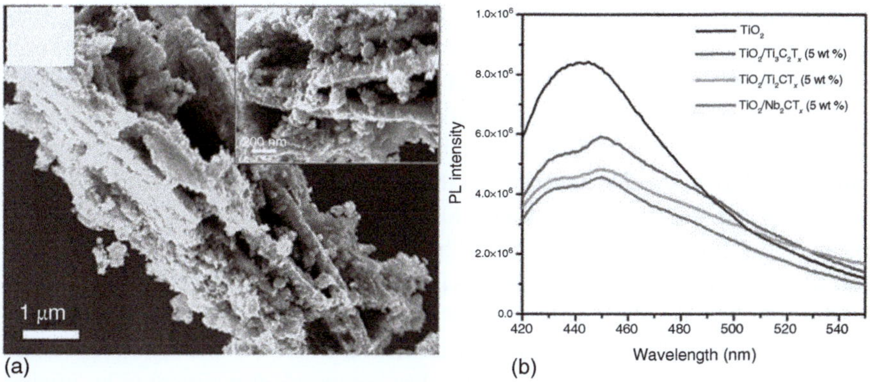

Figure 5.4 (a) SEM image of the $TiO_2/Ti_3C_2T_x$ composite Source: Reproduced with permission from Wang et al. [43]/© 2016, John Wiley & Sons, Inc; (b) the PL spectra of pure TiO_2 compared to TiO_2/MXene composites. Source: Wang et al. [43]/with permission of John Wiley & Sons.

prior to hydrothermal treatment at 160 °C for 8, 12, 16, and 32 hours. The acidic hydrothermal treatment transforms the titanium (Ti) atoms in MXene into hydrated ions of Ti^{3+} which are then oxidized to form the TiO_2.

Wang et al. [43] have also fabricated a TiO_2-incorporated MXene composite through a hydrothermal approach, with the two semiconductor materials aggregated together with good dispersion of TiO_2 nanoparticles in the $Ti_3C_2T_x$ MXene flakes as shown in Figure 5.4a, and the MXene acted as a cocatalyst and substituted the role of noble metals. The MXene composite with the grown rutile TiO_2 on the surface exhibited an enhanced photocatalytic performance due to the surpassed recombination rate of photogenerated charge carriers; this has been proven by the photoluminescence (PL) results which reflect the existence of charge recombination. As shown in Figure 5.4b, lower emission peaks were displayed with theTiO_2-MXene composites compared to the pure TiO_2 sample.

Peng et al. [44] reported that MXene can also be used as an efficient hole mediator for promoting the separation of photogenerated charges. The ternary Cu/TiO_2@MXene hybrid exhibited an exceptional photocatalytic activity for H_2 production owing to the synergistic effect of Cu nanoparticles which acted as a reduction co-catalyst and MXene which acted as a hole mediator. A p–n heterojunction of Cu_2O–TiO_2 was formed in the absence of MXene, and according to the band alignments the photogenerated electrons (e^-) were migrated from the Cu_2O CB to the CB of TiO_2, while the holes (h^+) are transferred from the TiO_2 VB to the Cu_2O VB. However, the presence of MXene promoted the charge separation efficiency by capturing the photogenerated holes excited by TiO_2. Moreover, the presence of MXene enabled the reduction of Cu_2O to metallic Cu due to the increased population of generated electrons (e^-).

5.2.3.2 Enhancing Light Harvesting

One of the most significant drawbacks of the TiO_2 material as a photocatalyst is its low ability for visible light harvesting; this is mainly attributed to its bandgap energy which is equal to 3.0–3.2 eV, making it UV-active only [22]. Shifting the TiO_2 optical properties of light absorption from UV toward visible light is considered the best technique to solve this issue. This could be achieved by coupling and incorporation of TiO_2 with other visible-active semiconductors such as MXenes. Due to their metallic nature, MXenes have no optical bandgap with an absorption edge approximately equal to 0, which is regarded as the key in narrowing the bandgap of TiO_2 and shifting it to the visible light range. Moreover, the darkish color of MXene promotes the absorption of visible light and hence improves the overall photocatalytic performance of TiO_2–MXene composites [45].

In the current development, Khan et al. [46] fabricated $Ti_3C_2/TiO_2/g\text{-}C_3N_4$ ternary nanocomposites for bi-reforming of methane into CO and H_2 under visible light in the presence of CO_2 gas. The ternary composite was prepared by mixing the three photocatalysts through a simple ultrasonication method as illustrated in Figure 5.5a. Pure TiO_2 and g-C_3N_4 samples exhibited UV–vis DRS spectra with absorption bands around the wavelengths of 400 and 460 nm, respectively, as depicted in Figure 5.5b. However, it is obvious that the photon's harvesting efficiency was significantly enhanced after coupling the TiO_2 and g-C_3N_4 with MXene (Ti_3C_2), leading to improved optical absorption in the visible light region. Similar trends were also observed and reported in another study by Ding et al. [48] who fabricated

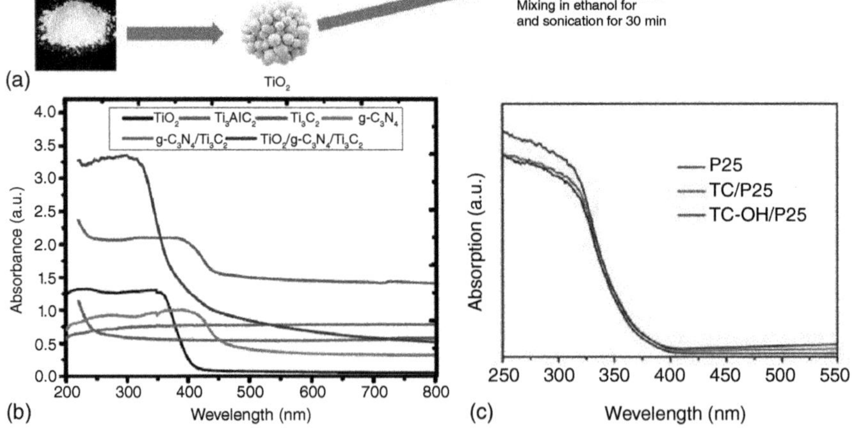

Figure 5.5 (a) The synthesis procedure of the $Ti_3C_2/TiO_2/g\text{-}C_3N_4$ ternary nanocomposite, (b) The UV–vis DRS spectrum of TiO_2, Ti_3AlC_2, Ti_3C_2, g-C_3N_4, g-C_3N_4/Ti_3C_2 and TiO_2/g-C_3N_4/Ti_3C_2. Source: Khan et al. [46]/with permission of American Chemical Society. (c) The UV–vis DRS spectrum of pure TiO_2 (P25) and its coupled composites of MXene and hydroxide-terminated MXene. Source: Ye et al. [47]/with permission of John Wiley & Sons.

the same ternary hybrids but used it for photo-degradation of RhB dye under visible light irradiation.

Ye et al. [47] investigated the effect of coupling MXene (Ti_3C_2) as a cocatalyst with a commercial type of TiO_2 (P25) through a simple physical mixing method for photocatalytic CO_2 reduction. Moreover, the effect of MXene surface termination was studied by synthesizing MXene with two different surface termination groups of hydroxyls ($-OH^-$) and fluorine (–F) groups. As shown in Figure 5.5c, the coupling of MXene (TC) with the TiO_2 (P25) exhibited a slight increase for the light absorbance in the visible range compared to pure TiO_2. However, more absorbance of light was displayed by the hydroxide-terminated MXene (TC-OH), resulting in enhanced photocatalytic CO_2 reduction into CO and CH_4 by 3 and 277 times, respectively.

5.2.3.3 Enhancing Reactant Adsorption

The MXene materials are known to have a large specific surface area, making them potential adsorbents for gases, liquids, and heavy metals [49]. In addition to boosting the separation of charge carriers and improving the light absorption in the TiO_2/MXene composites, the 2D heterostructure of MXene can maximize the contact area of TiO_2 composites with 0D, 1D, and 2D structures [50]. Owing to the good interaction between the surface termination groups of MXene and the reactant gas molecules, MXene can also provide adsorption active sites in the TiO_2/MXene composites which promote the reactant adsorption and hence improve the photocatalytic performance.

Qian et al. [51] reported that a one-step hydrothermally synthesized $TiO_2/Ti_3C_2T_x$ MXene composite displayed larger BET surface area ($14.72\,m^2\,g^{-1}$) compared to that of pure TiO_2 NSs ($10.197\,m^2\,g^{-1}$) and $Ti_3C_2T_x$MXene ($1.98\,m^2\,g^{-1}$). The highest ability for gas adsorption was obtained with the TiO_2/MXene composite compared to both pure materials, suggesting that more active sites were present by coupling MXene with TiO_2.

In another study [52], it was revealed that the BET specific surface area of TiO_2/MXene composites increases with the increase of MXene content up to 37%. The synthesized TiO_2/MXene composites with MXene contents of 2%, 7%, 30%, and 37% exhibited surface areas of 0.5, 0.9, 2.7, and $10.2\,m^2\,g^{-1}$ after hydrothermal treatment times of 28, 24, 20, and 16 hours, respectively. However, the best photoactivity for the degradation of methylene blue (MB) dye was achieved with a surface area of $2.7\,m^2\,g^{-1}$ obtained by 30% MXene, indicating that the photocatalytic enhancement cannot be merely determined by the photocatalyst surface area.

The type of the surface termination group in MXene can have a direct impact on the adsorption ability and photocatalytic performance of MXene as approved by Ye et al. [47]. The surface-alkalinized MXene exhibited higher adsorption of CO_2 molecules than the untreated MXene as shown in Figure 5.6a. Moreover, the product selectivity over the MXene/TiO_2 composite was shifted from CO to CH_4 production after the surface alkalization (–OH) treatment as depicted in Figure 5.6b. Sun et al. [53] have also reported that MXene with the–O terminal group exhibited higher results for photocatalytic H_2 production than those reported with MXene with the–F termination group.

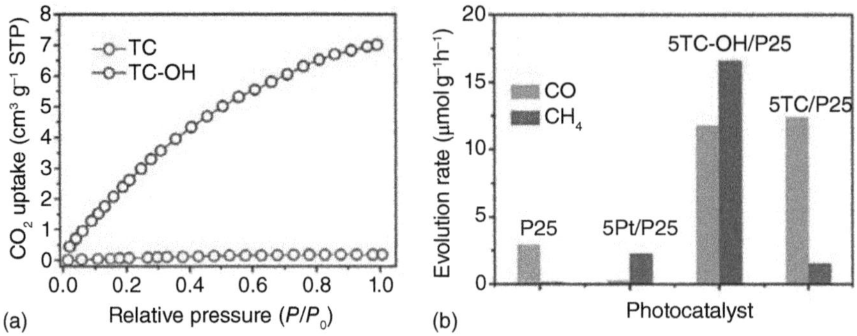

Figure 5.6 (a) The CO_2 gas uptake ability of MXene (TC) and MXene with −OH surface termination group; (b) the effect of MXene surface alkalization (−OH) treatment on the products selectivity over MXene/TiO_2 composite. Source: Ye et al. [47]/with permission of John Wiley & Sons.

5.3 TiC-Based TiO_2 Composite for CO_2 Reduction

In recent years, tremendous efforts have been devoted to solve the greenhouse gas effect and the ever-increasing demand on fossil fuels. Photocatalytic CO_2 reduction has been introduced as one of the most promising strategies for solving these issues [54]. Many photocatalysts have been investigated for the conversion of CO_2 gas into CO, CH_4, CH_3OH, and many other valuable chemicals. However, the low efficiency is considered the biggest challenge faced by most of the used photocatalysts due to the high recombination rate of photogenerated electron/hole pairs, high thermodynamic stability of the CO_2 gas, and the poor adsorption ability toward CO_2 molecules [55, 56]. Thus, the development of efficient and low-cost photocatalysts is highly needed. Recent research studies have approved that MXene/TiO_2 composites are good candidates in this field. The photocatalytic CO_2 reduction performances of the typical MXene/TiO_2 composite photocatalysts are summarized in Table 5.1.

In 2021, Khan et al. [57] synthesized a double-TiO_2 phase (anatase-rutile)/MXene composite for photocatalytic CO_2 reduction in the presence of water and under visible light irradiations as shown in Figure 5.7a. This composite was prepared by insitu growth of rutile TiO_2 nanoparticles over the MXene (Ti_3C_2) followed by coupling with the anatase TiO_2 through an ultrasonic approach. The $TiO_2A/Ti_3C_2/TiO_2R$ composite with 10 wt. % of Ti_3C_2/TiO_2R exhibited the highest production of both CO and CH_4, as shown in Figure 5.7b,c. The effect of reductant was also investigated by using $CO_2 + CH_4$ in a dry reforming process and $CO_2 + CH_4 + H_2O$ in a bi-reforming process for the production of CO and H_2 gases. However, the optimized $TiO_2A/Ti_3C_2/TiO_2R$ composite displayed the highest photocatalytic performance throughout all the investigated processes. This photocatalytic enhancement was attributed to the optoelectronic properties of MXene, which increased the light absorption and suppressed the high recombination rate of TiO_2. Moreover, the constructed 0D/2D/0D heterojunction facilitated the migration and transfer of electrons from the TiO_2 (anatase) to the TiO_2 (rutile) and then accumulated on the MXene (Ti_3C_2) surface which acted as an electron reservoir.

Table 5.1 Recent developments of MXene/TiO_2 composites for photocatalytic CO_2 reduction.

Photocatalyst(s)	Feed composition	Light source Power Wavelength Intensity	Reactor(parameters)	Product(s)(rate)	Year	References
$TiO_2A/Ti_3C_2/TiO_2R$	CO_2 gas,H_2O vapor 100 mg catalyst	Visible light 35 W Xe lamp $\lambda = 420$ nm $I = 20$ mW cm^{-2}	Stainless-steel reactor Vol. = 106 ml Temp. = 25 °C Press. = 0.1 bar Time = 4 h	CO 85 μmol h^{-1} g^{-1} CH_4 18 μmol h^{-1} g^{-1}	2021	[57]
CoAlLA-LDH/TiO_2/Ti_3C_2	CO_2 gas, H_2O vapor 100 mg catalyst	Visible light 35 W Xe lamp $\lambda = 420$ nm $I = 20$ mW cm^{-2}	Stainless-steel reactor Vol. = 106 ml Temp. = 25 °C Press. = 0.1 bar Time = 4 h	CO 46.32 μmol h^{-1} g^{-1} CH_4 31.02 μmol h^{-1} g^{-1}	2021	[58]
TiO_2/Ti_3C_2	CO_2 gas, 200 ml of 10% TEOA sol. 200 mg catalyst	500 W Hg lamp	Quartz reactor Vol. = 450 ml Temp. = 25 °C Time = 1 h	CO 120.83 μmol h^{-1} g^{-1} CH_4 15.91 μmol h^{-1} g^{-1}	2021	[59]
$TiO_2/C_3N_4/Ti_3C_2$	0.3 ml 2 M H_2SO_4 0.084 g$NaHCO_3$ 30 mg catalyst	Visible light 350 W Xe lamp $\lambda = > 400$ nm	Pyrex reactor Temp. = 25 °C Time = 1 h	CO 4.39 μmol h^{-1} g^{-1} CH_4 1.2 μmol h^{-1} g^{-1}	2020	[60]

(Continued)

Table 5.1 (Continued)

Photocatalyst(s)	Feed composition	Light source Power Wavelength Intensity	Reactor(parameters)	Product(s)(rate)	Year	References
TiO_2/g-C_3N_4/Ti_3C_2	CO_2 gas, CH_4 gas H_2O vapor 100 mg catalyst	Visible light 35 W Xe lamp $\lambda = 420$ nm $I = 20$ mWcm^{-2}	Stainless-steel reactor Vol. = 106 mL Temp. = 25 °C Press. = 0.1 bar Time = 4 h	CO 48.38 μmol h^{-1} g^{-1} H_2 83.2 μmol h^{-1} g^{-1}	2020	[46]
TiO_2/Ti_3C_2	CO_2 gas, DI H_2O 84 mg $NaHCO_3$, 3 ml 4 M HCl, 50 mg catalyst	Visible light 300 W Xe lamp	Pyrex reactor Vol. = 200 ml Temp. = 25 °C Time = 1 h	CH_4 0.22 μmol h^{-1} g^{-1}	2018	[61]
P25/ Ti_3C_2-OH	CO_2 gas, 3 ml H_2O, 50 mg catalyst	300 W Xe lamp	Reaction cell Press. = 70 kPa Temp. = 25 °C Time = 1 h	CO 11.74 μmol h^{-1} g^{-1} CH_4 16.61 μmol h^{-1} g^{-1}	2018	[47]

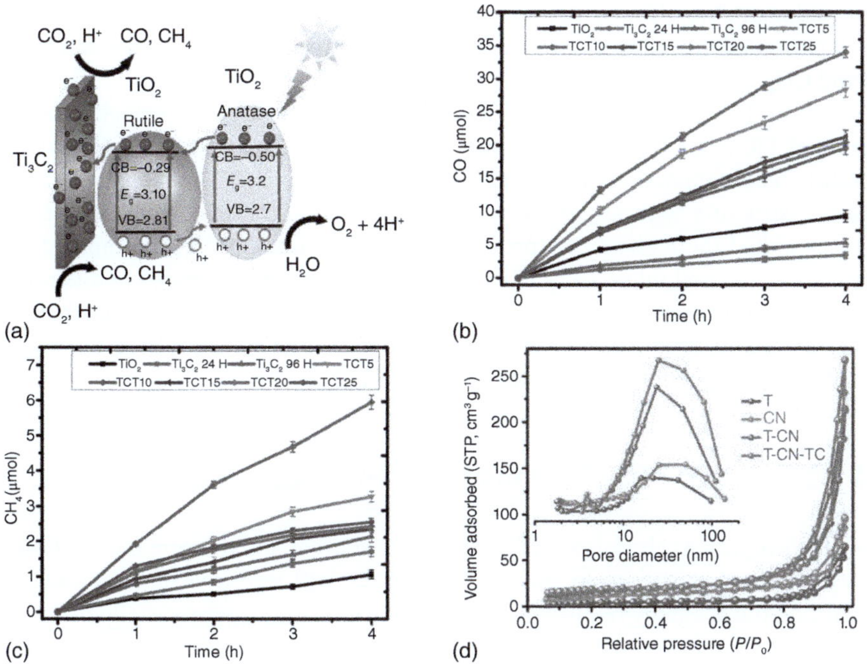

Figure 5.7 (a) The photocatalytic CO_2 reduction mechanism over the $TiO_2A/Ti_3C_2/TiO_2R$ composite, (b, c) The production rates of CO and CH_4 over pure TiO_2A, pure Ti_3C_2 and $TiO_2A/Ti_3C_2/TiO_2R$ composite with various TiO_2A contents. Source: Khan et al. [57]/with permission of Elsevier. (d) The N_2 physisorption isotherms and the corresponding pore size distribution curves of pure TiO_2, pure C_3N_4, TiO_2/C_3N_4, and $TiO_2/C_3N_4/Ti_3C_2$ composite. Source: He et al. [60]/with permission of Elsevier.

In another work, Khan et al. [57] have also fabricated a novel S-scheme photocatalyst of TiO_2/MXene composite with trimetallic CoAlLa-LDH for photocatalytic CO_2 reduction under visible light. Compared to the pure materials, the CoAlLa-LDH/TiO_2/Ti_3C_2 nanocomposite exhibited higher production rates of 46.32 and 31.02 $\mu mol\ g^{-1}\ h^{-1}$ for CO and CH_4, respectively. The enhanced photocatalytic performance was a result of the formed S-scheme heterojunction which provided extra pathways for the transfer of photo generated electron–hole pairs. Moreover, the presence of La^{3+} in the trimetallic LDH increased the surface-active sites for improved sorption of reactant molecules.

In a recent study conducted in 2020 by He et al. [60], it was revealed that the introduction of MXene (Ti_3C_2) into TiO_2/C_3N_4 nanosheet composites can effectively increase the adsorption of CO_2 gas molecules. Figure 5.7d shows the N_2 physisorption isotherms and the corresponding pore size distribution curves of the $TiO_2/C_3N_4/Ti_3C_2$ ternary composite compared to pure TiO_2, pure C_3N_4, and TiO_2/C_3N_4 composite. The ternary photocatalyst exhibited a pore volume of $0.37\ cm^3\ g^{-1}$ and an average pore diameter of 25.7 nm compared to $0.09\ cm^3\ g^{-1}$ and 22.0 nm, $0.13\ cm^3\ g^{-1}$ and 15.2 nm, and $0.30\ cm^3\ g^{-1}$ and 20.3 nm that of pure TiO_2, pure C_3N_4, and TiO_2/C_3N_4 composite, respectively. Coupling of these three semiconductors has also resulted in a construction of an S-scheme heterojunction which prompted the redox reaction for enhanced photocatalytic CO_2 reduction into CO and CH_4.

Low et al. [61] have designed a TiO_2/Ti_3C_2MXene composite through in situ growth of TiO_2 nanoparticles over the MXene surface by a simple calcination approach. The obtained composite (TT-550), calcined at 550 °C, displayed a rice crust-like structure with larger specific surface area. Moreover, the TT-550 composite exhibited the highest CH_4 production rate (0.22 μmol h^{-1} g^{-1}) compared to pure TiO_2 (TT0) and other TiO_2/Ti_3C_2 composites that were prepared at different calcination temperatures. The photocatalytic enhancement was mainly attributed to the MXene electron reservoir characteristic which greatly suppressed the high recombination rate of photogenerated charges in TiO_2.

5.4 TiC-Based TiO_2 Composite for Hydrogen Production

Fossil fuels represent almost 80% of our energy resources and the demand on these fuels is expected to rise up to 56% by 2040, which will lead to a severe shortage in fulfilling these energy requirements [22, 62]. Thus, looking for sustainable and renewable technologies for energy production and environmental remediation is essential in future. Hydrogen-based energy is one of the strongest competitors in this field. Hydrogen is the most abundant chemical element in the universe, representing 75% of normal matter by mass and over 90% by number of atoms. Hydrogen, a clean energy carrier, is also considered a clean alternative since the hydrogen direct combustion generates no pollutants or greenhouse gases and also because of its high energy yield (122 kJ g^{-1}), which is 2.75 times greater than that of hydrocarbon fuels [63, 64]. Nowadays, the H_2 gas is widely used as a raw material in chemical industries, food processing, hydrogenation processes, ammonia and methanol production, and energy production systems. Recently, a new technology known as photocatalytic water splitting for hydrogen production has attracted researchers' attention as a clean and sustainable strategy. However, most of the used photocatalysts still suffer from fast recombination rate of photogenerated charges and low visible light utilization capacity [65, 66], which limit their practical use. However, the TiO_2/MXene photocatalysts have shown great performance and gained a significant research interest in H_2 production.

In this perspective, a TiO_2/MXene (Ti_3C_2) nanocomposite was successfully prepared by Zhuang et al. [67] through an electrostatic self-assembly technique. As shown in Figure 5.8a, the 1D/2D nanocomposite with 3 wt% Ti_3C_2 exhibited excellent photocatalytic water splitting for hydrogen production with a rate of 6.979 μmol h^{-1} g^{-1} which is 3.8 times higher than that of pure TiO_2. Owing to the dark color of MXene, the TiO_2/Ti_3C_2 photocatalyst exhibited much stronger absorbance in the visible light range than the pure TiO_2 sample. Although the production of photogenerated electrons is not affected by the light absorption caused by MXene, the photothermal effect generated by the black MXene is favorable in the catalytic redox reactions [43, 70].

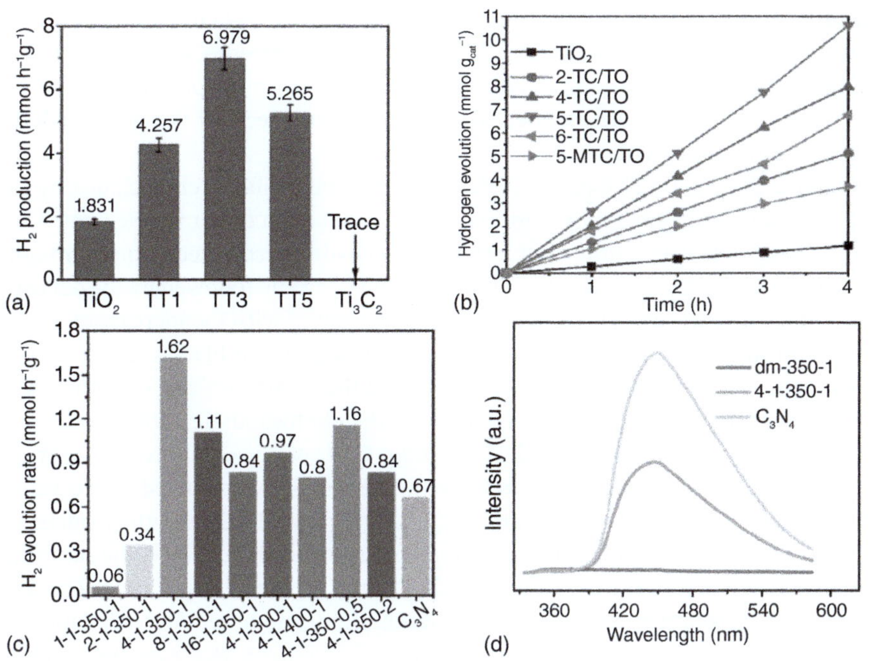

Figure 5.8 (a, b) The photocatalytic H_2 production over pure TiO_2 compared to different TiO_2/Ti_3C_2 composites. Source: (a) Zhuang et al. [67]/with permission of Elsevier. (b) Su et al. [68]/with permission of American Chemical Society. (c) The PL spectra of dm-350-1, 4-1-350-1, and g-C_3N_4, (d) The photocatalytic H_2 evolution rate over different samples. Source: Zhang et al. [69]/with permission of John Wiley & Sons.

In a similar study [68], a $TiO_2/Ti_3C_2T_x$ nanocomposite was synthesized through ultrasonic treatment for photocatalytic water splitting into hydrogen under UV light irradiation. However, the highest H_2 evolution of 2650 μmol h^{-1} g^{-1} was exhibited with the content of 5.0 wt% MXene as depicted in Figure 5.8b, and the achieved result was nine times higher than that of pure TiO_2. This superior photocatalytic enhancement was attributed to the electronic conductivity of monolayer MXene, which resulted in efficient separation and transfer of photogenerated charges. By using visible light, Wang et al. [43] have also conducted a study using the $TiO_2/Ti_3C_2T_x$ composite and revealed that the highest H_2 production through photocatalytic water splitting can be achieved with 5.0 wt% MXene ($Ti_3C_2T_x$) with the use of methanol as a sacrificial agent.

In a recent study conducted by Tian et al. [71], a porous metal organic framework known as UiO-66-NH_2TiO_2 was introduced to a MXene/TiO_2 composite for increasing the specific surface area and light absorption, and the novel Ti_3C_2/TiO_2/UiO-66-NH_2 composite displayed a BET surface area of 987.5 m^2 g^{-1}, compared to 4.2, 10.5, and 977.6 m^2 g^{-1} that of pure Ti_3C_2, Ti_3C_2/TiO_2, and UiO-66-NH_2, respectively. Furthermore, the ternary composite exhibited the highest pore volume (0.6 cm^3 g^{-1}) among all the tested samples. Although the addition of MOF was favorable as mentioned previously, the UiO-66-NH_2 MOF still suffers from a high recombination rate of photogenerated charges, which was greatly reduced by the introduction of Ti_3C_2 and furthermore with Ti_3C_2/TiO_2. Therefore, the superior photocatalytic performance with a H_2 production rate of 1980 μmol h^{-1} g^{-1} could be ascribed to the synergistic effects of the porous MOF and the conductive of MXene.

A unique 2D-2D-2D Ti_3C_2@TiO_2@MoS_2 photocatalyst has been designed by Li et al. [72] and investigated for photocatalytic H_2 production under visible light. The ternary composite was synthesized by a two-step hydrothermal technique through in situ growth of TiO_2 nanosheets on the conductive MXene followed by the deposition of MoS_2 nanosheets. The production of H_2 increased with the increase of MoS_2 content up to 15% due to the shielding effect caused by the black color of MoS_2 nanosheets. However, the investigations revealed the significant role of both Ti_3C_2 and MoS_2 as effective co-catalysts in improving the photocatalytic performance.

A binary mixture of delaminated Ti_3C_2 and g-C_3N_4 was used to prepare a series of d-Ti_3C_2/TiO_2/g-C_3N_4 nanocomposites [69]. The samples were mixed with certain mass ratios of 1 : 1, 2 : 1, 4 : 1, 8 : 1, and 16 : 1 for g-C_3N_4/d-Ti_3C_2 and then calcined at 300, 350, and 400 °C for 0.5, 1, and 2 hours. As shown inFigure 5.8c, the samples were denoted as *x-y-T-h* (where *x* is the g-C_3N_4 ratio, *y* is the d-Ti_3C_2 ratio, *T* is the calcination temperature, and *h* is the calcination time) and tested for their capability of photocatalytic hydrogen generation under visible light. However the highest H_2 evolution rate of 1.62 mmol h^{-1} g^{-1} was achieved with the 4-1-350-1 composite. The photocatalytic enhancement was attributed to the improved charge separation efficiency of the ternary nanocomposite, which was further confirmed with the PL analysis results as depicted in Figure 5.8d.

Table 5.2 summarizes the recent developments of MXene/TiO_2 composites for photocatalytic H_2 production with detailed information of the used light and the applied reaction parameters.

5.5 TiC-Based TiO_2 Composite for Degradation

The organic chemicals and dyes are considered the most dangerous environmental contaminants. The long-lasting toxicities of theses pollutants are effective even at low concentrations, making them high-risk contaminants that need immediate action. Many techniques have been employed for the removal of these pollutants including filtration, precipitation, and adsorption. However, the presence of these chemicals at low concentrations limits their applications; therefore, photocatalytic degradation has been introduced as an efficient and cost-effective technique for the capture and removal of these pollutants at ppb and ppm levels. MXene/TiO_2 composites proved to be excellent materials for photocatalytic degradation applications owing to the MXene wide adsorption wavelength ranging from the ultraviolet (UV) to infrared (IR) region [76]. Moreover, TiO_2 as a semiconductor has a good chemical stability, high photoactivity, abundancy, and low toxicity [77]. Table 5.3 summarizes the recent developments of MXene/TiO_2 composite for pollutant photodegradation.

Table 5.2 Recent developments of MXene/TiO_2 composites for photocatalytic H_2 production.

Photocatalyst(s)	Sacrificial agent	Light source Power Wavelength Intensity	Reactor Parameters	Production rate	Year	References
TiO_2/Ti_3C_2	CH_3OH	Visible light 300 W Xe lamp	Pyrex reaction cell Time = 1 h Catal. Wt. = 50 mg	6.979 μmol h^{-1} g^{-1}	2019	[67]
$TiO_2/Ti_3C_2T_x$	CH_3OH	UV light 200 WHg lamp λ = 285–325 nm	Quartz reactor Vol. = 65 ml Temp. = 25 °C Time = 1 h Catal. Wt. = 30 mg	2650 μmol h^{-1} g^{-1}	2019	[68]
Ti_3C_2/TiO_2/UiO-66-NH_2	Na_2S/Na_2SO_3	UV–visible light 350 W Xe lamp λ = 350–780 nm	CEL-SPH2N system Temp. = 5 °C Time = 1 h Catal. Wt. = 20 mg	1980 μmol h^{-1} g^{-1}	2019	[71]
$Ti_3C_2/TiO_2/MoS_2$	TEOA	Visible light 300 W Xe lamp I = 513 mW cm^{-2}	Pyrex reactor Temp. = 25 °C Time = 1 h Catal. Wt. = 10 mg	6425 μmol h^{-1} g^{-1}	2019	[72]
TiO_2/g-C_3N_4/Ti_3C_2	TEOA	Visible light 350 W Xe lamp λ = >420 nm	Pyrex reactor Vol. = 80 ml Time = 1 h	1620 μmol h^{-1} g^{-1}	2018	[69]

(Continued)

Table 5.2 (Continued)

Photocatalyst(s)	Sacrificial agent	Light source Power Wavelength Intensity	Reactor Parameters	Production rate	Year	References
$Mo_XS@TiO_2@Ti_3C_2$	TEOA	UV–visible light 300 W Xe lamp λ=> 200–1200 nm I = 180 mW cm^{-2}	Pyrex reactor Temp. = 25 °C Time = 1 h Catal. Wt. = 10 mg	10 505.8 μmol h^{-1} g^{-1}	2019	[73]
$TiO_2/Ti_3C_2/CoS_x$	CH_3OH	UV–visible light	Time = 1 h	950 μmol h^{-1} g^{-1}	2019	[74]
TiO_2/Ti_3C_2	None	UV light 300 W Xe lamp λ = 350 nm	Pyrex reactor Time = 1 h Catal. Wt. = 20 mg	526 μmol h^{-1} g^{-1}	2018	[75]
$Cu/TiO_2@Ti_3C_2T_x$	CH_3OH	Visible light 300 W Xe lamp	CEL-SPH2N system Temp. = 25 °C Time = 1 h Catal. Wt. = 20 mg	860 μmol h^{-1} g^{-1}	2018	[44]
$Cu_4/TiO_2@ Ti_3C_2T_x$	CH_3OH	Visible light 350 W Xe lamp	CEL-SPH2N system Time = 1 h Catal. Wt. = 20 mg	764 μmol h^{-1} g^{-1}	2018	[44]
$TiO_2/Ti_3C_2T_x$	CH_3OH	Visible light 200 W Hg lamp λ = >400 nm	Quartz reactor Temp. = 25 °C Time = 1 h Catal. Wt. = 40 mg	17.8 μmol h^{-1} g^{-1}	2016	[43]

Table 5.3 Recent developments of MXene/TiO_2 composites for pollutant photodegradation.

Photocatalyst(s)	Pollutant(s)	Light source Power Wavelength Intensity	Reaction conditions	Degradation efficiency	Year	References
$TiO_2/Ti_3C_2T_x$	Benzene	Near IR 100 W Philips lamp $I = 35\ mW\ cm^{-2}$	Flow = 150 ml min^{-1} Conc. = 100 μg m^{-3} Temp. = 25 °C Time = 5 h Catal. Wt. = 100 mg	57.6% 56.0%	2022	[78]
(001) TiO_2/Ti_3C_2	Methyl orange	Visible light 500 W Xe lamp	Temp. = 25 °C Catal. Wt. = 50 mg	96%	2021	[79]
g-$C_3N_4/Ti_3C_2/TiO_2$	Rhodamine B Tetracycline hydrochloride	Visible light 300 W Xe lamp $\lambda = < 420\ nm$	Vol. = 200 ml Conc. = 10 mg l^{-1} Temp. = 25 °C	96.04% 85.12%	2020	[80]
CdS@Ti_3C_2@TiO_2	Rhodamine B Methylene blue Sulfachloropyridazine Phenol	Visible light 300 W Xe lamp $\lambda = 400–1050\ nm$ $I = 300\ mW\ cm^{-2}$	Vol. = 200 ml Conc. = 20 mg L^{-1} Temp. = 25 °C Catal. Wt. = 50 mg	100% in 60 min 100% in 88 min 100% in 88 min 100% in 150 min	2019	[81]
In_2S_3/TiO_2@ $Ti_3C_2T_x$	Methyl orange	Visible light 300 W Xe lamp $\lambda = <420\ nm$ $I = 300\ mW\ cm^{-2}$	Vol. = 100 ml Conc. = 20 mg l^{-1} Time = 1 h Catal. Wt. = 60 mg	92.1%	2018	[82]
$Ti_2C/TiO_2/Ag$	Salicylic acid	UV light 150 W Hg lamp	Vol. = 400 ml Conc. = 100 μM Temp. = 18–20 °C Time = 3 h Catal. Wt. = 250 mg	97.1%	2019	[83]

Table 5.3 (Continued)

Photocatalyst(s)	Pollutant(s)	Light source Power Wavelength Intensity	Reaction conditions	Degradation efficiency	Year	References
$TiO_2/Ti_3C_2/CuO$	Methyl orange	UV light 175 W Hg lamp	Vol. = 100 ml Conc. = 20 mg L^{-1} Temp. = 10 °C Time = 80 min Catal. Wt. = 100 mg	99%	2017	[84]
$TiO_2/Ti_3C_2T_x$	Carbamazepine	UV light 100 W Hg lamp	Vol. = 10 ml Conc. = 5 mg l^{-1} Temp. = 25 °C Time = 4 h Catal. Wt. = 10 mg	98.67%	2018	[85]
(111) TiO_2/Ti_3C_2	Methyl orange Methylene blue	Visible light 500 W Xe lamp λ = <400 nm	Vol. = 200 ml Conc. = 20 mg l^{-1} Time = 150 min Catal. Wt. = 10 mg	75%	2017	[52]
(001) TiO_2/Ti_3C_2	Methyl orange	UV light 300 W Hg lamp	Vol. = 200 ml Conc. = 20 mg l^{-1} Time = 50 min Catal. Wt. = 10 mg	97.4%	2016	[77]
TiO_2/Ti_3C_2	Methyl orange	UV light 175 W Hg lamp	Vol. = 100 ml Conc. = 20 mg L^{-1} Temp. = 10 °C Time = 0.5 h Catal. Wt. = 100 mg	98%	2015	[86]

In 2022, Sergiienko et al. [78] fabricated MXene/TiO_2 composite for photocatalytic degradation of benzene with a simulated solar light atnear-IR range. The MXene/TiO_2 composite was synthesized through two strategies, the first one was conducted by the thermal treatment of TiO_2-based aqueous solutions at 450 °C for 2 hours, followed by mixing the yielded crystalline titania with different MXene amounts (1, 5, 10, 20, 30, 40, and 50 wt%) in a rotary evaporator containing 10 ml of IPA. The final product (I composite) was then obtained by the removal of the solvent at 60 °C. The second composite (II composite) was obtained through the direct mixing of the TiO_2-based aqueous solutions with the MXene amounts (same used in I composite) followed by the same procedure in the first strategy. Figure 5.9 a,b demonstrates the photocatalytic results of benzene removal by I composite and II composite, respectively. For I composite, the best result of 58.9% benzene removal was obtained with 5 wt% MXene compared to 57.6% that of pure TiO_2 while for II composite, the best result of 56.0% was obtained with 10 wt% MXene compared to 22.9% achieved by the unmodified TiO_2. Obviously, the photocatalytic

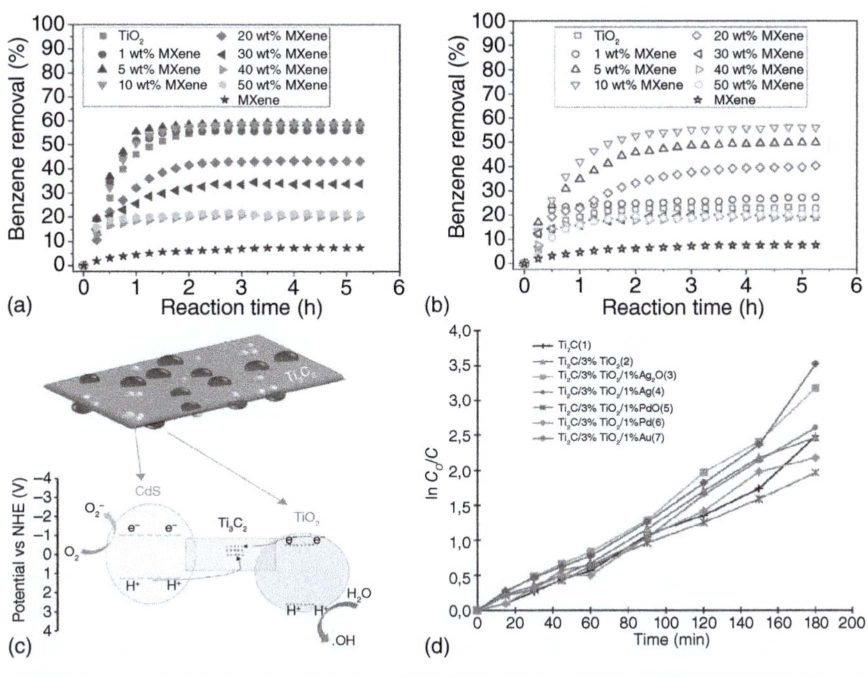

Figure 5.9 (a, b) The photocatalytic results of benzene removal over pure TiO_2 and pure MXene compared to type I composites and type II composites, respectively. Source: Sergiienko et al. [78]/with permission of Royal Society of Chemistry.(c) The proposed reaction mechanism over the Z-scheme CdS@Ti_3C_2@TiO_2 photocatalyst. Source: Liu et al. [81]/with permission of Royal Society of Chemistry. (d) The photodegradation performance of SA over different MXene photocatalysts. Source: Wojciechowski et al. [83]/with permission of American Chemical Society.

enhancement was attributed to the presence of MXene which promoted the optical absorption in the visible range.

A ternary g-C_3N_4/Ti_3C_2/TiO_2 NTA photocatalyst was successfully fabricated on Ti mesh by Diao et al. [80]. The composite was tested for photocatalytic degradation of rhodamine B (RhB) dye and tetracycline hydrochloride (TC-HCl) antibiotic under visible light irradiation. The pristine TNTAs exhibited the lowest photocatalytic performance for both RhB and TC-HCl due to its low response under visible light in addition to the high recombination rate of photogenerated charge carriers. However, the highest degradation rates of 96.04% and 85.12% for RhB and TC-HCl, respectively, were achieved with the TCN1 composite (TNTAs with 10 : 1, MXene/g-C_3N_4).

A novel ternary CdS@Ti_3C_2@TiO_2 composite was also synthesized through facile calcination of MXene (Ti_3C_2) followed by hydrothermal treatment of the obtained Ti_3C_2@TiO_2 product in a $CdCl_2$/Na_2S solution[81]. The Z-scheme photocatalyst was tested for photocatalytic degradation of four different organic pollutants; rhodamine B, methylene blue, sulfachloropyridazine, and phenol. Compared to pure MXene and CdS, the CdS@Ti_3C_2@TiO_2 composite exhibited efficient complete removal (100%) for the four contaminants after 60, 88, 88, and 150 minutes, respectively. The enhanced photodegradation performance was attributed to the suppressed recombination of electron–hole pairs due to the formed Z-scheme heterojunction as shown in Figure 5.9c.

In 2018, Wang et al. [82] synthesized a quasi-core–shell In_2S_3/TiO_2@MXene hybrid through a simple in situ hydrothermal method for the degradation for methyl orange (MO) dye under visible light irradiation. Compared to pure TiO_2, metallic $Ti_3C_2T_x$ and In_2S_3, the ternary hybrid exhibited a superior efficiency of 92.1% for the MO removal. This photocatalytic enhancement in fact was achieved as a result of the synergistic effect of the excellent electrical conductivity of the metallic MXene, the presence of visible light-responsive In_2S_3, and the formed type-II heterojunction structure which facilitated the separation and transfer of photogenerated charge carriers.

Wojciechowski et al. [83] compared the salicylic acid (SA) photodegradation ability of pure MXene with six composites consisting of the MXene modified by metal oxide (TiO_2, Ag_2O, and PdO) and metallic (Ag, Pd, and Au) nanoparticles. As shown in Figure 5.9d, the highest SA degradation efficiency of 97.1% was achieved with the Ti_2C/3% TiO_2/1% Ag nanocomposite under UV light. The key explanation for this photocatalytic enhancement was the oxidation of the metal with OH· radicals to form positive metal (M^+) centers attracting the photogenerated electrons that causes the charge recombination and reconstruction of M atoms.

Acknowledgment

The financial support of this work is provided by United Arab Emirates University (UAEU),UAE under Start up research grant vot no. 12N097.

References

1 Huskinson, B., Marshak, M.P., Suh, C. et al. (2014). A metal-free organic–inorganic aqueous flow battery. *Nature* 505 (7482): 195–198.

2 Alkhatib, I.I., Garlisi, C., Pagliaro, M. et al. (2020). Metal-organic frameworks for photocatalytic CO_2 reduction under visible radiation: a review of strategies and applications. *Catal. Today* 340: 209–224.

3 Bekun, F.V., Alola, A.A., and Sarkodie, S.A. (2019). Toward a sustainable environment: nexus between CO_2 emissions, resource rent, renewable and nonrenewable energy in 16-EU countries. *Sci. Total Environ.* 657: 1023–1029.

4 Ikreedeegh, R.R. and Tahir, M. (2021). A critical review in recent developments of metal-organic-frameworks (MOFs) with band engineering alteration for photocatalytic CO_2 reduction to solar fuels. *J. CO2 Util.* 43: 101381.

5 Goeppert, A., Czaun, M., Jones, J.-P. et al. (2014). Recycling of carbon dioxide to methanol and derived products–closing the loop. *Chem. Soc. Rev.* 43 (23): 7995–8048.

6 Tahir, M. and Amin, N.S. (2013). Advances in visible light responsive titanium oxide-based photocatalysts for CO_2 conversion to hydrocarbon fuels. *Energy Convers. Manage.* 76: 194–214.

7 Ikreedeegh, R. (2022). Recent developments of Fe-based metal organic frameworks and their composites in photocatalytic applications: fundamentals; synthesis and challenges. *Russ. Chem. Rev.* 91: 12.

8 Kabra, K., Chaudhary, R., and Sawhney, R.L. (2004). Treatment of hazardous organic and inorganic compounds through aqueous-phase photocatalysis: a review. *Ind. Eng. Chem. Res.* 43 (24): 7683–7696.

9 Tahir, B., Tahir, M., and Amin, N.S. (2015). Gold–indium modified TiO_2 nanocatalysts for photocatalytic CO_2 reduction with H_2 as reductant in a monolith photoreactor. *Appl. Surf. Sci.* 338: 1–14.

10 Sim, L.C., Leong, K.H., Saravanan, P., and Ibrahim, S. (2015). Rapid thermal reduced graphene oxide/Pt–TiO_2 nanotube arrays for enhanced visible-light-driven photocatalytic reduction of CO_2. *Appl. Surf. Sci.* 358: 122–129.

11 Nguyen, N.-T., Zheng, D.-D., Chen, S.-S., and Chang, C.-T. (2018). Preparation and photocatalytic hydrogen production of Pt-graphene/TiO_2 composites from water splitting. *J. Nanosci.* 18 (1): 48–55.

12 Shehzad, N., Tahir, M., Johari, K. et al. (2018). A critical review on TiO_2 based photocatalytic CO_2 reduction system: strategies to improve efficiency. *J. CO2 Util.* 26: 98–122.

13 Li, Q., Zong, L., Li, C., and Yang, J. (2014). Reprint of "photocatalytic reduction of CO_2 on MgO/TiO_2 nanotube films". *Appl. Surf. Sci.* 319: 16–20.

14 Ge, H., Xu, F., Cheng, B. et al. (2019). S-scheme heterojunction TiO_2/CdS nanocomposite nanofiber as H_2-production photocatalyst. *ChemCatChem* 11 (24): 6301–6309.

15 Al Jitan, S., Palmisano, G., and Garlisi, C. (2020). Synthesis and surface modification of TiO_2-based photocatalysts for the conversion of CO_2. *Catalysts* 10 (2): 227.

16 Zheng, Z., Huang, B., Qin, X. et al. (2009). Highly efficient photocatalyst: TiO_2 microspheres produced from TiO_2 nanosheets with a high percentage of reactive {001} facets. *Chem. Eur. J* 15 (46): 12576–12579.

17 Fang, W.Q., Zhou, J.Z., Liu, J. et al. (2011). Hierarchical structures of single-crystalline anatase TiO_2nanosheets dominated by {001} facets. *Chem. Eur. J.* 17 (5): 1423–1427.

18 Hajizadeh, H., Peighambardoust, S.J., Peighambardoust, S.H., and Peressini, D. (2020). Physical, mechanical, and antibacterial characteristics of bio-nanocomposite films loaded with Ag-modified SiO_2 and TiO_2 nanoparticles. *J. Food Sci.* 85 (4): 1193–1202.

19 Xu, J., Zhou, X., Gao, Z. et al. (2016). Visible-light-triggered drug release from TiO_2 nanotube arrays: a controllable antibacterial platform. *Angew. Chem. Int. Ed.* 55 (2): 593–597.

20 Diao, W., Cai, H., Wang, L. et al. (2020). Efficient photocatalytic degradation of gas-phase formaldehyde by Pt/TiO_2 nanowires in a continuous flow reactor. *ChemCatChem* 12 (21): 5420–5429.

21 Sokhandani, P., Babaluo, A.A., Rezaei, M. et al. (2013). Nanocomposites of PVC/TiO_2 nanorods: surface tension and mechanical properties before and after UV exposure. *J. Appl. Polym. Sci.* 129 (6): 3265–3272.

22 Ikreedeegh, R.R. and Tahir, M. (2021). Photocatalytic CO_2 reduction to CO and CH_4 using g-C_3N_4/RGO on titania nanotube arrays (TNTAs). *J. Mater. Sci.* 56 (34): 18989–19014.

23 Cheng, M., Yang, S., Chen, R. et al. (2017). Copper-decorated TiO_2 nanorod thin films in optofluidic planar reactors for efficient photocatalytic reduction of CO_2. *Int. J. Hydrogen Energy* 42 (15): 9722–9732.

24 Azam, M.U., Tahir, M., Umer, M. et al. (2019). Engineering approach to enhance photocatalytic water splitting for dynamic H_2 production using La_2O_3/TiO_2 nanocatalyst in a monolith photoreactor. *Appl. Surf. Sci.* 484: 1089–1101.

25 Shehzad, N., Tahir, M., Johari, K. et al. (2019). Fabrication of highly efficient and stable indirect Z-scheme assembly of $AgBr/TiO_2$ via graphene as a solid-state electron mediator for visible light induced enhanced photocatalytic H_2 production. *Appl. Surf. Sci.* 463: 445–455.

26 Marchal, C., Piquet, A., Behr, M. et al. (2017). Activation of solid grinding-derived Au/TiO_2photocatalysts for solar H_2 production from water-methanol mixtures with low alcohol content. *J. Catal.* 352: 22–34.

27 Yu, J., Wang, S., Low, J., and Xiao, W. (2013). Enhanced photocatalytic performance of direct Z-scheme g-C_3N_4–TiO_2 photocatalysts for the decomposition of formaldehyde in air. *Phys. Chem. Chem. Phys.* 15 (39): 16883–16890.

28 Giannakopoulou, T., Papailias, I., Todorova, N. et al. (2017). Tailoring the energy band gap and edges' potentials of g-C_3N_4/TiO_2 composite photocatalysts for NOx removal. *Chem. Eng. J.* 310: 571–580.

29 Hamandi, M., Berhault, G., Guillard, C., and Kochkar, H. (2017). Reduced graphene oxide/TiO_2 nanotube composites for formic acid photodegradation. *Appl. Catal. B* 209: 203–213.

30 Fujishima, A. and Honda, K. (1972). Electrochemical photolysis of water at a semiconductor electrode. *Nature* 238 (5358): 37–38.

31 Coleman, J.N., Lotya, M., O'Neill, A. et al. (2011). Two-dimensional nanosheets produced by liquid exfoliation of layered materials. *Science* 331 (6017): 568–571.

32 Abe, R. (2010). Recent progress on photocatalytic and photoelectrochemical water splitting under visible light irradiation. *J. Photochem. Photobiol. C* 11 (4): 179–209.

33 Zhao, Q.-N., Zhang, Y.-J., Duan, Z.-H. et al. (2021). A review on Ti_3C_2Tx-based nanomaterials: synthesis and applications in gas and humidity sensors. *Rare Met.* 40 (6): 1459–1476.

34 Huang, K., Li, C., Li, H. et al. (2020). Photocatalytic applications of two-dimensional Ti_3C_2 MXenes: a review. *ACS Appl. Nano Mater.* 3 (10): 9581–9603.

35 Pan, Z., Cao, F., Hu, X., and Ji, X. (2019). A facile method for synthesizing CuS decorated Ti_3C_2 MXene with enhanced performance for asymmetric supercapacitors. *J. Mater. Chem. A* 7 (15): 8984–8992.

36 Jawaid, A., Hassan, A., Neher, G. et al. (2021). Halogen etch of Ti_3AlC_2 MAX phase for MXene fabrication. *ACS Nano* 15 (2): 2771–2777.

37 Gong, K., Zhou, K., Qian, X. et al. (2021). MXene as emerging nanofillers for high-performance polymer composites: a review. *Composites Part B* 217: 108867.

38 Wei, Y., Zhang, P., Soomro, R.A. et al. (2021). Advances in the synthesis of 2D MXenes. *Adv. Mater.* 33 (39): 2103148.

39 Anasori, B., Lukatskaya, M.R., and Gogotsi, Y. (2017). 2D metal carbides and nitrides (MXenes) for energy storage. *Nat. Rev. Mater.* 2 (2): 1–17.

40 Sun, W., Shah, S., Chen, Y. et al. (2017). Electrochemical etching of Ti_2 AlC to Ti_2CT_X (MXene) in low-concentration hydrochloric acid solution. *J. Mater. Chem. A* 5 (41): 21663–21668.

41 Sreedhar, A. and Noh, J.-S. (2021). Recent advances in partially and completely derived 2D Ti_3C_2 MXene based TiO_2 nanocomposites towards photocatalytic applications: a review. *Sol. Energy* 222: 48–73.

42 Zhang, D., Yu, S., Wang, X. et al. (2022). UV illumination-enhanced ultrasensitive ammonia gas sensor based on (001) TiO_2/MXene heterostructure for food spoilage detection. *J. Hazard. Mater.* 423: 127160.

43 Wang, H., Peng, R., Hood, Z.D. et al. (2016). Titania composites with 2D transition metal carbides as photocatalysts for hydrogen production under visible-light irradiation. *ChemSusChem* 9 (12): 1490–1497.

44 Peng, C., Wei, P., Li, X. et al. (2018). High efficiency photocatalytic hydrogen production over ternary $Cu/TiO_2@Ti_3C_2Tx$ enabled by low-work-function 2D titanium carbide. *Nano Energy* 53: 97–107.

45 Biswal, L., Mohanty, R., Nayak, S., and Parida, K. (2022). Review on MXene/TiO_2 nanohybrids for photocatalytic hydrogen production and pollutant degradations. *J. Environ. Chem. Eng.* 107211.

46 Khan, A.A., Tahir, M., and Bafaqeer, A. (2020). Constructing a stable 2D layered Ti_3C_2 MXene cocatalyst-assisted $TiO_2/g\text{-}C_3N_4/Ti_3C_2$heterojunction for tailoring photocatalytic bireforming of methane under visible light. *Energy Fuels* 34 (8): 9810–9828.

47 Ye, M., Wang, X., Liu, E. et al. (2018). Boosting the photocatalytic activity of P25 for carbon dioxide reduction by using a surface-alkalinized titanium carbide MXene as cocatalyst. *ChemSusChem* 11 (10): 1606–1611.

48 Ding, X., Li, Y., Li, C. et al. (2019). 2D visible-light-driven $TiO_2@Ti_3C_2/g\text{-}C_3N_4$ ternary heterostructure for high photocatalytic activity. *J. Mater. Sci.* 54 (13): 9385–9396.

49 Peng, Q., Guo, J., Zhang, Q. et al. (2014). Unique lead adsorption behavior of activated hydroxyl group in two-dimensional titanium carbide. *J. Am. Chem. Soc.* 136 (11): 4113–4116.

50 Yang, J.-X., Yu, W.-B., Li, C.-F. et al. (2021). PtO nanodots promoting Ti_3C_2 MXene in-situ converted Ti_3C_2/TiO_2 composites for photocatalytic hydrogen production. *Chem. Eng. J* 420: 129695.

51 Qian, X., Wei, Y., Sun, M. et al. (2022). Heterostructuring 2D TiO_2 nanosheets in situ grown on $Ti_3C_2T_x$ MXene to improve the electrocatalytic nitrogen reduction. *Chin. J. Catal.* 43 (7): 1937–1944.

52 Peng, C., Wang, H., Yu, H., and Peng, F. (2017). (111) TiO_2-x/Ti_3C_2: synergy of active facets, interfacial charge transfer and Ti^{3+} doping for enhance photocatalytic activity. *Mater. Res. Bull.* 89: 16–25.

53 Sun, Y., Jin, D., Sun, Y. et al. (2018). $gC_3N_4/Ti_3C_2T_x$ (MXenes) composite with oxidized surface groups for efficient photocatalytic hydrogen evolution. *J. Mater. Chem. A* 6 (19): 9124–9131.

54 Ikreedeegh, R.R. and Tahir, M. (2021). Facile fabrication of well-designed 2D/2D porous $g\text{-}C_3N_4$–GO nanocomposite for photocatalytic methane reforming (DRM) with CO_2 towards enhanced syngas production under visible light. *Fuel* 305: 121558.

55 Zhou, Y., Abazari, R., Chen, J. et al. (2022). Bimetallic metal–organic frameworks and MOF-derived composites: recent progress on electro-and photoelectrocatalytic applications. *Coord. Chem. Rev.* 451: 214264.

56 Ikreedeegh, R.R. and Tahir, M. (2021). Indirect Z-scheme heterojunction of NH_2-MIL-125 (Ti) MOF/$g\text{-}C_3N_4$nanocomposite with RGO solid electron mediator for efficient photocatalytic CO_2 reduction to CO and CH_4. *J. Environ. Chem. Eng.* 105600.

57 Khan, A.A., Tahir, M., and Zakaria, Z.Y. (2021). Synergistic effect of anatase/rutile TiO_2 with exfoliated Ti_3C_2TR MXene multilayers composite for enhanced CO_2 photoreduction via dry and bi-reforming of methane under UV–visible light. *J. Environ. Chem. Eng.* 9 (3): 105244.

58 Ali Khan, A. and Tahir, M. (2021). Construction of an S-scheme heterojunction with oxygen-vacancy-rich trimetallic CoAlLa-LDscacH anchored on titania-sandwiched Ti_3C_2multilayers for boosting photocatalytic CO_2reduction under visible light. *Ind. Eng. Chem. Res.* 60 (45): 16201–16223.

59 Chen, L., Huang, K., Xie, Q. et al. (2021). The enhancement of photocatalytic CO_2 reduction by the in situ growth of TiO_2 on Ti_3C_2 MXene. *Catal. Sci. Technol.* 11 (4): 1602–1614.

60 He, F., Zhu, B., Cheng, B. et al. (2020). 2D/2D/0D $TiO_2/C_3N_4/Ti_3C_2$ MXene composite S-scheme photocatalyst with enhanced CO_2 reduction activity. *Appl. Catal. B* 119006.

61 Low, J., Zhang, L., Tong, T. et al. (2018). TiO_2/MXene Ti_3C_2 composite with excellent photocatalytic CO_2 reduction activity. *J. Catal.* 361: 255–266.

62 Chien, F., Kamran, H.W., Albashar, G., and Iqbal, W. (2021). Dynamic planning, conversion, and management strategy of different renewable energy sources: a sustainable solution for severe energy crises in emerging economies. *Int. J. Hydrogen Energy* 46 (11): 7745–7758.

63 Karadag, D., Köroğlu, O.E., Ozkaya, B. et al. (2014). A review on fermentative hydrogen production from dairy industry wastewater. *J. Chem. Technol. Biotechnol.* 89 (11): 1627–1636.

64 Kapdan, I.K. and Kargi, F. (2006). Bio-hydrogen production from waste materials. *Enzyme Microb. Technol.* 38 (5): 569–582.

65 Wang, L., Wang, W., Chen, Y. et al. (2018). Heterogeneous p–n junction CdS/Cu_2O nanorod arrays: synthesis and superior visible-light-driven photoelectrochemical performance for hydrogen evolution. *ACS Appl. Mater. Interfaces* 10 (14): 11652–11662.

66 Chen, Y., Wang, L., Wang, W., and Cao, M. (2017). Enhanced photoelectrochemical properties of $ZnO/ZnSe/CdSe/Cu_{2-x}Se$ core–shell nanowire arrays fabricated by ion-replacement method. *Appl. Catal., B* 209: 110–117.

67 Zhuang, Y., Liu, Y., and Meng, X. (2019). Fabrication of TiO_2 nanofibers/MXene Ti_3C_2 nanocomposites for photocatalytic H_2 evolution by electrostatic self-assembly. *Appl. Surf. Sci.* 496: 143647.

68 Su, T., Hood, Z.D., Naguib, M. et al. (2019). Monolayer $Ti_3C_2T_x$ as an effective co-catalyst for enhanced photocatalytic hydrogen production over TiO_2. *ACS Appl. Ener. Mater.* 2 (7): 4640–4651.

69 Zhang, M., Qin, J., Rajendran, S. et al. (2018). Heterostructured d-Ti_3C_2/TiO_2/g-C_3N_4 nanocomposites with enhanced visible-light photocatalytic hydrogen production activity. *ChemSusChem* 11 (24): 4226–4236.

70 Liu, W., Shen, J., Yang, X. et al. (2018). Dual Z-scheme g-$C_3N_4/Ag_3PO_4/Ag_2MoO_4$ ternary composite photocatalyst for solar oxygen evolution from water splitting. *Appl. Surf. Sci.* 456: 369–378.

71 Tian, P., He, X., Zhao, L. et al. (2019). Enhanced charge transfer for efficient photocatalytic H_2 evolution over UiO-66-NH_2 with annealed $Ti_3C_2T_X$ MXenes. *Int. J. Hydrogen Energy* 44 (2): 788–800.

72 Li, Y., Yin, Z., Ji, G. et al. (2019). 2D/2D/2D heterojunction of Ti_3C_2 MXene/MoS_2 nanosheets/TiO_2 nanosheets with exposed (001) facets toward enhanced photocatalytic hydrogen production activity. *Appl. Catal. B* 246: 12–20.

73 Li, Y., Ding, L., Liang, Z. et al. (2020). Synergetic effect of defects rich MoS_2 and $Ti3C_2$ MXene as cocatalysts for enhanced photocatalytic H_2 production activity of TiO_2. *Chem. Eng. J* 383: 123178.

74 Zhao, J.-H., Liu, L.-W., Li, K. et al. (2019). Conductive Ti_3C_2 and MOF-derived CoSx boosting the photocatalytic hydrogen production activity of TiO_2. *CrystEngComm* 21 (14): 2416–2421.

75 Li, Y., Deng, X., Tian, J. et al. (2018). Ti_3C_2 MXene-derived Ti_3C_2/TiO_2 nanoflowers for noble-metal-free photocatalytic overall water splitting. *Appl. Mater. Today* 13: 217–227.

76 Ng, V.M.H., Huang, H., Zhou, K. et al. (2017). Recent progress in layered transition metal carbides and/or nitrides (MXenes) and their composites: synthesis and applications. *J. Mater. Chem. A* 5 (7): 3039–3068.

77 Peng, C., Yang, X., Li, Y. et al. (2016). Hybrids of two-dimensional Ti_3C_2 and TiO_2 exposing {001} facets toward enhanced photocatalytic activity. *ACS Appl. Mater. Interfaces* 8 (9): 6051–6060.

78 Sergiienko, S.A., Tobaldi, D.M., Lajaunie, L. et al. (2022). Photocatalytic removal of benzene over $Ti_3C_2T_X$ MXene and TiO_2-MXene composite materials under solar and NIR irradiation. *J. Mater. Chem. C* 10 (2): 626–639.

79 Chen, Y., Li, X., Cai, G. et al. (2021). In situ formation of (001) TiO_2/Ti_3C_2 heterojunctions for enhanced photoelectrochemical detection of dopamine. *Electrochem. Commun.* 125: 106987.

80 Diao, Y., Yan, M., Li, X. et al. (2020). In-situ grown of g-$C_3N_4/Ti_3C_2/TiO_2$ nanotube arrays on Ti meshes for efficient degradation of organic pollutants under visible light irradiation. *Colloids Surf. A* 594: 124511.

81 Liu, Q., Tan, X., Wang, S. et al. (2019). MXene as a non-metal charge mediator in 2D layered CdS@ Ti_3C_2@TiO_2 composites with superior Z-scheme visible light-driven photocatalytic activity. *Environ. Sci. Nano* 6 (10): 3158–3169.

82 Wang, H., Wu, Y., Xiao, T. et al. (2018). Formation of quasi-core-shell In_2S_3/anatase TiO_2@metallic $Ti_3C_2T_X$ hybrids with favorable charge transfer channels for excellent visible-light-photocatalytic performance. *Appl. Catal. B* 233: 213–225.

83 Wojciechowski, T., Rozmysłowska-Wojciechowska, A., Matyszczak, G. et al. (2019). Ti_2C MXene modified with ceramic oxide and noble metal nanoparticles: synthesis, morphostructural properties, and high photocatalytic activity. *Inorg. Chem.* 58 (11): 7602–7614.

84 Lu, Y., Yao, M., Zhou, A. et al. (2017). Preparation and photocatalytic performance of Ti_3C_2/TiO_2/CuO ternary nanocomposites. *J. Nanomater.* 2017.

85 Shahzad, A., Rasool, K., Nawaz, M. et al. (2018). Heterostructural $TiO_2/Ti_3C_2T_X$ (MXene) for photocatalytic degradation of antiepileptic drug carbamazepine. *Chem. Eng. J* 349: 748–755.

86 Gao, Y., Wang, L., Zhou, A. et al. (2015). Hydrothermal synthesis of TiO_2/Ti_3C_2 nanocomposites with enhanced photocatalytic activity. *Mater. Lett.* 150: 62–64.

6

Titanium Carbide (TiC) MXene-based Graphitic Carbon Nitride Composites for Energy and Environment Applications

Abdelmoumin Y. Zerga[1] *and Muhammad Tahir*[2]

[1] *Universiti Teknologi Malaysia, School of Chemical and Energy Engineering, P.O. BOX 81310, Skudai Johor Bahru, UTM, Johor Bahru, Johor, Malaysia*
[2] *UAE University, Chemical and Petroleum Engineering Department, P.O. Box 15551, Sheik Khalifa Bin Zayed Street, Asharej, Al Ain, United Arab Emirates*

6.1 Introduction

The use of fossil resources as a fundamental power supply causes serious environmental issues. Its ephemeral usage is inextricably related to anthropogenic climate change. Furthermore, the degradation of fossil resources as a result of increased industrialization has become a worldwide danger to long-term economic progress. It has been emphasized that the difficulties with fossil fuel consumption and the negative environmental impacts are two interrelated concerns that require an immediate response. Using hydrogen to replace fossil fuels via energy recycling provides a safe, durable, and recyclable energy supply [1–5]. Photocatalytic hydrogen synthesis by solar energy collection has been a promising strategy for addressing the abovementioned difficulties. Furthermore, the use of photocatalytic water splitting in the formation of hydrogen energy has sparked scientific attention to improve its performance over other solar active catalysts in order to fulfil future power exigencies [6, 7].

The spontaneous production of hydrogen (H_2) via photocatalytic water splitting by semiconductors has received much interest. This approach of gathering and converting solar energy to chemical power could be deemed environmentally beneficial [8]. As a result, an optimal photocatalyst for H_2 generation must meet the following criteria: (i) the conduction band (CB) level should be greater than the hydrogen reduction reaction (H^+/H_2) value, and the valence band (VB) level should be greater than the water oxidation performance (H_2O/O_2) value, (ii) obtain as much solar light as possible, (iii) charged carriers should be shifted to the catalyst's surface without recombination, (iv) be stable under extreme circumstances, and (v) have a low heading expense [9–11].

Due to the appropriate construction, non-toxicity, and high reduction capacity, graphitic carbon nitride ($g\text{-}C_3N_4$) is regarded as an ideal element for hydrogen production in the visible light spectrum. Nonetheless, highly instinctive recombination

Titanium Carbide MXenes: Synthesis, Characterization, Energy, and Environmental Applications, First Edition. Edited by Muhammad Tahir.

of charged carriers significantly limits g-C_3N_4 photocatalytic efficacy. MXene is a new family of 2D metal nitrides, carbides, or carbonitrides with the regular formula of $M_{n+1}X_n$ (where X represents N and/or C, M stands for a transition metal, and $n = 1$, 2, or 3). It has recently demonstrated outstanding prospective application in several technical sectors, such as supercapacitors, photocatalysis, solar batteries, and different kinds of ion batteries, among which Ti_3C_2 is a prototype and the most studied one [12–15]. Due to its tunable terminal groups, metallic conductivity, layered structure, and simple morphological arrangement, MXene was chosen as a notable cocatalyst to drastically boost photocatalytic performance. Furthermore, impressive properties such as higher specific surface area, tunable terminal functional groups (OH, O, and F), exposed metallic sites, active sites, and outstanding electrical conductivity related to the creation of a Schottky junction make them an important cocatalyst to enhance the catalytic reaction and increase photocatalytic activity [16, 17].

In this chapter, recent development in MXene-based graphitic carbon nitride (g-C_3N_4)-based composite for photocatalytic solar energy application has been discussed. Initially, fundamentals and basic properties of MXenes for solar energy applications have been included. In the next stage, the recent development of TiC-based composites for photocatalytic applications has been discussed. Finally, conclusions and future recommendations have been included.

6.2 Principle of Photocatalysis for Using MXene/g-C_3N_4 Composites

A one-step thermal annealing procedure was used to create the g-C_3N_4/Ti_3C_2 hybrids. Generally, 3 g of melamine powder was distributed in 100 ml of deionized water and agitated for three hours at 60 °C. After cooling the suspension to room temperature, 0.5 g Ti_3C_2 powder was added to the resulting solution. The aforementioned suspension was then agitated for 30 minutes to evenly disseminate the Ti_3C_2 powder, proceeded by evaporation below 100 °C until dry. Further thermal processing at 550 °C for two hours under nitrogen at a rate of 5 °C min^{-1} yielded the g-C_3N_4/Ti_3C_2 composite. The g-C_3N_4/$Ti_3C_2T_x$ MXene was added to increase syngas generation with a rate of 51.24 μmol g^{-1} h^{-1} compared to that of a pure material of g-C_3N_4 with more optimized carrier dynamics. $Ti_3C_2T_x$ was synthesized via exfoliation with HF (hydrofluoric) as an etchant, whereas g-C_3N_4 was generated via thermal polymerization. In addition, the KOH-washing approach was used to prepare modified OH alkalized $Ti_3C_2T_x$ MXenes. As a cocatalyst, MXene is predominantly associated with the Schottky heterojunction. Because of its particular metallic qualities, MXene plays a part in photocatalytic water splitting by improving the efficiency of hybrid materials by creating spatial isolation and inhibiting the recombination of photogenerated electrons. The charge transfer process in the MXene-based composite, as shown in Figure 6.1, is accomplished by a variation in the work value of metallic MXenes and the semiconductor. When MXenes and semiconductors come into close proximity, significant contact charges occur between the metal–semiconductor interfaces, generating a Schottky barrier [19]. Photogenerated charges will be released from the semiconductor when exposed

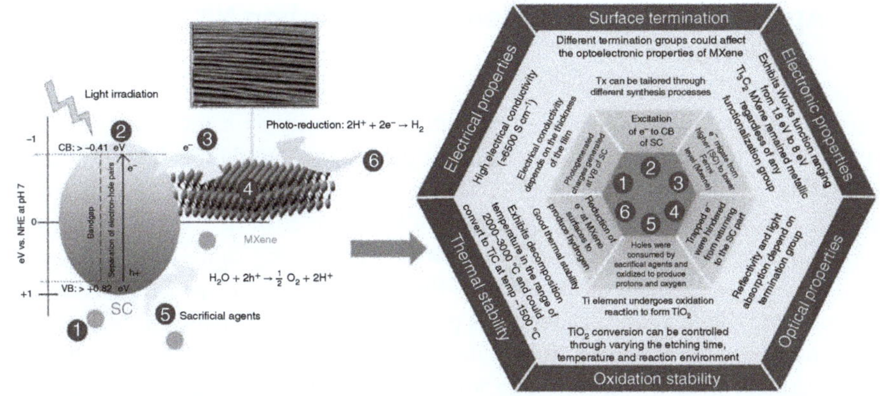

Figure 6.1 The fundamental system and features of the photocatalytic hydrogen production cycle using Mxene Ti_3C_2 as a cocatalyst are described. Source: Sherryna et al. [18]/with permission from American Chemical Society.

to light. In addition, electrons will be stimulated at the semiconductor's conduction band (CB), creating photogenerated holes at the valence band (VB). MXene acts as an electron acceptor, with a higher activity potential and a lower recombination level than its semiconductor partner [20]. Until the Fermi level is adjusted, the electron will move from the higher Fermi level semiconductor to the MXene. The substantial interfacial charges aid electron movement toward the MXene cocatalyst. In a subsequent process, the electrons will be gathered and reduced to form hydrogen. The Schottky barrier created by MXenes which, traps electrons and prevents collected electrons at the surfaces from moving back to the semiconductor component and recombining with the holes [21]. The Schottky barrier, which acts as an electron sink, enables optimal isolation of photogenerated charges and increases the photocatalytic performance of the photocatalyst. In principle, aspects such as photocatalyst activity, photoreactor structure, sacrificial agents applied, and light radiation demand should be considered when optimizing photocatalytic functionality. This is due to the concept that the effectiveness of the photocatalyst influences the process rate, but the aforementioned parameters may also impact the formation of products such as oxygen and hydrogen. On the other hand, sacrificial reagents serve an important function as hole scavengers in improving the isolation of photogenerated charges [22]. Varying sacrificial agents have variable adsorption capacities and considerable quantities of accessible protons. As a result, diverse photocatalysts may have varying degrees of efficiency when interacting with various sacrificial reagents. It has been revealed that using methanol as an electron supplier might feasibly produce hydrogen via methanol conversion. In addition, it has been noted that the number of carbon chains could assist in higher hydrogen production via alcohol conversion [23]. The optimum way to get increased photocatalytic activity is to enhance the photocatalyst's function through multiple engineering adjustments, such as producing a hybrid photocatalyst, incorporating metal-based materials, tailoring the morphology, and using engineering flaws. As a result, improving all of the interaction conditions could significantly increase the process rate.

6.3 Applications of TiC MXene-based Carbon Nitride for H_2 Evolution

Graphitic carbon nitride (g-C_3N_4) has gained popularity as a mono and hybrid composite for photocatalytic water splitting due to its abundance, low cost, and outstanding characteristics [24]. g-C_3N_4 has a bandgap of 2.7 eV, greater solar absorption, and a mesoporous surface area of less than $10\,m^2\,g^{-1}$ when compared to other traditional photocatalysts, all of which are strong signals for hydrogen production by photocatalysis. Therefore, the meso-porosity of g-C_3N_4 is one of the critical parameters that influence its photocatalytic performance. It was noted that photogenerated electron–hole pairs traveled smaller distances on meso-porous surfaces than in volume. The longer it takes for photogenerated charges to attain the photocatalyst surfaces, the more likely they recombine because of insufficient exciton energy. The redox process in the mesoporous photocatalyst, on the other hand, may proceed at the pore wall, minimizing the odds of charge coupling [25]. Although g-C_3N_4 has several benefits in the photocatalysis field, it has been observed that it has a substantial electrostatic potential barrier, which leads to greater electron and hole recombination rates. Furthermore, the low solar absorption properties of g-C_3N_4 reduce their photocatalytic activity for hydrogen generation. As a result, it was discovered that using a strongly metallic Ti_3C_2 MXene as a cocatalyst reduced the potential barrier of g-C_3N_4, allowing adept charge isolation and facilitating maximum photon absorption. Li et al. studied the photocatalytic performance of Ti_3C_2 MXene incorporated g-C_3N_4 in a sealed container photoreactor. The combination of g-C_3N_4/Ti_3C_2 MXene was synthesized using a one-step thermal annealing process with a temperature processing at 550 °C [26]. The hydrogen production increased up to $116.2\,\mu mol\,g^{-1}\,h^{-1}$, which is six times greater than that of raw g-C_3N_4. In terms of visual absorption, raw g-C_3N_4 exhibits strong absorption peak features in the 480 nm band, suggesting poor visual efficiency. Furthermore, the addition of Ti_3C_2 MXene effectively alters the absorption spectrum in the visible region of 450–700 nm. The addition of Ti_3C_2 MXene, which has metallic electronic band constructions, acts as an electron sink, preventing electron insertion back into g-C_3N_4 [27]. PL investigation indicated that photocarrier recombination was considerably slowed in the hybrid g-C_3N_4/Ti_3C_2 MXene, owing to the creation of the Schottky junction. Furthermore, it was discovered that the addition of 80 wt% Ti_3C_2 MXene significantly increases the photocatalytic potential of the hybrid composite for the below factors: (i) A stronger electrostatic connection between the two semiconductors results in a smaller electron transmission pathway, promoting electron migration from g-C_3N_4 to Ti_3C_2 MXene. (ii) The annealing operation converts the F termination in Ti_3C_2 MXene to an O termination, providing more significant merging impacts to attach the g-C_3N_4 to Ti_3C_2 MXene without using other surface-active agents [28]. (iii) The produced internal electric field at the metal–semiconductor contact efficiently separates electrons and holes, enhancing photocatalytic performance. (iv) The use of strongly conductivity Ti_3C_2 MXene increases photon absorption in the visible area.

Huang and his colleagues investigated the combination of two (2D) Ti_3C_2 MXene sheets with one-dimensional (1D) g-C_3N_4 [29]. The 1D/2D composite was created via the electrostatic self-assembly method. The delamination of 2D multilayer Ti_3C_2 MXene was initially accomplished by in situ HF etching of an MAX precursor using the HF/LiF etchant, resulting in a single-layer Ti_3C_2 MXene sheet [14]. The g-C_3N_4 was synthesized in a hollow cylindrical structure using the supramolecular self-assembly approach, accompanied by a carbonization step to produce a P-doped tubular g-C_3N_4 (PTCN). Including an appropriate number of Ti_3C_2 MXene sheets into PTCN results in a significant boost in hydrogen generation, with rates 5.3- and 3.0-fold more significant than in pure bulk g-C_3N_4 and PTCN, respectively. Furthermore, modifying g-C_3N_4 into a hollow cylindrical shape might give an orientated charge transfer pathway along the longitudinal dimension while enhancing the specific surface area, which could lower mass transfer tolerance, as Tong et al. suggested [30]. According to the BET surface area, changing the structure of bulk g-C_3N_4 into the hollow cylindrical shape significantly enhances the surface area from 2.3 to 17.4 $m^2\ g^{-1}$. In addition, Ti's access terminal metal locations from Ti_3C_2 MXene, integrating Ti_3C_2 MXene into the PTCN, might offer the composite with a large number of active catalytic locations and a higher redox reactivity than that of carbon materials [31]. As a result, Kang et al. demonstrated that g-C_3N_4 combined with Ti_3C_2 MXene hollow spheres produced 1.22-fold more hydrogen than the pure g-C_3N_4/Ti_3C_2 MXene composite [32]. Xu and colleagues investigated the potential of Ti_3C_2 MXene as a photoactivity amplifier more by altering the surface functional groups of Ti_3C_2 MXene using the plasma treatment approach [33]. The ultrathin 2D layered $Ti_3C_2T_x$ MXene was created initially by HF etching of the MAX precursor followed by intercalation in DMSO. The resultant $Ti_3C_2T_x$ MXene was then plasma-treated at a reactivity temperature of 500 °C in an N_2/H_2 environment. The thermal diffusion and ionization of H^+ from p-$Ti_3C_2T_x$ MXene plasma generation enable the distribution of Ti atoms to interact with O at the borders of the p-$Ti_3C_2T_x$ MXene sheets and form TiO bonds. Due to the decrease of TiC bonds in the p-$Ti_3C_2T_x$ MXene, these findings create a severely imperfect graphitic carbon shape. Furthermore, the impacts of H^+ ionization on the generation of Ti^{4+} cations, elimination of the F and OH functional groups, and increase in surface vacancies were discovered. This adds to the evidence that p-$Ti_3C_2T_x$ MXene underwent structural chemistry changes and had distinct photocatalytic capabilities compared to the native $Ti_3C_2T_x$ MXene. The incorporation of p-$Ti_3C_2T_x$ MXene nanosheets with a tailored surface functional group enhanced the specific surface region. In addition, it enlarged the space between the g-C_3N_4 layers, leading to an elongated carrier lifetime from 16.76 to 19.32 ns, which resulted in a 25.4-fold development in photocatalytic effectiveness over raw g-C_3N_4. It is also significant to note that the plasma-treated $Ti_3C_2T_x$ MXene hybrid outperforms the native $Ti_3C_2T_x$ MXene hybrid in terms of photocatalytic performance, with 2.4 times stronger process amplification. This increase in photoactivity was primarily due to the production of Ti^{4+} groups in the p-$Ti_3C_2T_x$ MXene structure, which acts as an electron trapping center with maximum segregation of electrons and holes. The interaction of produced Ti^{3+} with H^+ enhances the formation of hydrogen [34, 35].

The use of Ti_3C_2 MXene with enhanced electrical flexibility extends the spectrum of reactivity above 500 nm and reduces electron transfer resistance in the PCN. Furthermore, the involvement of Ti_3C_2 MXene in the PCN/Ti_3C_2 MXene composite is revealed by altering the volume of Ti_3C_2 MXene loading into 10, 20, and 40 mg, and varying the volume of Ti_3C_2 MXene loaded impacts the volume of hydrogen produced, confirming that Ti_3C_2 MXene does represent an essential function as a photoactivity amplifier. As a result, the composite of O-doped g-C_3N_4/Ti_3C_2 MXene exhibits a significant increase in hydrogen generation up to 25 124 μmol g^{-1} h^{-1}, with an AQE of 6.53% at 420 nm. The bulk g-C_3N_4 is thermally exfoliated at 500 °C before being protonated to create positively charged O-doped g-C_3N_4. The obtained O-doped g-C_3N_4 is electrostatically joined with Ti_3C_2 MXene to create an effective 2D/2D layered MX/HCN heterostructure. The enhanced engagement between Ti_3C_2 MXene and O-doped g-C_3N_4 is accomplished by changing the membrane charge of the g-C_3N_4, resulting in the formation of a layered multicharged MX/HCN hybrid. The electrostatic effects of Ti_3C_2 MXene and O-doped g-C_3N_4 are shown to be 3.32 and 5.19 eV, respectively, by the DFT computational study. The decreased activity capacity of Ti_3C_2 MXene causes electrons produced at the CB of HCN to move to the metallic Ti_3C_2 MXene nanosheets. In addition, the intimate interfacial interaction between them minimizes charge transfer resistance, allowing electrons to be transferred more easily. The hydrogen is produced when the collected electrons at the surfaces of Ti_3C_2 MXene interact with adsorbed water molecules. The development of the Schottky connection enhances the segregation of photogenerated carriers, which prolongs the durations in the redox process, resulting in an increase in the volume of hydrogen generation.

6.3.1 TiC/g-C_3N_4 Composites

Figure 6.2 a and b depicts the H_2-yield rates of O-doped g-C_3N_4 and MX/HCN samples with varying MXene masses when exposed to visible light. As demonstrated in Figure 6.2 a and b, bulk CN has poor photocatalytic H_2-generation (5366 μmol g^{-1} h^{-1}), but the OCN-2 sample has maximum H_2-production activity (13 745 μmol g^{-1} h^{-1}), which is nearly three times that of bulk CN. The protonated OCN-2 sample had a higher H_2-production efficiency (14 756 μmol g^{-1} h^{-1}) than OCN-2. Furthermore, coupling 0.5, 1, 3, and 5 wt% MXene with HCN improves the H_2-evolution efficiency. The proportion of MXene and the productivity of the MX/HCN combinations show an evident volcano-type trend, with the MX1/HCN sample demonstrating the optimum H_2-production rate (Figure 6.2 c and d). The stability of the MX1/HCN sample is also displayed in Figure 6.2e, and the photocatalyst is quite persistent after five cycles. The wavelength-dependent apparent quantum yield (AQY) values of the MX1/HCN sample based on H_2 generation illuminated by various monochrome lights are shown in Figure 6.2f, and the estimated AQY for 405 and 420 nm is up to 17.59% and 6.53%, correspondingly.

The hydrogen process efficiency of pure CN, PCN, and PCN-x composites deposited with various quantities of MXene ($x = 10$, 20, and 40 mg) were analyzed under a regular solar simulator to determine the impacts of the 2D MXene

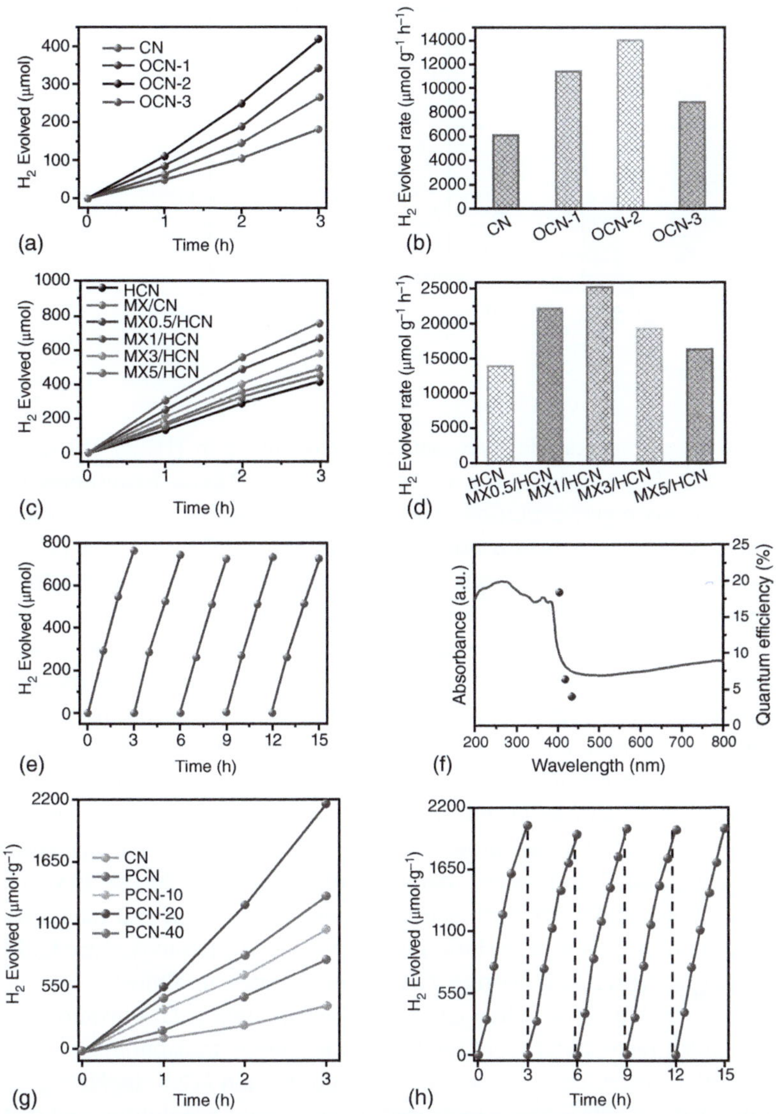

Figure 6.2 (a) Photocatalytic activities and (b) H_2-generation rates of O-doped g-C_3N_4 samples with various thermal processing intervals; (c) photocatalytic activities and (d) H_2-evolution rates of HCN and MX/HCN samples; (e) recycling durability studies and (f) AQY over MX1/HCN samples [35]. (g) CN, PCN, and PCN-x hydrogen production ($x = 10$, 20, and 40); (h) five hydrogen generation cycles utilizing the composite PCN-20 [34]. (i) Photocatalytic hydrogen production efficiency of the samples; (j) CNT3 composite durability evaluation for photocatalytic H_2 generation [26].

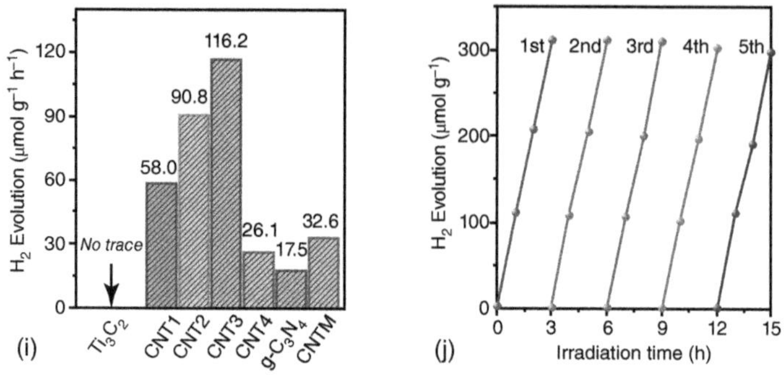

Figure 6.2 (*Continued*)

nanosheets on the photocatalytic hydrogen evolution reaction (HER) efficiency of the composites. To investigate the effects of the 2D MXene nanosheets on the photocatalytic hydrogen evolution reaction (HER) results of the composite materials, the hydrogen production effectiveness of pure CN, PCN, and PCN-x composites deposited with varying quantities of MXene ($x = 10$, 20, and 40 mg) was analyzed using a conventional solar simulator. As shown in Figure 6.2g, in the absence of a Pt co-catalyst, increasing the MXene content to 20 mg boosted PCN hydrogen production, while increasing the MXene content further reduced hydrogen production, implying that MXene played a crucial role in strengthening PCN hydrogen production efficiency. Under light irradiation for three hours, PCN-20 divided water into hydrogen at a frequency of roughly 2181 μmol g^{-1} with a quantum efficiency of 8.6% at 420 nm; this performance was significantly better than that of CN or PCN. Furthermore, after five cycles, PCN-20 still displayed consistent H_2 evolution (Figure 6.2h), demonstrating that PCN-20 can function as an excellent photocatalyst for hydrogen evolution. Furthermore, all materials' photocatalytic capabilities in water solution were evaluated by photocatalytic HER. As demonstrated in Figure 6.2i, pristine Ti_3C_2 demonstrates an almost negligible response to photocatalytic HER when exposed to visible light [12]. Furthermore, because of the quick recombination of photo-induced energy transporters, g-C_3N_4 has a relatively low visible light efficiency for H_2 production (17.5 μmol h^{-1} g^{-1}). It is fascinating that the hybrids have increased photocatalytic capabilities and that their process increases with increasing g-C_3N_4 content. The maximal HER product yield of 116.2 μmol h^{-1} g^{-1} achieved by the CNT3 sample (80 wt% g-C_3N_4) is more than six times greater than that of fundamental g-C_3N_4. When g-C_3N_4 content is enhanced considerably, photocatalytic HER activity deteriorates dramatically due to a reduction in active sites on Ti_3C_2 coated by abundant g-C_3N_4 and the presence of detached g-C_3N_4. The findings agree with hybrids' optical and photo-electron characteristics and pure materials. In contrast, a mechanical combination of g-C_3N_4 and Ti_3C_2 with an optimum g-C_3N_4 content (80 wt%) was developed and indicated as CNTM. The low HER output yield of the CNTM sample shows that the intimate contact in a hybrid is not built by direct interaction. Instead, the increased photocatalytic performance can be attributed to the creation of a heterojunction of g-C_3N_4 and Ti_3C_2, which has various advantages. The tightly combined composite has a larger photo-absorption region, resulting in more photo-induced charge transfer. Lastly, the many –O terminations adorned on the Ti_3C_2 substrate improve the hybrid's photocatalytic HER efficiency [36].

It is generally comprehensible that catalyst persistence and reliability have critical functions in usage. The findings of cyclic HER tests with CNT3 under similar circumstances are shown in Figure 6.2j. After five cycles, there is no discernible reduction in performance, confirming the CNT3 hybrid's outstanding stability.

6.3.2 TiC/Sensitizer/g-C_3N_4 Composite

Effective photocatalytic hydrogen evolution reaction (HER) at a cheap cost in the visible to near-infrared area represents a challenge. This study shows how to prepare organic–inorganic composites by depositing supramolecular aggregates of a chlorophyll derivative, zinc methyl 3-devinyl-3-hydroxymethyl-pyropheophorbide a (Chl), over the interface of $Ti_3C_2T_x$ MXene with a 2D form. This composite substance is a noble metal-free catalyst in the photocatalytic HER under white light irradiation. Chl replaces typical inorganic and polymer organic semiconductors like TiO_2, and g-C_3N_4 $Ti_3C_2T_x$ acts as a co-catalyst.

The illumination capacity, shape, charge transfer accuracy, and photocatalytic efficacy of various Chl/$Ti_3C_2T_x$ composition ratios are evaluated. After tuning, the optimum HER performance is found to be as high as $52 \pm 5\,\mu\text{mol}\,h^{-1}{g_{cat}}^{-1}$. The high HER activity is attributable to effective light absorption, accompanied by exciton transfer in Chl aggregates and the resulting charge isolation at the Chl/$Ti_3C_2T_x$ substrate. These composites are being evaluated as water-separating photocatalysts. Figure 6.3a depicts the HER performance of 3 mg $Ti_3C_2T_x$ with varying amounts of Chl. Visible-to-near infrared light is used to investigate the photocatalytic hydrogen generation of the Chl/$Ti_3C_2T_x$ composites (400 nm). The hydrogen production increases as the Chl level increases from 0.5% to 2%. The most incredible efficiency was obtained for the Chl/$Ti_3C_2T_x$ composite containing 2% Chl, with hydrogen generation attaining up to $20 \pm 2\,\mu\text{mol}\,h^{-1}\,g^{-1}$. When Chl adsorption increases, more photogenerated electrons are transported from Chl to $Ti_3C_2T_x$. When the volume of Chl increases by more than 2%, fewer stimulated electrons can be transported to $Ti_3C_2T_x$ in time. As a consequence of the increased mixing of electrons and holes formed by Chl upon light absorption, hydrogen generation decreases [38]. Figure 6.3b illustrates the HER values of composites comprising 60 g Chl and various $Ti_3C_2T_x$ masses. When the mass of $Ti_3C_2T_x$ is 1 mg, the maximal hydrogen production is $52 \pm 5\,\mu\text{mol}\,h^{-1}\,g^{-1}$. Hydrogen generation grows when $Ti_3C_2T_x$ concentration increases from 0.4 to 1 mg. This is due to photoelectrons being transmitted more simply from Chl to $Ti_3C_2T_x$ as $Ti_3C_2T_x$ grows and there are more active sites for HER. $Ti_3C_2T_x$ influences the light absorption of Chl, and the number of active sites decreases, causing a decrease in hydrogen production when the concentration is maintained over 1 mg. Figure 6.3c is a schematic representation of the Chl/$Ti_3C_2T_x$ composite for the HER. The stimulated condition is carried inside a Chl aggregate after being activated by visible-to-near infrared light.

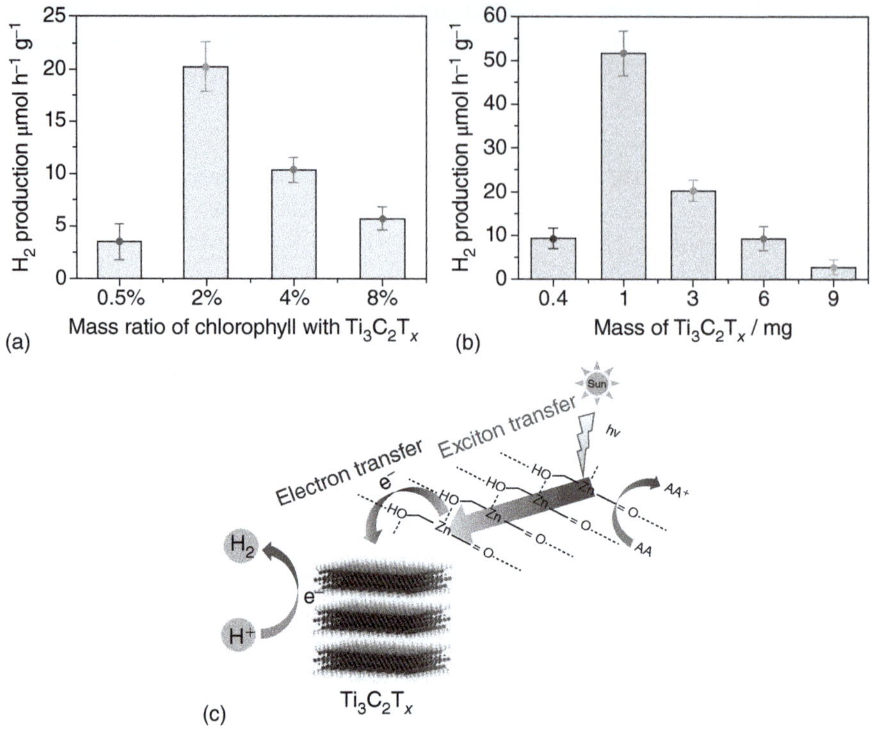

Figure 6.3 The hydrogen generation of (a) 3 mg $Ti_3C_2T_x$ with Chl at various mass ratios and (b) 60 μg Chl at various $Ti_3C_2T_x$ masses. (c) Under visible light irradiation, a schematic depiction of hydrogen generation for the Chl/$Ti_3C_2T_x$ composite is shown [37].

Charge separation happens when the excitonic state is attained at the $Ti_3C_2T_x$ contacting interface. Following the transmission of photogenerated electrons to $Ti_3C_2T_x$, the electrons interact with H^+ in an aqueous ascorbic acid (AA) mixture to produce hydrogen gas [37, 39].

6.3.3 TiC/Semiconductor/g-C_3N_4 Composite

Using semiconductor photocatalysts to transform clean and low energy intensity solar energy into sustainable and green H_2 with high potential density has recently attracted the curiosity and ambition of researchers [40–42]. Titanium dioxide (TiO_2) is a well-studied photocatalytic semiconductor photocatalyst with a wide range of applications in photocatalytic H_2 generation and contamination remediation. It has received a lot of attention due to its accessibility, high chemical consistency, low cost, and zero toxic effects [43–46]. Previous studies have identified g-C_3N_4 as a possible semiconductor for TiO_2. g-C_3N_4 is a typical 2D metal-free organic polymer semiconductor that was first used in photocatalysis by Wang et al. in 2009 [47]. It has piqued the interest of many researchers due to its visible light response, appropriate band edge (2.7 eV), more negative valence band position for photocatalytic H_2 evolution and photocatalytic degradation, low cost, and exceptional consistency [48].

Furthermore, the MXene-Ti_3C_2 contains carbons that might be integrated into TiO_2 as a suitable dopant, generating carbon-doped TiO_2 (C-TiO_2) without additional carbon precursors throughout the Ti_3C_2 oxidation phase. For example, Yang et al. discovered that a C-TiO_2/g-C_3N_4 photocatalyst with increased photocatalytic H_2 generation capability may be prepared by employing TiC as a carbon-doped TiO_2 precursor without additional carbon precursors. According to the aforementioned process, C-TiO_2/g-C_3N_4 heterojunction photocatalysts made using MXene-Ti_3C_2 as a precursor could be an effective photocatalyst for photocatalytic H_2 generation. Furthermore, the resulting C-TiO_2 may maintain the two-dimensional form of MXene, which is advantageous to photocatalytic activity [49]. To investigate the increased photocatalytic H_2 production efficiency of C-TiO_2/g-C_3N_4 composite photocatalysts, we evaluated H_2 evolution studies in 10 vol% TEOA solution under visible light exposure, and the amount of prepared H_2 was assessed every one hour utilizing gas chromatography with TCD. Figures 6.4 a and b demonstrate the average rate and time course of visible light photocatalytic H_2 production for the photocatalysts C-TiO_2, g-C_3N_4, and C-TiO_2/g-C_3N_4, respectively. Because of the in situ carbon-doping and improved absorption in the visible light region, the C-TiO_2 sample exhibits a clearly visible-light-response photocatalytic hydrogen production ability [53, 54]. The photocatalytic H_2 production activity is deficient for both pure g-CN and C-TiO_2 samples, with average hydrogen production rates of only 174 and 58 $\mu mol\,h^{-1}\,g^{-1}$, accordingly. This is due to the dissociation performance of their photogenerated carriers being extremely poor and there being insufficient functional electrons that can interact with H^+ to form H_2 in water [55]. C-TiO_2/g-C_3N_4 photocatalysts outperformed single g-CN and C-TiO_2 samples in photocatalytic H_2 production capability. The C-Ti/CN-10 photocatalyst has the greatest photocatalytic H_2 production capacity, with a rate of 1409 $\mu mol\,h^{-1}g^{-1}$, 8 and 24 times greater than that of the g-CN C-TiO_2 samples, correspondingly. The photocatalytic H_2 development data presented above clearly impact the coupling of g-C_3N_4 and C-TiO_2 samples. On the one hand, the coupling of g-C_3N_4 and C-TiO_2 increases its particular surface area and visible light absorption performance, which is advantageous to its photocatalytic active site and light consumption performance. Furthermore, a tight heterojunction can be produced between g-C_3N_4 and C-TiO_2 samples, which aids in the dissociation of photogenerated holes and electrons. Furthermore, as a reference, the P25/CN-10 (TiO_2/g-C_3N_4) sample was manufactured using the same procedure as that of the C-Ti/CN-10 photocatalyst. The photocatalytic H_2 production findings of the P25/CN-10 and the C-Ti/CN-10 are shown in Figure 6.4c. Remarkably, the addition of P25 can considerably enhance g-photocatalytic g-C_3N_4's H_2 efficiency, which is consistent with earlier research [44]. Nonetheless, the photocatalytic H_2 production rate of the P25/CN-10 sample is 810 $\mu mol\,g^{-1}\,h^{-1}$, which is much lower than that of C-Ti/CN-10 under the same circumstances. This demonstrates the importance of carbon-doped TiO_2 that can be stimulated by visible light in enhancing the photocatalytic H_2 generation efficiency of C-TiO_2/g-C_3N_4. Cycling photocatalytic tests were performed on the g-CN and C-Ti/CN-10 samples to assess their practicability and durability, as shown in Figure 6.4d. The photocatalytic H_2

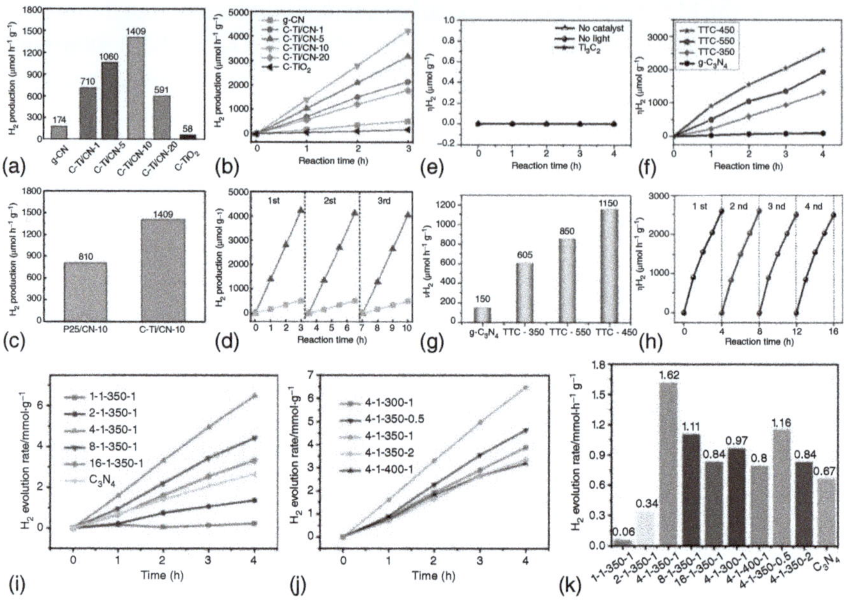

Figure 6.4 (a) Photocatalytic H_2 generation level and (b) time course hydrogen production graphs for various samples under visible light exposure, (c) H_2 production rate comparison for P25/CN-10 and C-Ti/CN-10 samples, and (d) recycling photocatalytic H_2 generation tests for g-CN and C-Ti/CN-10 materials [50]. (e) Control experiments for hydrogen evolution without light, without catalyst, and with only the Ti_3C_2 catalyst; (f) photocatalytic hydrogen efficiency; (g) average hydrogen evolution speed of g-C_3N_4, TTC-350, TTC-450, and TTC-550; and (h) hydrogen evolution stability test over standard TTC heterojunction [51]. (i) The photocatalytic H_2 evolution rate of the samples with different g-C_3N_4/d-Ti_3C_2 mass ratios, (j) the photocatalytic H_2 evolution rate of the samples with g-C_3N_4/d-Ti_3C_2 mass ratio of 4/1 but calcined at different temperatures or for different soaking times, (k) comparison of the photocatalytic H_2 evolution rate of all the samples [52].

generation activity of C-Ti/CN-10 stays constant without apparent deactivation following three continuous cycling photocatalytic H_2 generation trials, demonstrating improved photocatalytic consistency [50].

According to Vu Quang Hieu et al., 2021, through photocatalytic hydrogen production processes from water with triethanolamine (TEOA) as a sacrificial agent, the photocatalytic efficiency of as-prepared samples was investigated. Control studies show that no H_2 gas is produced in the absence of a photocatalyst, light exposure, or in the existence of pristine Ti_3C_2 MXene (Figure 6.4e) [51]. In contrast, combining Ti_3C_2 and g-C_3N_4 results in a considerable increase in H_2 generation performance (Figure 6.4f). Based on the promoting influence of Ti_3C_2 MXene and in situ TiO_2 on the dissociation of photogenerated charge carriers, all three composites at various processing temperatures (TTC-350, TTC-450, and TTC-550) demonstrated better photocatalytic performance than pure g-C_3N_4. The basic composite generated at 450 °C (TTC-450) had the most significant figure for H_2 generation, around 2592 μmole g^{-1}, which was significantly greater than that of TTC-550 (approximately 1935 μmol g^{-1}) and TTC-350 (1320 μmol g^{-1}). In addition, the overall H_2 production frequency was calculated and shown in Figure 6.4g. The amount of H_2 generation in TTC-450 was very significant, at 1150 μmol h^{-1} g^{-1}, nearly eight times that of raw g-C_3N_4. On the other hand, TTC-550 and TTC-350 had lower figures, at 850 and 650 μmol h^{-1} g^{-1}, correspondingly. The difference in H_2 production rates can be clarified by the fact that the TTC-450 sample comprises an ideal amount of components (TiO_2, Ti_3C_2, and g-C_3N_4), resulting in the highest performance. However, at lower temperatures, the TiO_2 phase composition of the TTC-350 sample is insufficient for synthesizing produced electrons, resulting in a decreased hydrogen generation yield. Furthermore, the extra Ti_3C_2 inhibits light from attaining the TiO_2 layer. Nevertheless, in the TTC-550 sample, the majority of the Ti_3C_2 tended to transform into the TiO_2 layer, resulting in the absence of metallic Ti_3C_2 components and the retention of H_2 production in the heterostructure. The TTC composite's photostability was tested under the same conditions (Figure 6.4h). After four runs, there is no significant decrease in H_2 generation performance for the typical TTC sample, indicating that the TTC composite can be employed as a sustainable photocatalyst for H_2 generation.

Based on Mengyuan Zhang et al., 2018, Figure 6.4i shows the photocatalytic efficiency of raw g-C_3N_4 and all heterostructured-Ti_3C_2/TiO_2/g-C_3N_4 nanocomposites, whereas dm-350-1 is removed since its H_2 generation cannot be observed [52]. The composite's capacity to produce H_2 increases initially and then diminishes as the proportion of g-C_3N_4 increases. With a g-C_3N_4/d-Ti_3C_2 mass ratio of 4/1, it had the highest H_2 output of 324.2 mol after 4 hours of processing, which is similar to or even higher than that of the other Ti_3C_2-based and carbon/TiO_2/g-C_3N_4 hetero-structured composite photocatalysts. However, under the same conditions, only 133.3 mol of H_2 can be identified in pristine g-C_3N_4. This large increase in output is due to enhanced charge segregation performance, as demonstrated by EIS and photocurrent intensity. In addition, the photocatalytic characteristics of samples with the same g-C_3N_4/d-Ti_3C_2 mass ratio of 4/1 but carbonized at various temperatures for one hour or carbonized for one hour at multiple temperatures are also evaluated to discover the optimum carbonized factors. The experimental findings are shown in Figure 6.4j, demonstrating that under the other process circumstances, the d-Ti_3C_2/TiO_2/g-C_3N_4 nanocomposite would no longer have such high photocatalytic capability. A histogram in Figure 6.4k allows for a visual evaluation of the photocatalytic activity of several samples.

6.4 Conclusions

In this chapter, the development of the TiC MXene-based graphitic carbon nitride composites has been explained. We examined the efficiency of previously published TiC MXene-based nanohybrids and nanocomposites in this chapter. The explanation for the improved photocatalytic activity in the presence of different TiC MXene-based g-C_3N_4 is as follows: (i) changing the surface chemistry of TiC MXene-based photocatalysts by reduction/oxidation and surface functional group substitution enhances their characteristics. (ii) Photocatalysts based on

TiC MXene with heterojunctions improve bandgap structure, increase charge carrier separation, and minimize electron–hole pair recombination. In general, the disadvantages of g-C_3N_4 were addressed when coupled with TiC Mxenes. As a result, the photocatalytic performance was achieved.

Acknowledgment

The financial support of this work is provided by United Arab Emirates University (UAEU), UAE under Start up research grant vot no. 12 N097.

References

1 Höök, M. and Tang, X. (2013). Depletion of fossil fuels and anthropogenic climate change—a review. *Energy Policy* 52: 797–809.

2 Yan, T., Liu, H., and Jin, Z. (2020). g-C_3N_4/α-Fe_2O_3 supported zero-dimensional Co_3S_4 nanoparticles form S-scheme heterojunction photocatalyst for efficient hydrogen production. *Energy Fuels* 35 (1): 856–867.

3 Umer, M., Tahir, M., Usman Azam, M. et al. (2019). Synergistic effects of single/multi-walls carbon nanotubes in TiO_2 and process optimization using response surface methodology for photo-catalytic H_2 evolution. *J. Environ. Chem. Eng* 7 (5).

4 Fan, W.K. and Tahir, M. (2021). Recent trends in developments of active metals and heterogenous materials for catalytic CO_2 hydrogenation to renewable methane: a review. *J. Environ. Chem. Eng.* 9 (4).

5 Katsumata, H., Ando, H., Suzuki, T., and Kaneco, S. (2015). Highly efficient photocatalytic hydrogen production over PdS@CdS+ZnS(en)0.5 photocatalyst under visible light irradiation. *Ind. Eng. Chem. Res.* 54 (13): 3532–3535.

6 Kim, Y., Coy, E., Kim, H. et al. (2021). Efficient photocatalytic production of hydrogen by exploiting the polydopamine-semiconductor interface. *Appl. Catal., B* 280.

7 Morshedy, A.S., Abd El Salam, H.M., El Naggar, A.M.A., and Zaki, T. (2020). Hydrogen production and in situ storage through process of water splitting using mono/binary metal–organic framework (MOF) structures as new chief photocatalysts. *Energy Fuels* 34 (9): 11660–11669.

8 Le, V.T., Doan, V.D., Le, T.T.N. et al. (2021). Efficient photocatalytic degradation of crystal violet under natural sunlight using Fe_3O_4/ZnO nanoparticles embedded carboxylate-rich carbon. *Mater. Lett.* 283.

9 Dao, M.U., Le, H.S., Hoang, H.Y. et al. (2020). Natural core-shell structure activated carbon beads derived from *Litsea glutinosa* seeds for removal of methylene blue: facile preparation, characterization, and adsorption properties. *Environ. Res.* 198: 110481.

10 Le, V.T., Doan, V.D., Tran, V.A. et al. (2020). Cu/Fe_3O_4@carboxylate-rich carbon composite: one-pot synthesis, characterization, adsorption and photo-Fenton catalytic activities. *Mater. Res. Bull.* 129.

11 Liu, F., Zhou, A., Chen, J. et al. (2017). Preparation of Ti_3C_2 and Ti_2C MXenes by fluoride salts etching and methane adsorptive properties. *Appl. Surf. Sci.* 416: 781–789.

12 Ran, J., Gao, G., Li, F.-T., and Ma, T.-Y. (2016). Ti_3C_2 MXene co-catalyst on metal sulfide photo-absorbers for enhanced visible-light photocatalytic hydrogen production. *Nat. Commun.* 8: 13907.

13 Alhabeb, M., Maleski, K., Anasori, B. et al. (2017). Guidelines for synthesis and processing of two-dimensional titanium carbide ($Ti_3C_2T_x$ MXene). *Chem. Mater.* 29 (18): 7633–7644.

14 Ghidiu, M., Lukatskaya, M.R., Zhao, M.Q. et al. (2014). Conductive two-dimensional titanium carbide 'clay' with high volumetric capacitance. *Nature* 516 (7529): 78–81.

15 Anasori, B., Lukatskaya, M.R., and Gogotsi, Y. (2017). 2D metal carbides and nitrides (MXenes) for energy storage. *Nat. Rev. Mater.* 2 (2).

16 Khazaei, M., Ranjbar, A., Arai, M. et al. (2017). Electronic properties and applications of MXenes: a theoretical review. *J. Mater. Chem. C* 5 (10): 2488–2503.

17 Lu, Y., Fan, D., Wang, Y. et al. (2021). Surface patterning of two-dimensional nanostructure-embedded photothermal hydrogels for high-yield solar steam generation. *ACS Nano* 15 (6): 10366–10376.

18 Sherryna, A. and Tahir, M. (2021). Role of Ti_3C_2 MXene as prominent Schottky barriers in driving hydrogen production through photoinduced water splitting: a comprehensive review. *ACS Appl. Energy Mater.* 4 (11): 11982–12006.

19 Choi, J., Kim, Y.J., Cho, S.Y. et al. (2020). In situ formation of multiple schottky barriers in a Ti_3C_2 MXene film and its application in highly sensitive gas sensors. *Adv. Funct. Mater.* 30 (40).

20 Khan, A.A., Tahir, M., and Bafaqeer, A. (2020). Constructing a stable 2D layered Ti_3C_2 MXene cocatalyst-assisted TiO_2/g-C_3N_4/Ti_3C_2 heterojunction for tailoring photocatalytic bireforming of methane under visible light. *Energy Fuels* 34 (8): 9810–9828.

21 Tahir, M. and Tahir, B. (2020). 2D/2D/2D O-C_3N_4/Bt/$Ti_3C_2T_x$ heterojunction with novel MXene/clay multi-electron mediator for stimulating photo-induced CO_2 reforming to CO and CH_4. *Chem. Eng. J.* 400.

22 Guzman, F., Chuang, S.S.C., and Yang, C. (2013). Role of methanol sacrificing reagent in the photocatalytic evolution of hydrogen. *Ind. Eng. Chem. Res.* 52 (1): 61–65.

23 Wang, J., Yang, P., Cao, B. et al. (2015). Photocatalytic carbon–carbon bond formation with concurrent hydrogen evolution on the Pt/TiO_2 nanotube. *Appl. Surf. Sci.* 325: 86–90.

24 Zhang, J., Zhu, Q., Wang, L. et al. (2020). g-C_3N_4/CoAl-LDH 2D/2D hybrid heterojunction for boosting photocatalytic hydrogen evolution. *Int. J. Hydrogen Energy* 45 (41): 21331–21340.

25 Chen, X., Yu, T., Fan, X. et al. (2007). Enhanced activity of mesoporous Nb_2O_5 for photocatalytic hydrogen production. *Appl. Surf. Sci.* 253 (20): 8500–8506.

26 Li, J., Zhao, L., Wang, S. et al. (2020). In situ fabrication of 2D/3D g-C_3N_4/Ti_3C_2 (MXene) heterojunction for efficient visible-light photocatalytic hydrogen evolution. *Appl. Surf. Sci.* 515.

27 Jhon, Y.I., Seo, M., and Jhon, Y.M. (2017). First-principles study of a MXene terahertz detector. *Nanoscale* 10 (1): 69–75.

28 Zhang, L., Yang, J., Xie, T. et al. (2020). Boosting visible-light-driven photocatalytic activity of $BiPO_4$ via constructing Schottky junction with Ti_3C_2 MXene. *Mater. Design* 192.

29 Huang, K., Li, C., Zhang, X. et al. (2021). Self-assembly synthesis of phosphorus-doped tubular g-C_3N_4/Ti_3C_2 MXene Schottky junction for boosting photocatalytic hydrogen evolution. *Green Energy Environ.*

30 Tong, Z., Yang, D., Sun, Y. et al. (2016). Tubular g-C_3N_4 isotype heterojunction: enhanced visible-light photocatalytic activity through cooperative manipulation of oriented electron and hole transfer. *Small* 12 (30): 4093–4101.

31 Li, Z. and Wu, Y. (2019). 2D early transition metal carbides (MXenes) for catalysis. *Small* 15 (29): e1804736.

32 Kang, J., Byun, S., Kim, S. et al. (2020). Design of three-dimensional hollow-sphere architecture of $Ti_3C_2T_x$ MXene with graphitic carbon nitride nanoshells for efficient photocatalytic hydrogen evolution. *ACS Appl. Energy Mater.* 3 (9): 9226–9233.

33 Xu, F., Zhang, D., Liao, Y. et al. (2019). Synthesis and photocatalytic H_2-production activity of plasma-treated $Ti_3C_2T_x$ MXene modified graphitic carbon nitride. *J. Am. Ceram. Soc.* 103 (2): 849–858.

34 Haotian, X., Xiao, R., Huang, J. et al. (2021). In situ construction of protonated g-C_3N_4/Ti_3C_2 MXene Schottky heterojunctions for efficient photocatalytic hydrogen production. *Chin. J. Catal.* 42 (1): 107–114.

35 Lin, P., Shen, J., Yu, X. et al. (2019). Construction of Ti_3C_2 MXene/O-doped g-C_3N_4 2D-2D Schottky-junction for enhanced photocatalytic hydrogen evolution. *Ceram. Int.* 45 (18): 24656–24663.

36 An, X., Wang, W., Wang, J. et al. (2018). The synergetic effects of Ti_3C_2 MXene and Pt as co-catalysts for highly efficient photocatalytic hydrogen evolution over g-C_3N_4. *Phys. Chem. Chem. Phys.* 20 (16): 11405–11411.

37 Li, Y., Chen, X., Sun, Y. et al. (2020). Chlorosome-like molecular aggregation of chlorophyll derivative on $Ti_3C_2T_x$ MXene nanosheets for efficient noble metal-free photocatalytic hydrogen evolution. *Adv. Mater. Interfaces* 7 (8).

38 Shao, M., Shao, Y., Chai, J. et al. (2017). Synergistic effect of 2D Ti2C and g-C3N4 for efficient photocatalytic hydrogen production. *J. Mater. Chem. A* 5 (32): 16748–16756.

39 Xie, X., Zhang, N., Tang, Z.-R. et al. (2018). $Ti_3C_2T_x$ MXene as a Janus cocatalyst for concurrent promoted photoactivity and inhibited photocorrosion. *Appl. Catal., B* 237: 43–49.

40 Ran, J., Jaroniec, M., and Qiao, S.Z. (2018). Cocatalysts in semiconductor-based photocatalytic CO_2 reduction: achievements, challenges, and opportunities. *Adv. Mater.* 30 (7).

41 Chen, S., Takata, T., and Domen, K. (2017). Particulate photocatalysts for overall water splitting. *Nat. Rev. Mater.* 2 (10).

42 Zheng, X. and Zhang, L. (2016). Photonic nanostructures for solar energy conversion. *Energy Environ. Sci.* 9 (8): 2511–2532.

43 Gu, W., Lu, F., Wang, C. et al. (2017). Face-to-face interfacial assembly of ultrathin g-C3N4 and anatase TiO_2 nanosheets for enhanced solar photocatalytic activity. *ACS Appl. Mater. Interfaces* 9 (34): 28674–28684.

44 Tan, Y., Shu, Z., Zhou, J. et al. (2018). One-step synthesis of nanostructured g-C_3N_4/TiO_2 composite for highly enhanced visible-light photocatalytic H_2 evolution. *Appl. Catal., B* 230: 260–268.

45 Zhang, Q. and Gao, L. (2003). Preparation of oxide nanocrystals with tunable morphologies by the moderate hydrothermal method: insights from rutile TiO_2. *ACS Appl. Mater. Interfaces.*

46 Qinghong Zhang, L.G. and Guo, J. (2000). Effects of calcination on the photocatalytic properties of nanosized TiO_2 powders prepared by $TiCl_4$ hydrolysis. *Appl. Catal. B.*

47 Wang, X., Maeda, K., Thomas, A. et al. (2009). A metal-free polymeric photocatalyst for hydrogen production from water under visible light. *Nat. Mater.* 8 (1): 76–80.

48 Yu, H., Shi, R., Zhao, Y. et al. (2017). Alkali-assisted synthesis of nitrogen deficient graphitic carbon nitride with tunable band structures for efficient visible-light-driven hydrogen evolution. *Adv. Mater.* 29 (16).

49 Yang, C., Qin, J., Xue, Z. et al. (2017). Rational design of carbon-doped TiO_2 modified g-C_3N_4 via in-situ heat treatment for drastically improved photocatalytic hydrogen with excellent photostability. *Nano Energy* 41: 1–9.

50 Han, X., An, L., Hu, Y. et al. (2020). Ti_3C_2 MXene-derived carbon-doped TiO_2 coupled with g-C_3N_4 as the visible-light photocatalysts for photocatalytic H_2 generation. *Appl. Catal., B* 265.

51 Hieu, V.Q., Lam, T.C., Khan, A. et al. (2021). TiO_2/Ti_3C_2/g-C_3N_4 ternary heterojunction for photocatalytic hydrogen evolution. *Chemosphere* 285: 131429.

52 Zhang, M., Qin, J., Rajendran, S. et al. (2018). Heterostructured d-Ti_3C_2/TiO_2/g-C_3N_4 nanocomposites with enhanced visible-light photocatalytic hydrogen production activity. *ChemSusChem* 11 (24): 4226–4236.

53 Park, Y., Kim, W., Park, H. et al. (2009). Carbon-doped TiO_2 photocatalyst synthesized without using an external carbon precursor and the visible light activity. *Appl. Catal., B* 91 (1–2): 355–361.

54 Ren, W., Ai, Z., Jia, F. et al. (2007). Low temperature preparation and visible light photocatalytic activity of mesoporous carbon-doped crystalline TiO_2. *Appl. Catal., B* 69 (3–4): 138–144.

55 Han, X., Xu, D., An, L. et al. (2019). Ni-Mo nanoparticles as co-catalyst for drastically enhanced photocatalytic hydrogen production activity over g-C_3N_4. *Appl. Catal., B* 243: 136–144.

7

Titanium Carbide MXene-Based MOF Composites for Energy and Environment Applications

Animesh T. Partho[1], Muhammad Tahir[2], and Naveen Kumar[3]

[1] *Universiti Teknologi Malaysia, School of Chemical and Energy Engineering, Department of Chemical Engineering, 81310 UTM, Skudai, Johor Bahru, Johor, Malaysia*
[2] *UAE University, Chemical and Petroleum Engineering Department, Sheik Khalifa Bin Zayed Street, Asharej, Al Ain, United Arab Emirates*
[3] *Maharshi Dayanand University, Department of Chemistry, Rohtak 124001, Haryana, India*

7.1 Introduction

The energy crisis and the resulting environmental problems caused by excessive fossil fuel use force academics to investigate clean and renewable energy alternatives. Hydrogen is a "green" alternative to fossil fuels due to its high heat value and lack of CO_2 emissions. Hydrogen (H_2) generation by photocatalytic water splitting can help meet the massive energy demands of the future civilization. Besides, H_2 is a clean and efficient form of energy [1, 2]. However, significant overpotential and slow kinetics must be overcome in order to accelerate hydrogen reaction rates by the use of effective catalysts. Although precious metal elements are the most active catalysts, their high cost and scarcity preclude their use in large-scale water photocatalysis technology. As a result, producing a substantial amount of hydrogen by photocatalysis of water necessitates the development of low-cost, high-catalytic activity non-precious metal-based materials [3–5].

Growing urbanization and industrialization has led to a significant increase in worldwide water usage over the past years, which has resulted in a significant increase in the volume of waste water released into the environment, leading water bodies to become contaminated. More attention is being paid to water shortages and pollution. Pollutants from human activities, such as industrial pollution, agricultural pollution, and residential pollution, are the primary contributors of water contamination [6, 7]. As a primary pollutant source, dyes are utilized in a wide range of sectors, including leather, textiles, cosmetics, and paper-making. Toxic, cancer-causing, mutagenic, and allergenic organic contaminants are ubiquitous [8–10].

Titanium Carbide MXenes: Synthesis, Characterization, Energy, and Environmental Applications, First Edition. Edited by Muhammad Tahir.

Due to the excellent optoelectronic properties, environmental friendliness, and low cost, a developing family of 2D ternary transition metal carbide materials have sparked substantial research interest in the energy storage, adsorption, and conversion domains. Numerous investigations revealed that MXenes might act as efficient co-catalysts in photocatalytic H_2 generation, degradation, and could possibly be a viable choice for noble-metal catalysts [11, 12]. MXenes have an abundance of fascinating mechanical, electrical, magnetic, and electrochemical capabilities due to their diverse chemistry [13]. High flexibility, along with the two-dimensional shape and layered structure, enable them to create composites with other materials, allowing for the complementary integration of the exceptional qualities of other materials. Not only MXenes but also MXene-based composites have sparked tremendous research interest, as they offer great potential for a variety of applications, including photocatalytic hydrogen generation [14, 15].

Metal–organic frameworks (MOFs) have been explored for photocatalytic hydrogen evolution in recent years because of their high specific surface area, tunable bandgap, changeable crystalline structures, and adjustable chemistry and functionality. Furthermore, MOFs have outstanding physical qualities that can be employed for adsorption processes in environmental applications. Based on prior research, Al-based MOF was used as an adsorbent for synthetic dyes in another investigation [16, 17]. However, the crystalline architecture of MOFs inherently introduces structural flaws such as electron–hole recombination centers and low electrical conductivity, limiting their efficiency in photocatalytic water splitting [18, 19]. Numerous efforts have been undertaken to solve the problems, including the inclusion of precious metal nanoparticles, naturally occurring nanoparticles, and polyoxometalates into MOFs. Apparently, overcoming the disadvantages remains a task. However, it has been determined that combining with other semiconductors for future development of hydrogen generation is promising [20, 21].

TiC MXenes are permissive materials that can be used with MOFs for photocatalytic hydrogen generation. Apart from charge carrier usage and separation, transport efficiency has a significant effect on photocatalytic HER activity. Unfortunately, MOFs have a low conductivity, which works against the HER reaction [22, 23]. TiC MXenes are characterized by a strong electrical conductivity and a high degree of stability. Notably, conductive TiC MXenes may improve charge carrier transport efficiency and allow for fine-tuning of catalytic performance in MXene/MOF multicomponent catalyst systems [23]. Besides photocatalytic water splitting, few reports have been published on photodegradation using MXene-based MOF composites.

This chapter provides an outline of current developments in MXenes and MOFs for photocatalytic H_2 production, degradation, and CO_2 reduction applications. In the main stream, the characteristics and performance of previously reported TiC MXene-based MOF nanocomposites and nanohybrids are discussed. Finally, we have concluded this chapter by describing the problems and prospects for future study in the field of TiC MXene-based MOF composites and hybrids, in order to stimulate further investigation on MXenes.

7.2 Overview of MXenes and MOFs for Photocatalytic Applications

7.2.1 Overview of MXenes

2D transition metal carbide MXenes, like graphene, have a common formula of $M_{n+1}X_nT_x$. M, A, and X stand for transition metal, IIIA or IVA elements, and N or C, correspondingly, and may be used to produce these materials by the exfoliating of an "A" layer from a "MAX" predecessor [24, 25]. In 2011, Naguib et al. [26] brought unprecedented 2D transition metal carbides, nitrides, and carbon nitrides (MXenes) to the family. This development has lately sparked off yet another surge of interest for the research of advanced ideas and their potentially useful areas of application. The standard formula for MXenes is $M_{n+1}X_nT_x$, where M is an initial transition metal (including such Ti, V, Sc, Zr, Ta, Cr, Mo, Hf, Nb, and Mn), X is either carbon or nitrogen, T is a surface termination component (such as oxygen, hydroxyl, or fluorine), and n is either 1, 2, or 3. The production of surface terminations is caused when the metals that make up an MXene's chemically active outer layer are exposed to the environment. As illustrated in Figure 7.1, depending on their atomic lattices and composition, MXenes can be classified as ordered mono-M, ordered double-M, or solid-solution M elements [28, 29].

When compared to unfunctionalized MXenes, the surface-functionalized MXenes have additional flexibility. The mechanical characteristics of MXenes can be enhanced by generating composite materials with polymeric materials or nanotubes. Furthermore, by covalently attaching the surface-functionalized MXenes to polymers or carbon nanotubes, the mechanical characteristics of MXenes can be enhanced [30–32]. To describe the electronic characteristics, MXenes are composed

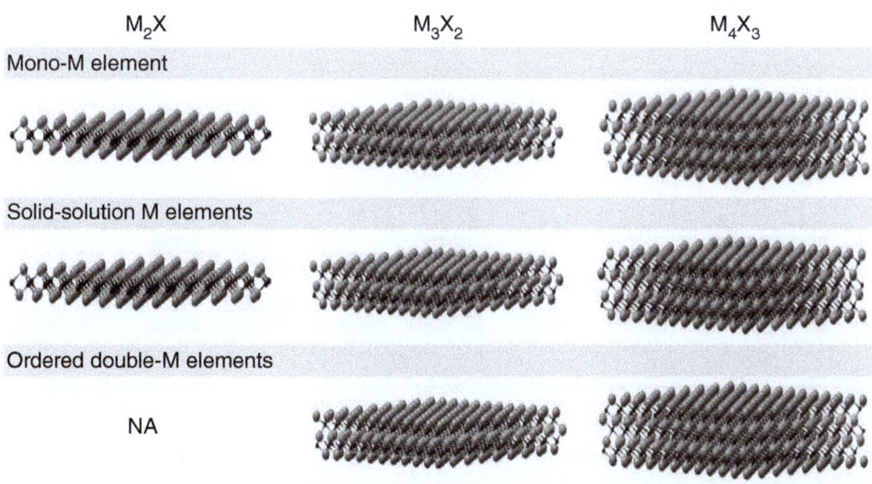

Figure 7.1 Typical layered structures of MXenes. The metal atom lattice can be classified as Mono-M element, Solid-solution M elements, and Ordered double-M elements. Source: Anasori et al. [27]/Springer Nature.

of a wide range of materials, and their electronic properties range from metallicity to topological insulativity, depending on their manufacture and debonding processes. The conductivity of MXenes is strongly dependent on their production and debonding methods. MXenes can be used for capacitive deionization, as adsorbents, membranes, for photocatalysis, and electrochemical segregation [24, 30, 32]. Because of their large surface area and various functional groups such as –OH and –O, these compounds are excellent candidates for water treatment. They also exhibit impressive catalytic efficiency against a variety of hazardous contaminants [28, 33, 34].

Three primary phases are involved in the photocatalytic water splitting process: (i) light uptake by semiconductors to form electron–hole pairs; (ii) charge separation and transport to the surface; and (iii) surface reactivity for oxidation and reduction [35, 36]. Photocatalytic effectiveness depends on the kinetics of all three of the aforementioned stages [35–37].

In the MXene class, there are over 70 distinct forms of MAX ternary carbides and nitrides that have been identified [27, 29]. More than 30 different types of MXenes have been discovered experimentally to date, and conceptual forecasts suggest that the number will continue to increase [29, 38, 39]. A brief timeline of the progress in MXenes is presented in Figure 7.2.

7.2.2 Overview of MOF

The coordinating polymers were found in the 1950s [40], and further study by Johnson [41], Drumel [42], and Férey [43] established their existence. Nevertheless, significant interest in these substances was sparked in the late 1990s by Omar Yaghi, a prominent material scientist, and his colleagues. Yaghi and his colleagues have created the stiffest coordinated networks using metallic ions and organic linkers. They began by substituting multidentate carboxylate ligands for the 4,4′-bipyridine moieties and dubbed the resultant neutralized frameworks MOFs [44, 45]. Among these, MOF-5 has been extensively explored in the literature, and its layout has

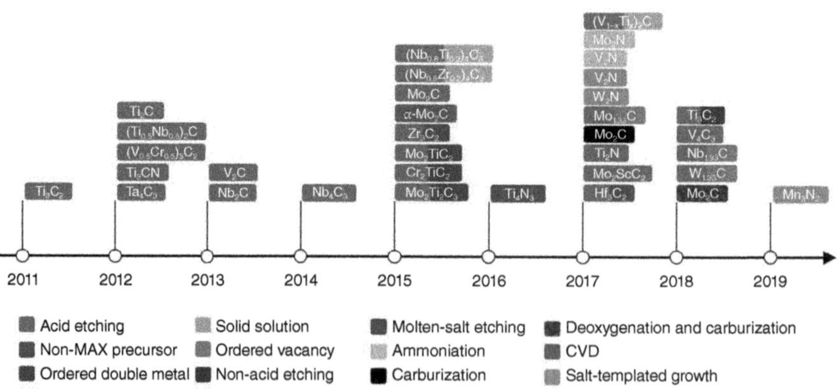

Figure 7.2 Chronology of MXene investigation. Source: Jiang et al. [28]/with permission of Elsevier.

contributed in the discovery of potential ideas in MOF design and tuning. Subsequently, as a result of these structures' unique and extraordinary qualities, a rapid expansion in publications on the creation of novel materials and their uses was seen. Li et al. [45] described the challenges and opportunities extensively in their recent study. MOFs, also referred as porous coordination polymers (PCPs), are a comparatively new family of crystalline porous materials that are built using metal ions/clusters and organic linkers [45, 46]. MOFs have demonstrated tremendous potential applications in a variety of fields, such as gas sorption and separation, catalysis, luminescence, biomedicine, proton conductivity, and sensor design due to their fascinating and immense configurations, high crystallinity, lasting porosity, extremely high surface area, tunable features, and highly customizable framework topology [45–49]. Numerous outstanding studies on the production and uses of metals, metal oxides, and porous carbons produced from MOFs have been published in recent years. MOF research has risen to prominence as a hot topic of research in chemistry and materials science during the last decade. Among all MOF applications, heterogeneous catalysis was first investigated in 1994 [50] and continues to attract the interest of researchers due to the chemical mutability, bespoke pore structures, and huge, easily available inner surface areas. The original heterogeneous catalysis has been progressively implemented in various types of catalytic reactions (Figure 7.3), most notably photocatalysis and electrocatalysis [45].

7.2.3 Structure and Properties of MXenes

Earth-abundant and non-toxic elements make up MXenes, which are 2D substances with interesting properties like (i) rich surface chemistry, (ii) effective metallic hydroxide sites, (iii) higher electronic conductivity, (iv) resistance to corrosion, (v) superior chemical resilience, and (vi) preferred opto-electronic features can be achieved by changing the size of the particles, the spacing between layers, and the number of layers [23, 51, 52]. The structural formula for MXenes is $M_{n+1}AX_n$

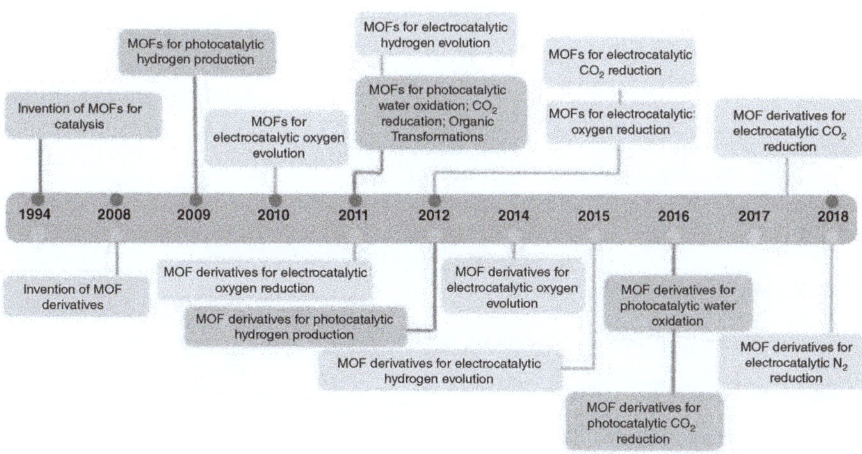

Figure 7.3 The trend of MOFs in catalysis. Source: Li et al. [45]/with permission of Elsevier.

(n = 1–3), where M is a transition metal element, A is a cluster of early transition III/IV elements, and X is carbon or nitrogen [23, 53–55]. In the formula $M_{n+1}X_nT_x$, which shows how many terminations there are, the strong metallic ties between M and A are broken and replaced by weaker bonds of (–O, –F, or –OH, for example), which are called T_x in the scheme. Figure 7.4 illustrates the layout of MX stacks in a hexagonal alternative option stacking arrangement with the A layer in between [55]. Following the etching process, there are three types of MXenes: M_2X, M_3X_2, and M_4X_3. They all have the same hexagonal closed-pack crystal structure, with the X atoms in the center of the MX octahedrons. M_2X phase: It has honeycomb lattices of transition metals in a honeycomb pattern on the top. M_3X_2 and M_4X_3 phase: They have out-of-plane arrangements, and the transition metals are on the outer layer and the other atoms are in the middle [23, 56]. The M_2X phase is made up of three-layer sheets, where X (C or N) is between the two layers of M (early transition metal). Most transition metals have a co-ordination number of 6, which means they have six chemical bonds with the X atoms next to them and the bonded functional groups at the surface termination. The M_2X phase has two hollow spots, with either no "X" atom or one atom in between the interlayers of transition metals [23, 57].

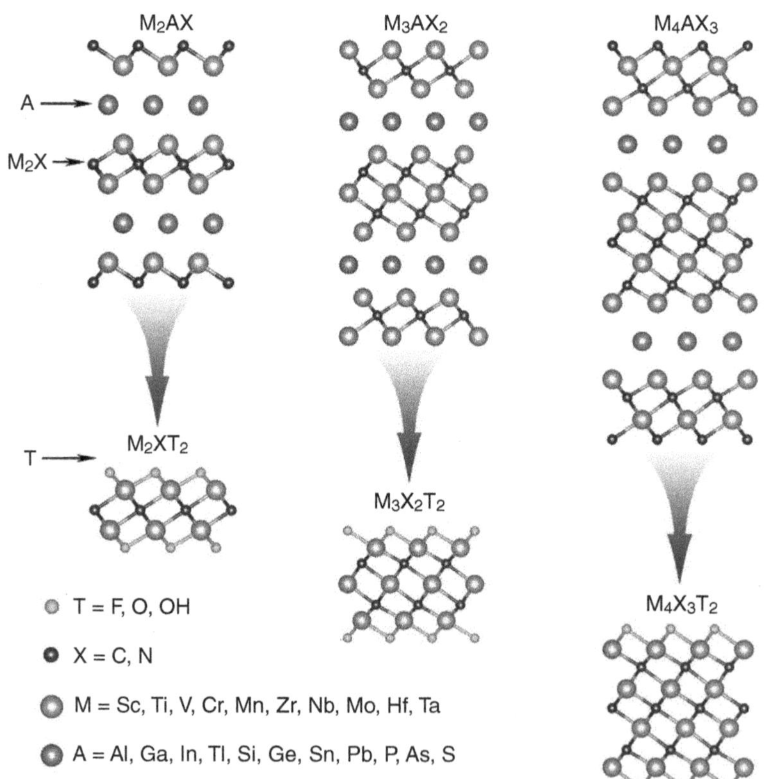

Figure 7.4 Structures of different MAX phases and derived MXene monosheets. Source: Peng et al. [55]/with permission of Elsevier.

Face-centered cubic stacking is how the M_3C_2 and M_4C_3 phases are made. Not to mention, the correct atomic ordering of phases is very essential for the sustainability of MXenes [23, 54, 55].

To describe the properties, MXenes have a high Young's modulus, a high electric and thermal conductivity, and a customizable bandgap. The hydrophilic surfaces of MXenes, which have high metallic conductivities, are different from those of most 2D materials, including graphene [58–60]. Lastly, their properties and performances may be tuned through (i) composition (as such, solid solution formation and different transition metals "M" and "X" elements), (ii) surface functionalization (through chemical and thermal treatments) and structure/morphology alterations [27, 61–65].

Mechanical characteristics of MXene may vary significantly based on the surface terminations. Bai et al. [66] discovered that the connection between the O terminations and Ti atoms is greater in Ti_2C and Ti_3C_2 than in F- or OH-terminated MXenes. Additionally, the O-terminated MXenes exhibit a high rigidity, which is attributable to the greater bonding strength between Ti and O than between Ti and OH or Ti and F. Magnuson et al. [67] also discovered that the surface groups remove charge from Ti—C and weaken the Ti—C bonds. They discovered that the Ti—C bond is longer in $Ti_2C\text{-}T_x$ than in $Ti_3C_2\text{-}T_x$, which they propose may alter the materials' elastic characteristics. Additionally, the researchers note that altering the binding strength may be employed to maximize the flexibility. Zha et al. [68] indicated that because of their increased structural rigidity, O-terminated MXenes should be the top option for structural materials, super-capacitors, and other applications. Furthermore, Yorulmaz et al. [69] observed that when the concentration of the transition metal M increases, the MXene becomes stiffer. For the nitrides, the aforementioned findings could not be reproduced. The great majority of carbide-based MXenes are mechanically stable, while nitride-based MXenes are not [64].

As described in a study by Jiang et al. [28], non-terminated MXenes are usually metallic with a high density of states (DOS) near the Fermi region owing to the presence of an outer layer of transition metallic elements. The DOS is driven by transition metal d-electrons around the Fermi surface, while the p-electrons of the X atoms form energy bands about −3 to −5 eV below the Fermi surface [70, 71]. It is claimed that the outer transition metal layers of MXenes have a greater influence on their electrical characteristics than the inner transition metal layers [72–74]. As a result, surface terminations that are bonded to the outer layer transition metal atoms have the potential to drastically modify the electronic properties including the band structures. The electronegative termination accepts one or two electrons from the outer transition metal layers and creates a new energy band below the Fermi surface, resulting in the decrease in the DOS concentration at the Fermi surface. Due to the fact that OH- and F- groups can only take one electron, they display identical effects on the MXenes' electronic configuration. However, an O-group may accept two electrons and hence has a distinct impact. Electronic [75] and thermal [76] transport in MXenes is similarly influenced by the surface functional groups. Electronic transmission is greatest for F-terminated MXenes, but surface functionalization with O atoms significantly reduces electronic transmission [75]. To date, the electronic characteristics of terminated MXenes have been predicted theoretically or partially experimentally, ranging from highly conductive metallic [77–79], semiconducting [74, 80], to topologically insulating and very insulating states [81]. With the exception of surface terminations, MXenes' energy structure may be adjusted by stoichiometry tailoring [82], doping [83], stresses [84], crystal lattice symmetry [85], and an external electric field [86].

From the description of Papadopoulou et al. [64], in terms of optics, MXenes have displayed a variety of intriguing properties during the last several years. Optical transparency, plasmonic behavior, and efficient photothermal conversion are examples of these. MXenes' capacity to interact with light in a variety of ways has significantly impacted the science community. Again, the surface terminations are critical in defining the optical characteristics of a material [64, 87]. Halim et al. [88] show that changing the nature of the transition metal may modify the optical characteristics of MXenes. They found that the Nb-based films showed an increase in absorption toward the infrared spectral range. This behavior was markedly different from that of the MAX phase Nb_2AlC, which was not found for the Ti-based films. MXenes have a greater capacity for light absorption than carbon nanotubes, as shown by Li et al. [89] (CNTs). In addition, Jiang et al. [90] in 2018 examined the optical non-linearity of $Ti_3C_2T_x$ using excitation sources with various wavelengths (800, 1064, 1550, and 1800 nm).

7.2.4 Structure and Properties of MOFs

As described in the publication by Jiao et al. [91], MOFs are composed of two components: secondary building blocks (clusters or metal ions) and organic molecules that connect the former to produce essentially periodic porous structures. Numerous combinations of these structural components result in a vast array of MOFs. The many functionalities of MOFs are based on their various crystalline forms. MOFs are synthesized by coupling metals or metal clusters (sometimes referred to as secondary building units, SBUs) with organic linkers, the majority of which are carboxylic acid or nitrogen-containing ligands (Figure 7.5). Figure 7.5 illustrates many sample SBUs and organic linkers. One distinguishing characteristic of MOFs over other porous materials such as zeolites and carbons is their tunability. Due to the fact that the constructions of MOFs are established by the topography of SBUs as well as the size and shape of the organic linkers, MOFs can be tuned to a certain extent by carefully selecting SBUs and connectors to reach the required pore size, structure, and functionality for specific uses. The MOF structures may theoretically be anticipated using the metals and organic linker building components. Recently, over 20 000 MOF formations were believed to have been recorded [91, 92], and the number continues to increase by the thousands each year. MOFs with specified structures enable the investigation of structure–property interactions, which is crucial for the rational design of novel MOFs with intended functionalities. From the pore surfaces of MOF materials, active sites may be created. For instance, organic bonds may directly integrate functional groups such as pyridyl and amine that are capable of detecting certain small molecules. Latest advancements in multivariate

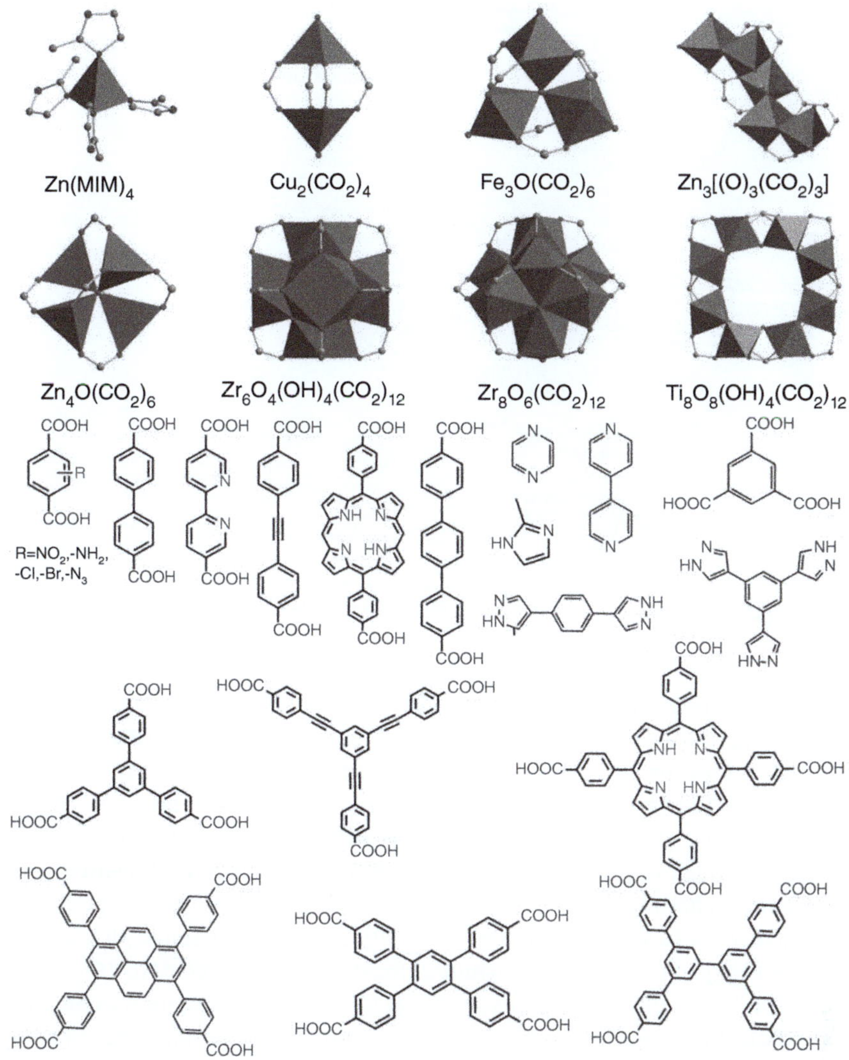

Figure 7.5 Some representative SBUs and organic linkers of MOFs. Source: Jiao et al. [91]/with permission of Elsevier.

MOFs enable the rapid introduction of numerous functional sites onto pore walls. Modifications made after the SBUs or organic linkers are synthesized might include functional groups that are inappropriate with MOF production. Metal ions/clusters and organic linkers in MOFs with a range of electrical, magnetic, and optical properties may be precisely tailored to meet the requirements of a particular purpose. Additionally, the pore space of MOFs may be occupied by a variety of functional hosts to perform a variety of duties. MOFs may tolerate a wide variety of species, including single metal atoms, nanoparticles, metal complexes, organic dyes, polyoxometalates,

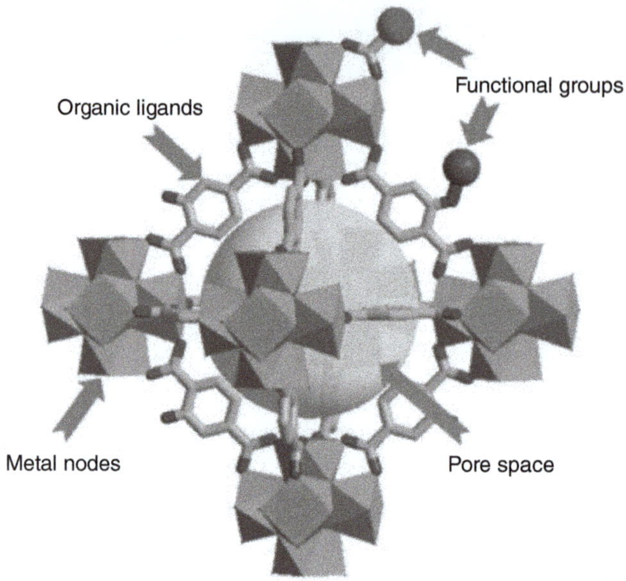

Figure 7.6 Illustration of the construction elements of functional MOFs. Source: Jiao et al. [91]/with permission of Elsevier.

polymers, and tiny enzymes [91, 93–96]. The presence of a guest species greatly expands the applications that are not possible with a single MOF [91] (Figure 7.6).

Further described by Jiao et al. [91] stating, even though a substantial majority of novel compounds have been described during the last two decades, MOFs often exhibit poor stability in terms of water, acid/base, thermal, and mechanical properties. Nonetheless, MOF resilience is critical for a variety of useful situations. Numerous efforts have been undertaken and great advancement has been achieved in subsequent years to solve this difficulty [97–99]. Several water-stable MOFs have been intensively explored to date, including the chromium-based MIL-101 series [100], zeolitic imidazolate frameworks (ZIFs) [101] or metal azolate frameworks (MAFs) [102], the zirconium-based carboxylates [103–105], the aluminum-based carboxylates [106], and the pyrazole-based MOFs [107]. Additionally, novel ways for improving the water/moisture stability of MOFs have been explored lately via the creation of hydrophobic surfaces or interfaces. Nguyen and Cohen devised an effective technique for protecting the moisture-sensitive MOF by incorporating medium-to-long alkyl groups into it [108]. Yang et al. reported the development of a series of fluorinated MOFs (FMOFs) with exceptional water stability due to their super-hydrophobicity [109]. Thermal stability of MOFs is especially critical for applications requiring greater temperatures, such as gas phase reactions at elevated temperatures. Thermal stability is typically governed by the strength of the connection between the node and the linker, and the number of linkers attached to each metal node. MOFs with excellent thermal stability are often synthesized using high valence metal ions such as Ln^{3+}, Al^{3+}, Zr^{4+}, and Ti^{4+}. While the majority of MOFs are stable between 350 and 400 °C, a few, such as MIL-53 [106] and UiO-66 [104], are stable over 500° Celsius. Thermal stability is often the only form of stability examined for novel MOFs since it is a good predictor of resistance to other stressors [91].

Jiao et al. [91] discuss the pore characteristics and surface area in their article mentioning that persistent and highly ordered porosity is one of the most distinguishing characteristics of MOFs and serves as the foundation for their many functionalities. The majority of MOFs created to date are microporous (2 nm pore diameter) in nature and have shown excellent adsorption capabilities for a variety of gases, including hydrogen and carbon dioxide. While substantial micropore volumes and large surface area are desired for many purposes, such tiny holes cannot accommodate massive items or molecular functions nor do they allow for rapid mass diffusion and transfer, limiting their utility in storage, separation, catalysis, and drug delivery. For some of the more recent applications, such as catalysis and drug delivery, mesoporous MOFs with pore diameters of 2–50 nm and macroporous MOFs with pore diameters greater than 50 nm are preferable [92, 110]. MIL-100(Cr) and MIL-101(Cr) are two very well-known cage-type meso-MOFs. MIL-100(Cr) was synthesized and crystallized for the first time in 2004 by Frey et al., while MIL-101(Cr) was synthesized and crystallized by the same group in 2005 [100]. Numerous ways for fabricating and adjusting MOFs with high pore diameters have been shown to be successful. Increasing the length of ligands is an obvious and efficient strategy to do this [111]. IR-MOF-74-XI currently holds the record for the biggest pores in MOFs, with an astounding pore diameter [96]. Along with pore diameter, the Brunauer–Emmett–Teller (BET) and Langmuir surface areas of MOFs are critical for their functionalities [112]. NU-110 has the greatest surface area of any material ever known (7140 $m^2\ g^{-1}$) [113]. Calculations show that the theoretical maximum limit for MOF surface areas is roughly 14 600 $m^2\ g^{-1}$ [113], but determining how much of this can be achieved experimentally remains as a difficulty [91].

7.2.5 Photocatalytic Mechanism

The photocatalytic mechanism is described as follows: the photocatalyst is made up of two distinct bands: the empty conduction band and the occupied valence band. It all starts with optical absorption, and when the energy of the photon is more than or equivalent to the bandgap, the photocatalytic process produces electronic transition and forms e^--h^+ pairs, which are subsequently stimulated to the CB, while the holes are left behind. VB holes cause the oxidation process, which is catalyzed by the CB electrons and results in H_2. The VB's peak band edge needs to be more positive than O_2/H_2O's redox potential, while the CB's lowest band edge must be more negative than the redox potential of H^+/H_2 (Figure 7.7) [35–37, 114, 115].

On the other hand, organic substances can be broken down by advanced oxidation process, which includes photocatalytic degradation. In comparison to other advanced oxidation processes, this one is more successful because of the low cost and ease with which semiconductors can mineralize a wide range of organic molecules [116–118]. One theory suggests that the process depicted in Figure 7.8 is responsible for dye photocatalytic degradation. Because of UV light, electrons in a catalyst's conduction band are elevated. An electron–hole couple is formed

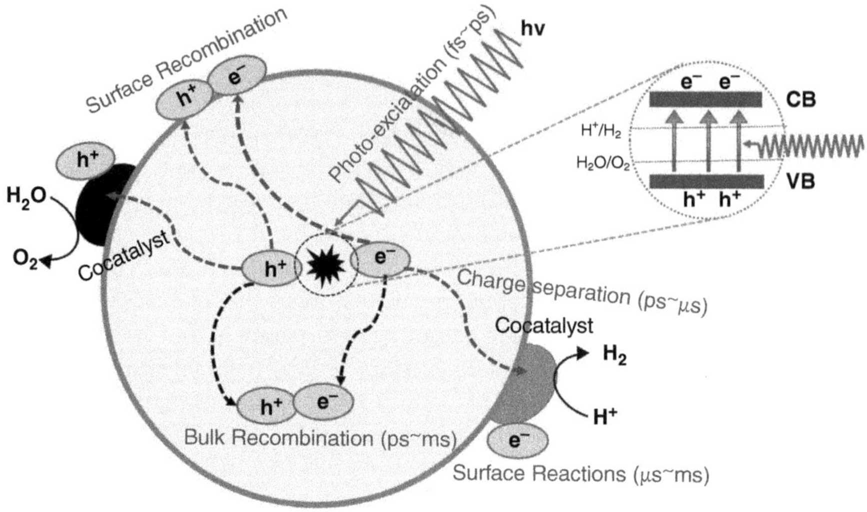

Figure 7.7 Basic mechanism of photocatalytic water-splitting for the semiconductor. Source: Alfaifi et al. [35]/Public Domain CC BY 4.0.

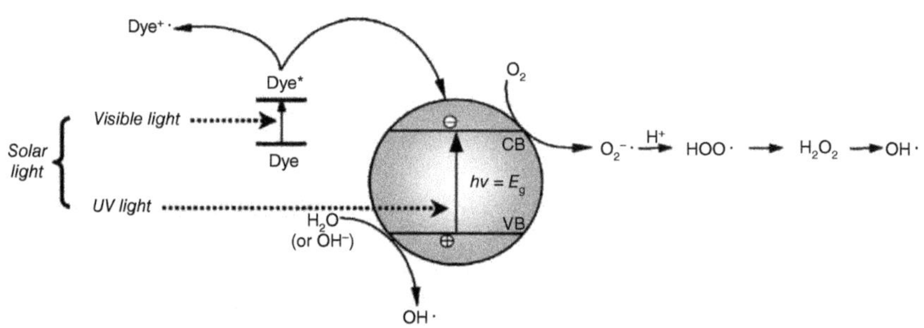

Figure 7.8 Illustration of the generation of oxidative species in a photocatalytic study. Source: Rauf et al. [118]/with permission of Elsevier.

as a result [119]. A redox reaction can occur on the catalyst surface between $e^-{}_{cb}$ and $h^+{}_{vb}$ and other species present there [116, 117]. ˙OH radicals are formed in most instances by the reaction of $h^+{}_{vb}$ with surface-bound H_2O, while superoxide radical anions of oxygen are formed by reacting with O_2[120]. This reaction keeps the electron and hole from combining, as they would have done in the first stage. ˙OH and O_2 single bond radical dot generated in the way described above can react with the dye to make other species, causing the dye to degrade [116, 118, 119].

7.3 Photocatalytic Hydrogen Production

7.3.1 TiC Supported Nanohybrids

TiC MXene has received significant interest in photocatalysis recently due to its strong electrical mobility for effective carrier segregation [121, 122]. Many studies have shown that TiC is an excellent co-catalyst for photocatalysis, based on the enhanced activity of TiO_2, $ZnIn_2S_4$, Bi_2WO_6, and other Ti_3C_2-modified semiconductors [121, 123]. Attributed to the prevalence of functional groups like –OH, –F, and –O on the surface of Ti_3C_2, semiconductors are often modified by electrostatic adsorption. The organic ligand in MOFs can be easily coordinated with metals because of the MOF's structure [122–124]. The organic ligand in MOFs is therefore extremely likely to pair with Ti in Ti_3C_2, generating a different interaction mechanism [123, 124]. The photocatalytic efficiency of MOFs in combination with Ti_3C_2 may be affected in this way [122]. Some researchers have reported the efficiency of MXene-based nanohybrids recently [123, 124].

Li et al. [125] coupled Ti_3C_2 MXene to a Ti-based MOF (denoted as MIL-NH_2). Ti_3C_2 is thought to perform two purposes: as a source of Ti for in situ development of MIL-NH_2 and as a co-catalyst for photocatalytic hydrogen generation. They obtained very tiny MIL-NH_2 particles under the laboratory conditions. Ti_3C_2@MIL-NH_2 hybrids were generated using an in situ growing process (mole proportions of Ti_3C_2 to MIL-NH_2 are 1 : 1.2, 1 : 1.6, 1 : 2.0, and 1 : 2.4). Ti_3C_2@MIL-NH_2-1.2, Ti_3C_2@MIL-NH_2-1.6, Ti_3C_2@MIL-NH_2-2.0, and Ti_3C_2@MIL-NH_2-2.4 were used to label the materials. The resulting hybrid (Ti_3C_2@MIL-NH_2) exhibits a substantial improvement in H_2 generation under simulated sunlight, with an H_2 generation rate of 4383.1 $\mu mol\,h^{-1}\,g^{-1}$, which is far greater than that of any other MOF-based composite to date. The improved photocatalytic activity is attributable to Ti_3C_2's effective carrier dissociation capability and the close interfacial proximity formed by the coupling of the –NH_2 group in MIL-NH_2 with the Ti on the surface of Ti_3C_2. As seen in Figure 7.9 a,b, Ti_3C_2 has a characteristic accordion-like stacked MXene configuration [25], while pure MIL-NH_2 exhibits an uneven nanoparticle shape that aggregates. Ti_3C_2@MIL-NH_2 significantly inhibits MIL-NH_2 self-aggregation. MIL-NH_2 is uniformly immobilized on the surface of Ti_3C_2 due to the in situ growth.

Figure 7.9d shows the XRD patterns for all subjects. The MIL-NH_2 diffraction peaks are compatible with those reported earlier, and the coexistence of MIL-NH_2 and Ti_3C_2 diffraction peaks suggests that the Ti_3C_2@MIL-NH_2 hybrids were successfully prepared. The assessed BET surface area was reported as 389.5, 7.1, 53.7, 69.2, 83.9, and 112.4 $m^2\,g^{-1}$ for MIL-NH_2, Ti_3C_2 MXene, Ti_3C_2@MIL-NH_2-1.2, Ti_3C_2@MIL-NH_2-1.6, Ti_3C_2@MIL-NH_2-2.0, and Ti_3C_2@MIL-NH_2-2.4, respectively. The progressive increase in BET surface area for the composites as MIL-NH_2 concentration increases further implies that MIL-NH_2 is formed.

It is generally documented that increased carrier segregation and transfer efficiency are always advantageous for photocatalytic process actuation. As a result, the photocatalytic H_2 generation over MIL-NH_2 and Ti_3C_2@MIL-NH_2 hybrids is evaluated using a 300-W Xe lamp equipped with an AM 1.5 filter. As seen in Figure 7.10a,b, unadulterated MIL-NH_2 has very little activity toward H_2 evolution. By comparison, the Ti_3C_2@MIL-NH_2 hybrids demonstrate significantly increased activity. To be precise, the optimized Ti_3C_2@MIL-NH_2-1.6 exhibits the maximum efficiency (4383.1 $\mu mol\,h^{-1}\,g^{-1}$), which is almost six times that of MIL-NH_2 and five

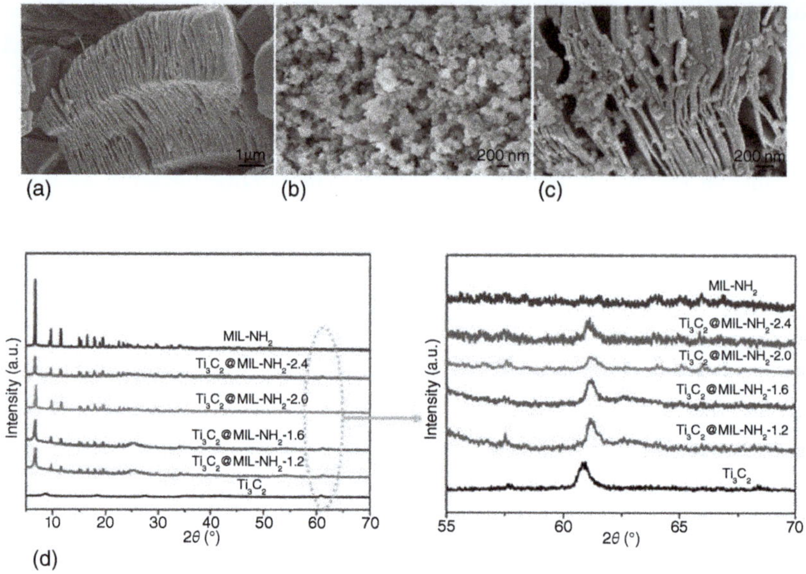

Figure 7.9 SEM images of (a) Ti_3C_2 MXene, (b) MIL-NH_2, (c) Ti_3C_2@MIL-NH_2-1.6 and (d) XRD patterns of Ti_3C_2, MIL-NH_2, and Ti_3C_2@MIL-NH_2 composites. Source: Adapted from Li et al. [125]/with permission from Elsevier.

times that of a physical combination of MIL-NH_2 and Ti_3C_2. Ti_3C_2@MIL-NH_2-1.6 exhibits a 3.140% apparent quantum efficiency (AQE).

Furthermore, the stability of Ti_3C_2@MIL-NH_2-1.6 was investigated. The photocatalytic H_2 generation rate over Ti_3C_2@MIL-NH_2-1.6 does not significantly decrease after four cycles, showing its good photostability. Additionally, the XRD pattern, SEM picture, and N_2 adsorption/desorption isotherms of Ti_3C_2@MIL-NH_2-1.6 after four cycles exhibit no discernible variation from those of the fresh sample, showing the composites' resilience. Ti_3C_2@MIL-NH_2-1.6 (4383.1 $\mu mol\,h^{-1}\,g^{-1}$) has much higher photocatalytic H_2 generation activity than other typical photocatalytic MOFs (Figure 7.10c). This enhanced performance can be attributed to the following: first, Ti_3C_2's outstanding electronic conductivity enables faster charge transfer; second the presence of Ti_3C_2 acts as a co-catalyst, lowering the potential of water reduction; and, most notably, third, both the in situ growth of MIL-NH_2 on Ti_3C_2. Furthermore, the coordination between N and Ti enables charge transfer. To demonstrate the critical role of connectivity between Ti and N in functionality, MIL-125-NH_2 is substituted with MIL-125 employing terephthalic acid as the organic ligand rather than bdc-NH_2. The photocatalytic H_2 generation rate indicates that the resulting Ti_3C_2@MIL exhibits much less activity (139.1 $\mu mol\,h^{-1}\,g^{-1}$) than Ti_3C_2@MIL-NH_2 (626.3 $\mu mol\,h^{-1}\,g^{-1}$) (Figure 7.10d).

Zhao et al. [126] described the synthesis of a ternary TiO_2–Ti_3C_2–CoS_x hybrid using a straightforward solvothermal technique (Figure 7.11). The porous morphology of the ZIF-67-templated CoS_x enhanced not only the morphology of TiO_2 nanoparticles but also the segregation and utilization effectiveness of charge

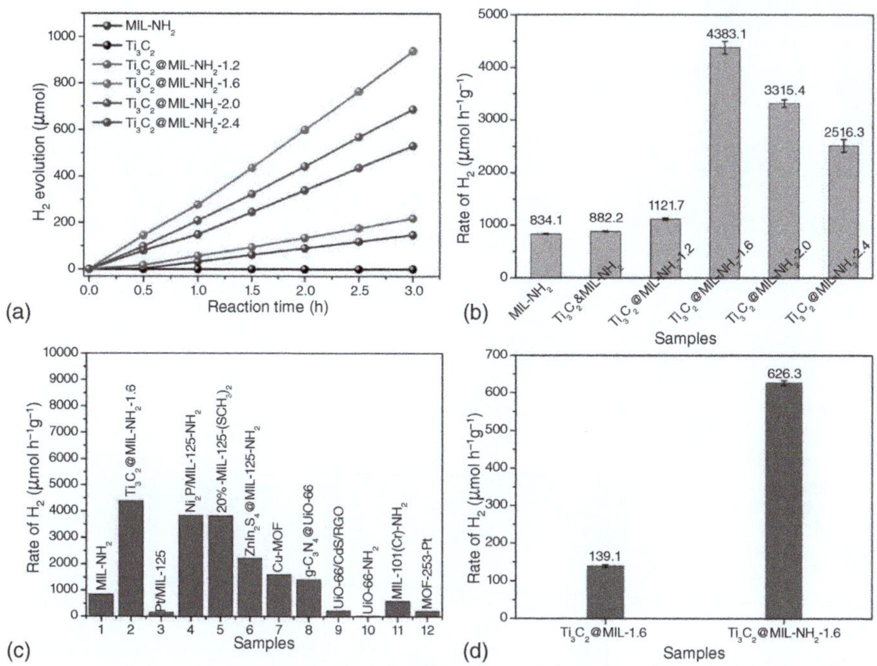

Figure 7.10 (a) Time course of photocatalytic H_2 production over MIL-NH_2, Ti_3C_2 MXene, and Ti_3C_2@MIL-NH_2 composites, (b) Photocatalytic H_2 production rate over MIL-NH_2, Ti_3C_2& MIL-NH_2, and Ti_3C_2@MIL-NH_2 composites, (c) Comparison on the activity of photocatalytic H_2 production over MIL-NH_2, Ti_3C_2@MIL-NH_2-1.6 and other representative photocatalytic MOFs and (d) Rate of photocatalytic H_2 production over Ti_3C_2@MIL and Ti_3C_2@MIL-NH_2 under UV irradiation. Source: Li et al. [125]/Elsevier.

transfer. Additionally, the conductive Ti_3C_2 MXenes might boost the heterostructure system's conductivity, hence improving the transfer of photo-generated carriers. As a result, the produced TiO_2–Ti_3C_2–CoS_x heterostructure shows significant improvements in photocatalytic activity. The samples created with varying mole ratios of CoS_x were designated as TC_x ($x = 0.2$, 0.5, 1, 2, and 5), whereas those produced with varying mass fractions of Ti_3C_2 and 1% CoS_x were designated as TM_xC1 ($x = 0.2, 0.5, 1,$ and 2, 5).

SEM and TEM were used to determine the morphology of the produced samples. As reported, pure TiO_2 had a smooth round block shape with a radius of about 450 nm. In contrast to the pure TiO_2, the TC1 sample revealed a well-dispersed nanoparticle morphology owing to the presence of porous ZIF-derived CoS_x. Importantly, it was indicated that TiO_2–CoS_x particles were incorporated inside Ti_3C_2 nanosheets, resulting in the rough block structure constituted of tiny nanoparticles. The transmission electron microscopy (TEM) technique was used to further characterize the microstructures of as-formed TiO_2–Ti_3C_2–CoS_x samples. As reported, TiO_2 agglomerates, while uniformly scattered TiO_2 nanoparticles with a diameter smaller than 20 nm form when CoS_x is introduced. The Ti_3C_2 nanosheets, which worked as a scaffold for the creation of the TiO_2–CoS_x heterostructure,

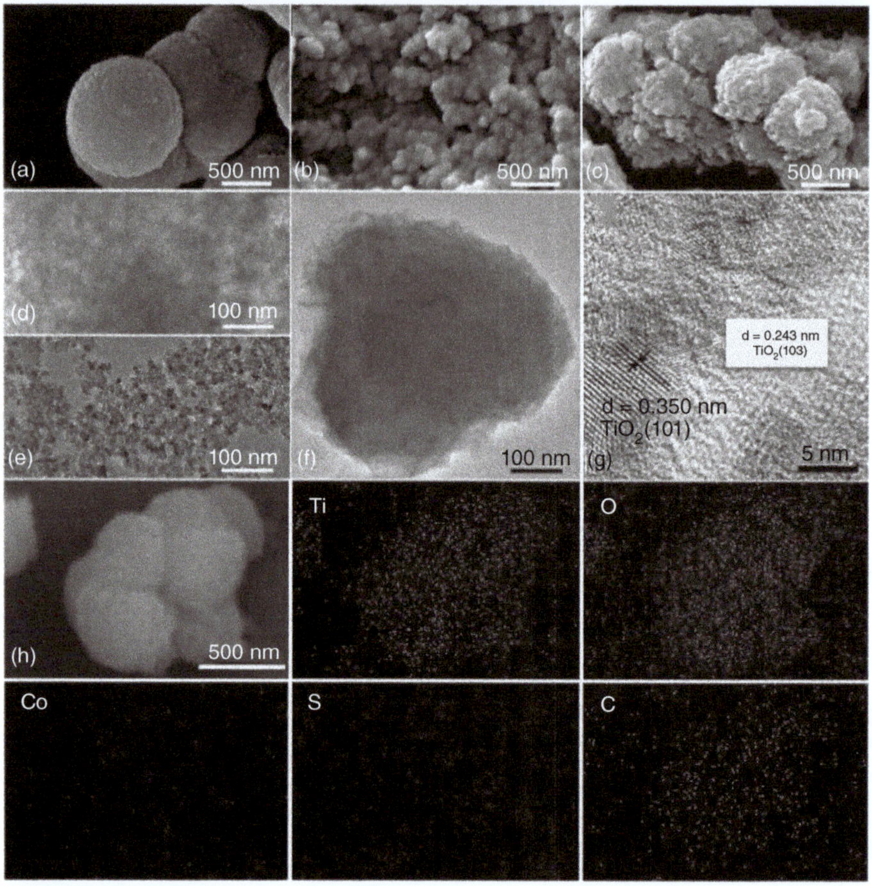

Figure 7.11 The SEM and TEM of images (a, d) TiO_2, (b, e) TC1, (c, f) TM0.5C1 samples, (g) The HRTEM image and (h) Element mapping of TM0.5C1. Source: Zhao et al. [126]/with permission from Royal Society of Chemistry.

were coupled to create a porous morphology, as verified by the evident nanosheet morphology of the TiO_2–Ti_3C_2–CoS_x sample's TEM picture. Furthermore, it benefited charge carrier transportation. Additionally, the high-resolution transmission electron microscopy (HRTEM) technique was used to investigate the phase structure of the as-prepared heterostructure. The lattice gap of the TiO_2 nanoparticle was found to be 0.35 nm, which matched to the (101) crystal plane, and the unique fringe space of 0.243 nm was attributed to the (103) crystal plane of TiO_2. The component distributions of the TiO_2–Ti_3C_2–CoS_x sample (TM0.5C1) were described, and a homogeneous distribution of Ti, O, C, Co, and S components was found. Under UV irradiation, the photocatalytic efficiency of the as-prepared TiO_2-ZIF-67 templated TiO_2–CoS_x and TiO_2–Ti_3C_2–CoS_x photocatalysts was investigated using methanol as the sacrificial donor. Figure 7.12a compares the photocatalytic hydrogen generation capability of TiO_2–CoS_x with that of various molar amounts of CoS_x. TiO_2 in its purest form has a hydrogen generation activity of 0.14 mmol h^{-1} g^{-1}. In comparison,

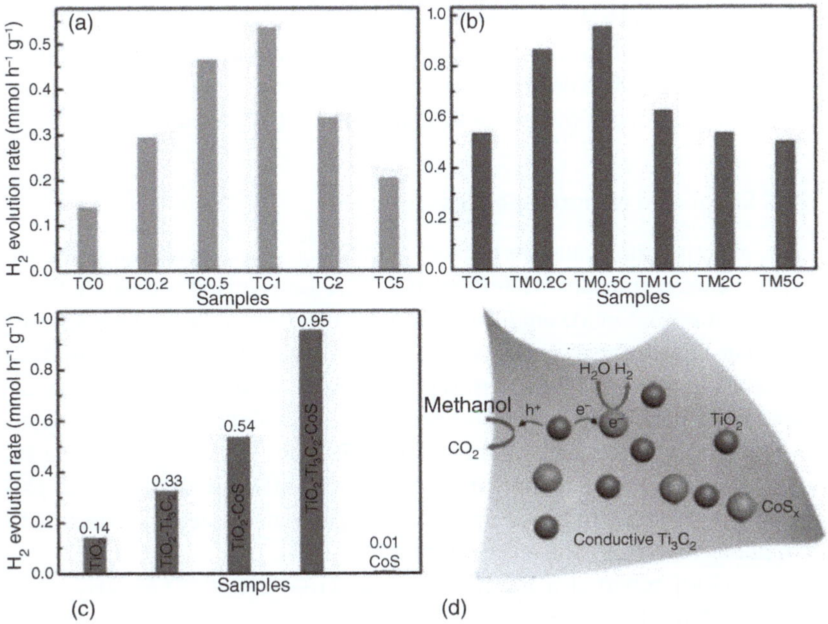

Figure 7.12 Influence of the content of CoS_x on the photocatalytic activity of (a) TiO_2–CoS_x, the photocatalytic H_2-production activities of (b) TiO_2–Ti_3C_2–CoS_x samples with different content of Ti_3C_2 and 1% CoS_x. The comparison results of the photocatalytic activity of (c) TiO_2, TM0.5, TC1, TM0.5C1, and CoS_x, and (d) Schematic illustrating the mechanism of the HER. Source: Zhao [126]/with permission of Royal Society of Chemistry.

adding merely 1% ZIF-67-templated CoS_x resulted in a substantial enhancement in photocatalytic activity. The hydrogen generation rates of TC_x increased when the CoS_x concentration was increased. Once the molar ratio of CoS_x approached 1%, the ideal amount of H_2 generation was 0.54 mmol h^{-1} g^{-1}, which was 2.8 times that of pure TiO_2. Increased CoS_x concentration caused a decrease in the H_2-generation rate, while the hydrogen production activity remained greater than that of pure TiO_2. Likewise, the impact of Ti_3C_2 presence on TiO_2–Ti_3C_2 suggested that the best photocatalytic performance occurred at a Ti_3C_2 concentration of 0.5 wt%, which should be attributable to Ti_3C_2's high conductivity.

Additionally, as seen in Figure 7.12b, the use of conductive Ti_3C_2 may improve charge carrier transport efficiency, hence increasing photocatalytic H_2 generation activity to 0.95 mmol h^{-1} g^{-1} for the best TiO_2–Ti_3C_2–CoS_x system. When the concentration of black Ti_3C_2 was increased further, the sample's color darkened, reducing optical absorption in TiO_2 and resulting in a reduction in photocatalytic performance. To provide a more complete picture of the influence of ZIF-67-derived CoS_x and Ti_3C_2 on the photocatalytic performance of TiO_2, a full comparison of pure TiO_2 and the as-synthesized heterostructure samples is shown in Figure 7.12c. The presence of porous CoS_x generated from ZIF-67 and conductive Ti_3C_2 boosted the photocatalytic HER rate from 0.14 to 0.54 and 0.33 mmol h^{-1} g^{-1}, correspondingly. Surprisingly, co-loading of CoS_x and Ti_3C_2 had a synergistic impact on increasing this value to 0.95 mmol h^{-1} g^{-1}. Importantly, the TiO_2, TC1, and TM0.5C1 have shown excellent durability throughout the photocatalytic hydrogen evolution process, with no discernible drop in the rate of H_2 evolution for the TiO_2, TC1, and TM0.5C1 over the course of five cycles in 15 hours under UV irradiation.

7.3.2 TiC Supported Nanocomposites

MXenes are effective photocatalysts because of their high Fermi levels, superior conductivity, and carrier transport characteristics [127, 128]. It is also possible to create heterostructures out of MXene/semiconductor heterostructures, and a Schottky junction will be built at the interface between the two materials, which might facilitate the flow of photo-generated electrons [129]. When charge-carrier transport may be efficiently facilitated, interfacial architecture is often done by selecting appropriate components and reducing interfacial imperfections [128, 129]. Coupling TiC MXenes, which have good conductivity, with porosity MOFs with fully filled photoactive sites is predicted to speed up the photo-generated electron transport for long-term photocatalytic hydrogen generation [127–129].

Tian et al. [130] reported a novel composite Ti_3C_2/TiO_2/UiO-66-NH_2 which was investigated in detail for the production of photocatalytic hydrogen. The photoresponsive photocatalyst UiO-66-NH_2 was examined in their study as a typical MOF, and annealed $Ti_3C_2T_x$ MXenes was utilized as an electron acceptor to create pathways for interfacial charge transfer. To do this, the $Ti_3C_2T_x$ MXenes were first annealed in nitrogen to form TiO_2 layers which was denoted as TCA. Following that, UiO-66-NH_2 was electrostatically adsorbed onto the surfaces of annealed $Ti_3C_2T_x$ using a one-step hydrothermal process, as shown in Figure 7.13.

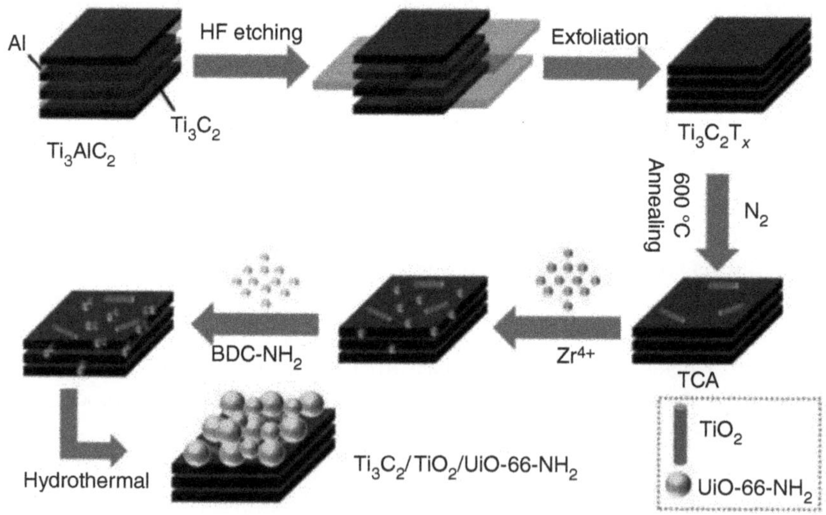

Figure 7.13 A proposed chemical pathway for the production of Ti_3C_2/TiO_2/UiO-66-NH_2 is shown schematically. Tian et al. [130]/with permission of Elsevier.

Ti_3C_2/TiO_2/UiO-66-NH_2 showed superior photocatalytic HER performance as compared to $Ti_3C_2T_x$/UiO-66-NH_2 and pure UiO-66-NH_2. From this vantage point, the TCA not only served as a platform for the UiO-66-NH_2 but also as co-catalysts by acting as a storage for the charge carriers, hence extending their lifespan. As-prepared samples were subjected to X-ray diffraction (XRD) analysis to determine the crystalline phase. The disappearance of the Ti_3AlC_2 greatest diffraction pattern at $2\theta = 39°$ showed the removal of Al layers and the development of 2D $Ti_3C_2T_x$ MXenes after HF etching. Additionally, the peaks corresponding to the Ti_3AlC_2 (002) and (004) planes widened and migrated to lower degrees, indicating an increase in the c-lattice parameter. The XRD patterns of pure UiO-66-NH_2, Ti_3C_2/TiO_2/UiO-66-NH_2, and $Ti_3C_2T_x$/UiO-66-NH_2 are reported. It can be observed that the addition of TCA had no discernible effect on the basic crystal structure of UiO-66-NH_2. Additionally, no Ti_3C_2 or TiO_2 diffraction peak was seen in Ti_3C_2/TiO_2/UiO-66-NH_2, which was presumably due to the low TCA loading [30].

SEM was used to examine the morphologies of as-prepared samples. Pristine UiO-66-NH_2, $Ti_3C_2T_x$, and TCA were also included for comparison. Typically, $Ti_3C_2T_x$ displayed accordion-like structures during HF etching, indicating that Al layers were successfully removed from Ti_3AlC_2. Additionally, TCA exhibited a layered structure comparable to that of pure $Ti_3C_2T_x$, and no TiO_2 can be observed owing to its low concentration. UiO-66-NH_2 was found to be firmly wrapped around $Ti_3C_2T_x$ owing to the very hydrophilic surfaces of pure $Ti_3C_2T_x$, resulting in severe agglomeration of UiO-66-NH_2. In comparison, Ti_3C_2/TiO_2/UiO-66-NH_2 exhibited a more homogeneous attachment of UiO-66-NH_2 to the surface of TCA with more accessible reactive sites. The elemental arrangement of Ti_3C_2/TiO_2/UiO-66-NH_2 was determined using the matching EDS imaging spectrum and a piece of TCA as the test target. The elements C, N, O, F, Ti, and Zr were identified, and all the elements were evenly distributed in Ti_3C_2/TiO_2/UiO-66-NH_2, demonstrating that UiO-66-NH_2 was properly disseminated on the surface of TCA.

The Brunauer–Emmett-Teller (BET) surface area of $Ti_3C_2T_x$, TCA, UiO-66-NH_2, Ti_3C_2/TiO_2/UiO-66-NH_2, and Ti_3C_2/UiO-66-NH_2 was reported as 4.2, 10.5, 977.6, 987.5, and 1019.5 $m^2\ g^{-1}$ respectively. The photocatalytic generation of H_2 was carried out using an aqueous solution comprising 0.1 M Na_2S and 0.1 M Na_2SO_3 as a sacrificial agent for 4 hours under artificial sunlight illumination ($350 < \lambda < 780$ nm). Without illumination or catalysts, no H_2 was measured, indicating that the majority of H_2 was generated by the photocatalytic process. Unadulterated $Ti_3C_2T_x$ had a lower photocatalytic H_2-production rate than TCA, which may be attributed to the presence of TiO_x surface oxides. Additionally, the quantity of H_2 created by $Ti_3C_2T_x$/UiO-66-NH_2 was around 3665 μmol g^{-1}, which was greater than the amount produced by pure UiO-66-NH_2, owing to $Ti_3C_2T_x$'s high electrical conductivity and effective charge transport mechanism. The quantity of H_2 produced by Ti_3C_2/TiO_2/UiO-66-NH_2 increased to 7840 μmol g^{-1} with the addition of TCA. This improved improvement was ascribed not only to the enhanced light collecting capacity and higher specific surface area but also to the synergistic impact of Ti_3C_2, TiO_2, and UiO-66-NH_2. To validate the synergistic impact shown in Ti_3C_2/TiO_2/UiO-66-NH_2, another sample TiO_2/UiO-66-NH_2 was evaluated by loading TiO_2 onto UiO-66-NH_2. The absence of H_2 indicated a high recombination efficiency for photo-induced electron–hole pairs, which might have been caused by the weak binding force between TiO_2/UiO-66-NH_2 during the impregnation step. The mean recorded H_2 evolution rates are shown in Figure 7.14b. The addition of TCA significantly increased the H_2 evolution rates of Ti_3C_2T/TiO_2/UiO-66-NH_2, achieving a HER performance of 1980 $\mu mol\,h^{-1}\,g^{-1}$, 1.5 times and 2.1 times that of $Ti_3C_2T_x$/UiO-66-NH_2 and unadulterated UiO-66-NH_2, respectively. Subsequently, the stability of Ti_3C_2/TiO_2/UiO-66-NH_2 was tested further in recycling photocatalytic tests during a 12-hour period. As a result of the structure's resilience, only a minor decrease in HER activity was identified after three cycles with a 4-hour intermission over Ti_3C_2/TiO_2/UiO-66-NH_2, as shown in Figure 7.14c. Table 7.1 shows different types of Ti_3C_2-based MOF composites and their performance for photocatalytic hydrogen production. It was observed that MXenes-based MOF ternary composites are more efficient for hydrogen production due to more efficient charge carrier separation.

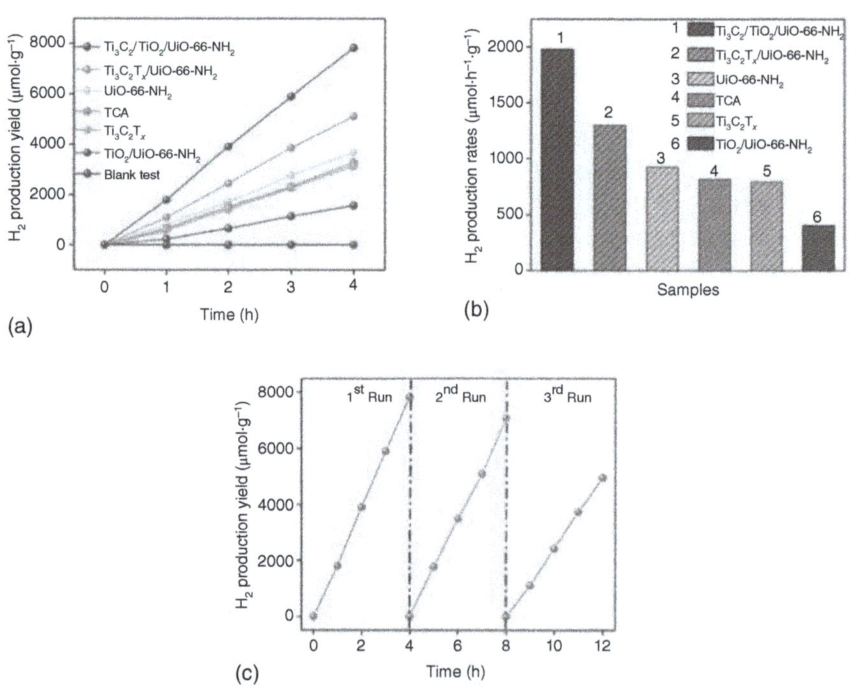

Figure 7.14 (a) Photocatalytic H_2 production yield, (b) Average H_2 evolution rates obtained by the as-prepared samples, and (c) Recyclability test of Ti_3C_2/TiO_2/UiO-66-NH_2 for photocatalytic hydrogen generation under UV–visible light irradiation ($350 < \lambda < 780$ nm). Source: Tian et al. [130]/with permission of Elsevier.

Table 7.1 Photocatalytic hydrogen evolution rate investigation by using TiC MXene-based MOF Composites.

Photocatalyst	Bandgap, Pore Size (nm) and BET (m^2/g)	Sacrificial agent	Cut-off filter	Illumination source	HER	References
Ti_3C_2@MIL-NH_2	B.g.-2.4 eV, P.S.-n.a., BET-69.2	TEOA	1100 nm > λ > 300 nm	300 W Xe lamp	4383.1 μmol h^{-1} g^{-1}	[125]
TiO_2–Ti_3C_2–CoS_x	—	Methanol	—	300 W Xe lamp	0.95 mmol h^{-1} g^{-1}	[126]
Ti_3C_2/TiO_2/UiO-66-NH_2	B.g.-n.a., P.S.-n.a., BET-987.5	Na_2S and Na_2SO_3	780 nm > λ > 350 nm	300 W Xe lamp	1980 μmol h^{-1} g^{-1}	[130]
$Ti_3C_2T_x$/UiO-66-NH_2	B.g.-n.a., P.S.-n.a., BET-1019.5	Na_2S and Na_2SO_3	780 nm > λ > 350 nm	300 W Xe lamp	1320 μmol h^{-1} g^{-1}	[130]

7.4 Photocatalytic Degradation Application

Ti_3C_2 MXene has components that are efficient for photocatalysis; however, it is often not utilized directly for photocatalysis, which impedes its usage in photocatalytic water purification. MXene and MOFs have excellent application potential in water treatment, yet there are few publications that mix the two [131, 132].

Long et al. [133] reported the solvothermal method to grow MOF (Fe) in situ on Ti_3C_2 MXene to construct a composite Schottky photocatalytic adsorbent. Depending on the outcomes of the experiments, a dye-removing mechanism depending on the success of adsorption and the synergistic Fenton effect at a Schottky knot is proposed. It could be used to get rid of and break down pollution in water. Under the same circumstances (170 °C, 24 hours), the obtained Ti_3C_2 MXene can be added to the antecedent of the synthesizing of NH_2-MIL-88B, followed by the addition of 1.5 ml NaOH (2 mol l^{-1}) after the MXene has been added to the predecessor. Various MXene/NH_2-MIL-88B nanocomposites were created by varying the quantity of MXene used in the composite. The MXene dosage may be changed to 50, 100, or 150 mg. There are three different samples in this series, each with a different mass ratio of MXene to NH_2-MIL-88B. The three samples are called 0.125-MXene/NH_2-MIL-88B, 0.25-MXene/NH_2-MIL-88B, and 0.375-MXene/NH_2-MIL-88B. Figure 7.15 depicts a simplified procedure.

It was determined by SEM that different materials had varied shapes and that monomer and solvothermal processes had different shapes. A huge block architecture can be seen in Figure 7.16, with $Ti_3C_2T_x$ MXene presenting an accordion-like layered structure (Figure 7.16b). Figure 7.16c is NH_2-MIL-88B, which shows the hexagonal prism's regular shape. The dissolution rate of NH_2-MIL-88B will improve as the amount of MXene increases, as shown in Figure 7.16e, largely because

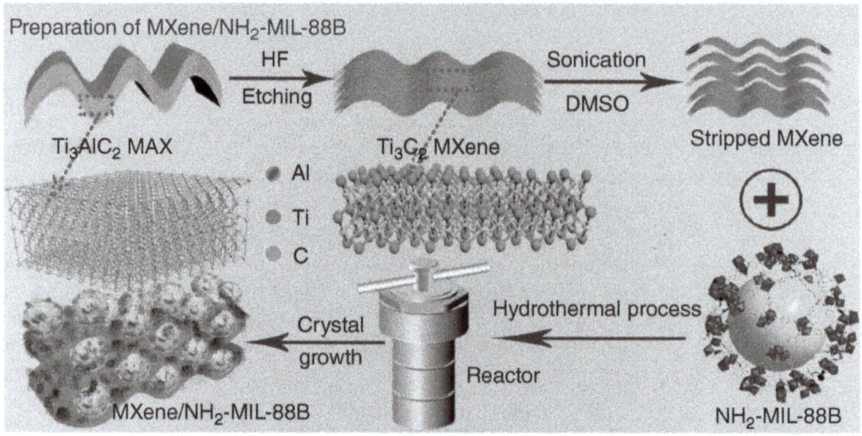

Figure 7.15 Preparation of MXene/NH_2-MIL-88B. Source: Long et al. [133]/with permission of Elsevier.

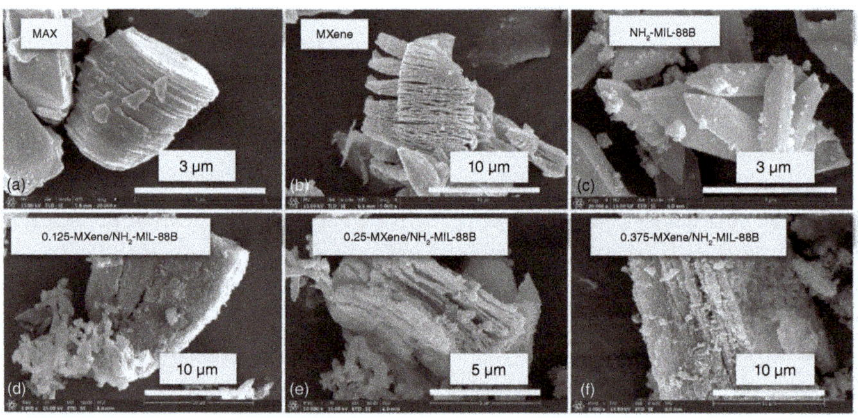

Figure 7.16 SEM images of different materials. Source: Long et al. [133]/with permission of Elsevier.

MXene offers additional NH_2-MIL-88B growth platforms so that the produced NH_2-MIL-88B may be disseminated on the surface of MXene, and the phenomenon of agglomeration is lessened. However, as seen in Figure 7.16f, too much MXene will interfere with the crystallization of NH_2-MIL-88B, leading to an imperfect structure.

Mapping is used to examine the distribution of various components on the surface of the material, as shown in Figure 7.17. As can be seen, the C and O atoms are equally spaced out. Fe and N elements are equally dispersed around the dense Ti elements, showing that NH_2-MIL-88B has grown uniformly on the surface of Ti_3C_2 MXene.

In the laboratory test, the recorded activity is depicted in Figure 7.18, and different materials degrade 100 ml MB (50 ppm) at various dosages. As reported by the research group, using 10 mg of material, MXene, NH_2-MIL-88B, 0.125-MXene/

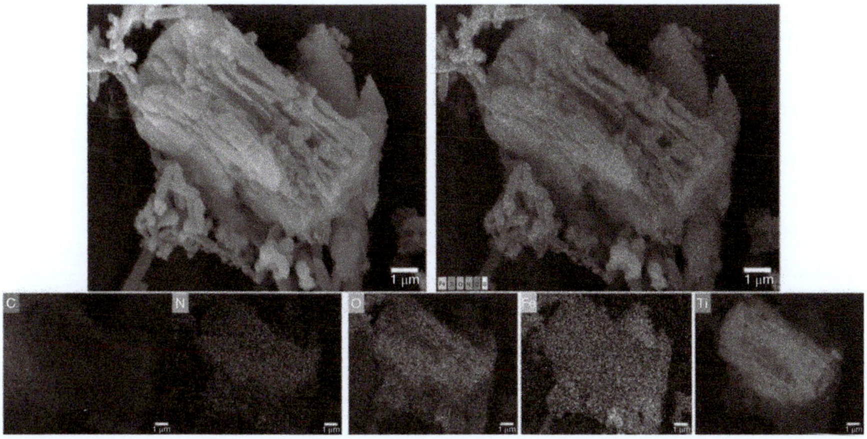

Figure 7.17 SEM image and mapping image of 0.25-MXene/NH_2-MIL-88B. Source: Long et al. [133]/with permission from Elsevier.

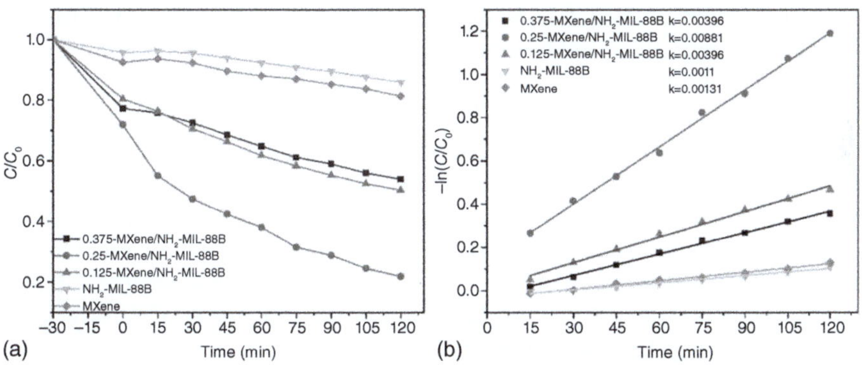

Figure 7.18 (a) The photocatalytic degradation efficiency of different materials, (b) corresponding quasi-first-order kinetic curve, and pseudo-first-order rate constant. Source: Long et al. [133]/with permission of Elsevier.

NH_2-MIL-88B, 0.25-MXene/NH_2-MIL-88B, and 0.375-MXene/NH_2-MIL-88B adsorbed 7.5%, 4.3%, 20%, 28.2%, and 22.7%, accordingly; after 1 hour of light condition, they degraded by 4.5%, 3.2%, 18.3%, 33.8%, and 12.4%; after 2 hours of light condition, the final removal rate reached 18.7%, 14.2%, 49.7%, 78.2%, and 46%. 100 MB's overall removal percentage and removal effect of 391 mg g^{-1} can be seen to be the highest. Few studies have shown that a smaller dose can effectively remove higher concentrations of MB than similar studies. There are many uses for composite materials in the near future.

The Fenton effect should be considered due to the MOF's presence of Fe. After dark treatment, add 50 μl H_2O_2 (30%) and monitor its deterioration under light settings. Fenton effect intensity has been assessed. The composite material's photocatalytic efficiency has been considerably enhanced with the addition of H_2O_2. In 30 minutes, the clearance rate of 0.25-MXene/NH_2-MIL-88B reaches

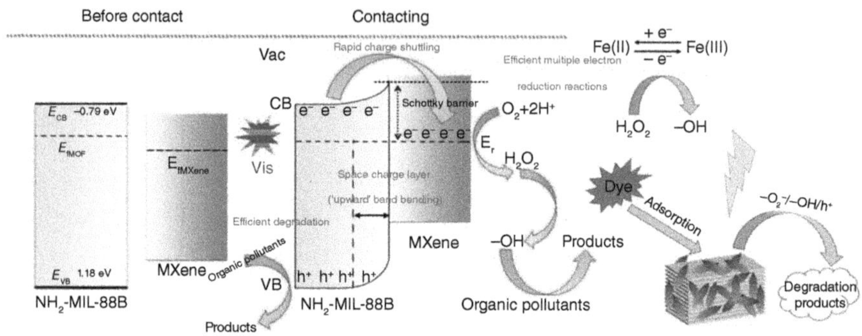

Figure 7.19 Illustration of the 0.25-MXene/NH_2-MIL-88B photocatalytic elimination process. Source: Long et al. [133]/with permission of Elsevier.

96.8%, which is a noteworthy impact. The degradation rate constant $k = 0.10468$ for 0.25-MXene/NH_2-MIL-88B is 1.19 fold that of 0.125-MXene/NH_2-MIL-88B, 2.5 times that of 0.375-MXene/NH_2-MIL-88B, 2.47 times that of MXene, and 1.65 fold that of NH_2-MIL-88B, which is 11.88 times that when H_2O_2 is not added. This is a significant difference. Photocatalysis may be accelerated by the Fenton impact, and the material exhibits a high selectivity in a short period of time.

The nanocomposite 0.25-MXene/NH_2-MIL-88B's photodegradation capability was evaluated over four cycles. Each cycle lasts 120 minutes, and 5 ml of water is taken every 30 minutes for UV detection throughout this time. The photocatalyst should be collected at the conclusion of each cycle, and the experiment is repeated. The clearance rate can still be as high as 95% in the first two cycles, but after that, it begins to decline. MB adsorbed on too little material may have had an effect on the composite's architecture under low-light circumstances, causing it to lose strength. But even after four cycles, a little quantity is still consumed each time compared to the same sort of treatment method which still provides some cost benefits.

On the basis of prior experimental results and the study of interface interactions between Ti_3C_2 MXene and NH_2-MIL-88B, a putative photocatalytic degradation pathway has been postulated. Figure 7.19 depicts a potential mechanism for this phenomenon.

7.5 Photocatalytic CO_2 Reduction Application

In the past, there have been numerous studies conducted on Ti_3C_2 MXene-based composites and MOF-based composites for the photocatalytic production of solar fuels. For instance, MOFs are materials that have HOMO and LUMO separated by an energy bandgap, indicating its capability to generate charge carriers upon light irradiation. Even though pristine MOFs are reported to show photocatalytic activity for solar fuel production [134, 135], but their efficiency is very low due to the fast recombination of charge carriers. On the other hand, due to the metallic properties of Ti_3C_2 MXene, they can act as a co-catalyst to form a Schottky junction with other

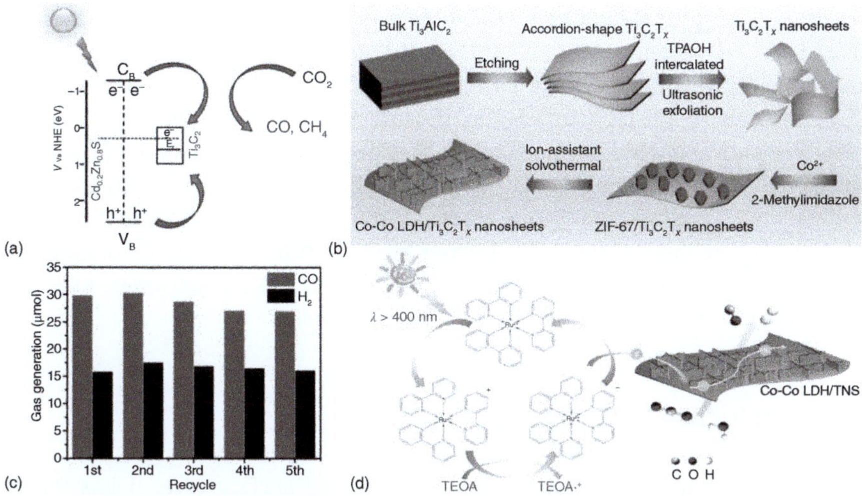

Figure 7.20 (a) Scheme of Schottky junction between the CdZnS@Ti_3C_2 nanocomposite for CO_2 reduction. Source: Saeed et al. [136]/with permission of Elsevier., (b) Schematic diagram of the preparation of MOF derived Co–Co LDH/$Ti_3C_2T_x$ nanosheets, (c) Photocatalytic stability over Co–Co LDH/$Ti_3C_2T_x$ for the CO_2 reduction reaction, (d) Mechanism of the CO_2 reduction over the nanocomposite with $[Ru(bpy)_3]Cl_2$ photosensitizer and TEOA sacrificial agent. Source: Chen et al. [137]/with permission of Elsevier.

photo-active materials to effectively trap photogenerated electrons. It was observed that the Schottky junction between the CdZnS/Ti_3C_2 nanocomposite (Figure 7.20a) successfully ameliorated the photocatalytic reduction of CO_2 to form CH_4 and CO due to a highly efficient charge carrier separation [136]. Ti_3C_2 MXene-based MOF composites present a number of advantageous synergies. First, MOFs which possess high surface area and porosity are able to host MXenes and prevent the restacking and agglomeration of MXene layers. The composites are also able to show improved stability due to the synergistic effects between the surface terminal groups of MXenes and functionalities of MOFs [138]. Not only that, a built-in electric field between the MXene and MOF known as the Schottky junction is induced, where efficient separation, rapid mobility, and transportation of charge carriers are promoted [139]. Ti_3C_2 MXene-based MOF composite can function as attractive photocatalysts for CO_2 reduction and H_2 production via water splitting. However, there are only a handful of studies conducted over the composite for solar fuel production.

Chen et al. [137] prepared a Ti_3C_2 composite with MOF-derived Co–Co LDH nanosheets as shown in Figure 7.20b for visible light CO_2 reduction in the presence of $[Ru(bpy)_3]Cl_2$ photosensitizer. The bulk Ti_3AlC_2 MAX was subjected to etching to remove the Al layers, followed by ultrasonic exfoliation to obtain $Ti_3C_2T_x$ nanosheets (TNS). Then, in situ growth of ZIF-67 was conducted on the $Ti_3C_2T_x$ nanosheets. Upon successful loading of ZIF-67, the nanocomposite was subjected to solvothermal treatment, whereby Co—Co LDH/TNS nanosheets were obtained. Pristine $Ti_3C_2T_x$ nanosheets did not show any photocatalytic CO_2 reduction activity.

However, upon optimal loading of 15 mg TNS on the MOF-derived Co–Co LDH, a maximum CO generation rate of 1.25×10^4 μmol h^{-1} g^{-1} was achieved. The nanocomposite also exhibited an excellent stability, maintaining good photocatalytic performance up to five cycles as illustrated in Figure 7.20c. Figure 7.20d depicts the photocatalytic mechanism of the Co–Co LDH/TNS nanocomposite in the presence of a photosensitizer. Upon visible light irradiation, the photosensitizer is excited, but is subsequently quenched by the TEOA electron donor to form $[Ru(bpy)_3]Cl_2^-$ reduced state. The electrons are then transmitted to the nanocomposite, whereby the electrons are rapidly migrated to the Co active sites for the reduction of CO_2 to CO.

7.6 Conclusion and Outlook

Due to the fact that TiC MXene offers an increased surface area, it has found widespread application as a co-catalyst in the field of photocatalysis. The intriguing features of TiC MXenes encourage the usage of MOFs combined with TiC MXenes. We examined the efficiency of previously published TiC MXene-based MOF nanohybrids and nanocomposites in this chapter. The excellent performances observed in the reported studies were due to their improved characteristics when combined. Thus, common problems of inefficient photocatalysts including fast recombination rate of electron–hole pairs and low charge-carrier separation were found to be resolved, resulting in high efficiency when exposed to illumination.

Nonetheless, issues like as scalability hampered the deployment of TiC with MOFs in actual applications. It is critical to understand how to effectively utilize the co-catalyst function of TiC MXenes as a guide for the preparation of high-efficiency TiC MXene-based photocatalysts. The surface termination groups of TiC MXenes have a significant effect on the characteristics and charge transfer mechanism of TiC MXene-based MOF composites. Concerning MXene in the realm of photocatalysis, the study on the integration of theory and experiment is rather limited and is not yet complete. Likewise, present research on the surface chemistry and interfacial carrier dynamics of TiC MXene-based photocatalysts is limited, and there are disagreements among the published findings. Thus, more research is necessary to elucidate the interfacial charge transfer process, which will considerably aid in the manufacture of a TiC MXene-based photocatalyst and its related reactions. By preparing TiC MXene-based MOF photocatalysts, the photocatalytic activity can be effectively enhanced due to various synergistic effects such as boosted stability and formation of built-in electric field known as the Schottky junction. Despite promising findings from TiC-based MOF composites for solar fuel production, there are still very limited studies conducted. Hence, it is recommended to further explore TiC with various MOFs for photocatalytic CO_2 reduction and water-splitting reactions.

In summary, TiC MXene-based MOF composites and hybrids warrant more attention and investigation. Without a doubt, the present findings are encouraging. However, more study will aid the scientific community in gaining a better understanding of its scalability and industrial application.

Acknowledgment

The authors would like to extend their deepest appreciation to United Arab Emirates University (UAEU) for the financial support of this work under start-up research grant (grant # 12N097).

References

1 Zhang, B., Zhang, S.-X., Yao, R. et al. (2021). Progress and prospects of hydrogen production: opportunities and challenges. *J. Electron. Sci. Technol.* 19 (2): 100080.

2 Suman, S. (2018). Hybrid nuclear-renewable energy systems: a review. *J. Cleaner Prod.* 181: 166–177.

3 Xie, J., Yang, C., Duan, M. et al. (2018). Amorphous NiP as cocatalyst for photocatalytic water splitting. *Ceram. Int.* 44 (5): 5459–5465.

4 Liu, M., Zhou, L., Luo, X. et al. (2020). Recent advances in noble metal catalysts for hydrogen production from ammonia borane. *Catalysts* 10 (7).

5 Zong, H., Qi, R., Yu, K., and Zhu, Z. (2021). Ultrathin Ti_2NT_x MXene-wrapped MOF-derived CoP frameworks towards hydrogen evolution and water oxidation. *Electrochim. Acta* 393: 139068.

6 Rajasulochana, P. and Preethy, V. (2016). Comparison on efficiency of various techniques in treatment of waste and sewage water–a comprehensive review. *Res. -Effic. Technol.* 2 (4): 175–184.

7 Briffa, J., Sinagra, E., and Blundell, R. (2020). Heavy metal pollution in the environment and their toxicological effects on humans. *Heliyon* 6 (9): e04691.

8 Carmen, Z.; Daniela, S., *Textile organic dyes-characteristics, polluting effects and separation/elimination procedures from industrial effluents-a critical overview. Organic Pollutants Ten Years After the Stockholm Convention*IntechOpenRijeka: 2012; Vol. 3.

9 Ardila-Leal, L.D., Poutou-Piñales, R.A., Pedroza-Rodríguez, A.M., and Quevedo-Hidalgo, B.E. (2021). A brief history of colour, the environmental impact of synthetic dyes and removal by using laccases. *Molecules* 26 (13): 3813.

10 Tchounwou, P.B., Yedjou, C.G., Patlolla, A.K., and Sutton, D.J. (2012). Heavy metal toxicity and the environment. *Mol. Clin. Environ. Toxicol.* 133–164.

11 Qin, T., Wang, Z., Wang, Y. et al. (2021). Recent progress in emerging two-dimensional transition metal carbides. *Nano-Micro Lett.* 13 (1): 183.

12 Sharma, S.K., Kumar, A., Sharma, G. et al. (2022). MXenes based nano-heterojunctions and composites for advanced photocatalytic environmental detoxification and energy conversion: a review. *Chemosphere* 291: 132923.

13 Han, F., Luo, S., Xie, L. et al. (2019). Boosting the yield of MXene 2D sheets via a facile hydrothermal-assisted intercalation. *ACS Appl. Mater. Interfaces* 11 (8): 8443–8452.

14 Han, X., An, L., Hu, Y. et al. (2020). Ti_3C_2 MXene-derived carbon-doped TiO_2 coupled with g-C_3N_4 as the visible-light photocatalysts for photocatalytic H_2 generation. *Appl. Catal., B* 265: 118539.

15 Gao, Y., Wang, L., Zhou, A. et al. (2015). Hydrothermal synthesis of TiO_2/Ti_3C_2 nanocomposites with enhanced photocatalytic activity. *Mater. Lett.* 150: 62–64.

16 Liu, X., Shan, Y., Zhang, S. et al. (2022). Application of metal organic framework in wastewater treatment. *Green Energy Environ.*

17 Lin, S., Song, Z., Che, G. et al. (2014). Adsorption behavior of metal–organic frameworks for methylene blue from aqueous solution. *Microporous Mesoporous Mater.* 193: 27–34.

18 Liu, S., Zhang, C., Sun, Y. et al. (2020). Design of metal-organic framework-based photocatalysts for hydrogen generation. *Coord. Chem. Rev.* 413: 213266.

19 Reddy, C.V., Reddy, K.R., Harish, V.V.N. et al. (2020). Metal-organic frameworks (MOFs)-based efficient heterogeneous photocatalysts: synthesis, properties and its applications in photocatalytic hydrogen generation, CO_2 reduction and photodegradation of organic dyes. *Int. J. Hydrogen Energy* 45 (13): 7656–7679.

20 Reddy, D.A., Kim, Y., Gopannagari, M. et al. (2021). Recent advances in metal–organic framework-based photocatalysts for hydrogen production. *Sustainable Energy Fuels* 5 (6): 1597–1618.

21 Wang, Q., Gao, Q., Al-Enizi, A.M. et al. (2020). Recent advances in MOF-based photocatalysis: environmental remediation under visible light. *Inorg. Chem. Front.* 7 (2): 300–339.

22 Wu, H., Almalki, M., Xu, X. et al. (2019). MXene Derived Metal–Organic Frameworks. *J. Am. Chem. Soc.* 141 (51): 20037–20042.

23 Sharma, K., Hasija, V., Patial, S. et al. (2022). Recent progress on MXenes and MOFs hybrids: structure, synthetic strategies and catalytic water splitting. *Int. J. Hydrogen Energy.*

24 Hong, W., Wyatt, B.C., Nemani, S.K., and Anasori, B. (2020). Double transition-metal MXenes: atomistic design of two-dimensional carbides and nitrides. *MRS Bull.* 45 (10): 850–861.

25 Gogotsi, Y. and Anasori, B. (2019). *The Rise of MXenes*, vol. 13, 8491–8494. ACS Publications.

26 Naguib, M., Kurtoglu, M., Presser, V. et al. (2011). Two-dimensional nanocrystals produced by exfoliation of Ti_3AlC_2. *Adv. Mater.* 23 (37): 4248–4253.

27 Anasori, B., Lukatskaya, M.R., and Gogotsi, Y. (2017). 2D metal carbides and nitrides (MXenes) for energy storage. *Nat. Rev. Mater.* 2 (2).

28 Jiang, X., Kuklin, A.V., Baev, A. et al. (2020). Two-dimensional MXenes: from morphological to optical, electric, and magnetic properties and applications. *Phys. Rep.* 848: 1–58.

29 De, S., Acharya, S., Sahoo, S. et al. (2022). From 0D to 3D MXenes: their diverse syntheses, morphologies and applications. *Mater. Chem. Front.* 6 (7): 818–842.

30 Su, T., Ma, X., Tong, J. et al. (2022). Surface engineering of MXenes for energy and environmental applications. *J. Mater. Chem. A* 10: 10265–10296.

31 Dayananda, D., Reddy, P.L., Deshmukh, K. et al. (2022). MXene-based flexible polymer composites as high dielectric constant materials. *Mxenes Composites* 725–758.

32 Monastyreckis, G., Mishnaevsky, L. Jr., Hatter, C. et al. (2020). Micromechanical modeling of MXene-polymer composites. *Carbon* 162: 402–409.

33 Jeon, J., Yang, Y., Choi, H. et al. (2020). MXenes for future nanophotonic device applications. *Nanophotonics* 9 (7): 1831–1853.

34 Shi, Z., Khaledialidusti, R., Malaki, M., and Zhang, H. (2021). MXene-based materials for solar cell applications. *Nanomaterials* 11 (12): 3170.

35 Alfaifi, B. Y.; Ullah, H.; Alfaifi, S.; Tahir, A. A.; Mallick, T. K., *Photoelectrochemical Solar Water Splitting: From Basic Principles to Advanced Devices.* 2018.

36 Maeda, K. (2011). Photocatalytic water splitting using semiconductor particles: history and recent developments. *J. Photochem. Photobiol., C* 12 (4): 237–268.

37 Shkrob, I.A. and Sauer, M.C. (2004). Hole scavenging and photo-stimulated recombination of electron–hole pairs in aqueous TiO_2 nanoparticles. *J. Phys. Chem. B* 108 (33): 12497–12511.

38 Frey, N.C., Wang, J., Vega Bellido, G.I. et al. (2019). Prediction of synthesis of 2D metal carbides and nitrides (MXenes) and their precursors with positive and unlabeled machine learning. *ACS Nano* 13 (3): 3031–3041.

39 Pan, J., Lany, S., and Qi, Y. (2017). Computationally driven two-dimensional materials design: what is next? *ACS Nano* 11 (8): 7560–7564.

40 Knobloch, F.W. and Rauscher, W.H. (1959). Coordination polymers of copper(II) prepared at liquid-liquid interfaces. *J. Polym. Sci.* 38 (133): 261–262.

41 Johnson, J.W., Jacobson, A.J., Butler, W.M. et al. (1989). Molecular recognition of alcohols by layered compounds with alternating organic and inorganic layers. *J. Am. Chem. Soc.* 111 (1): 381–383.

42 Drumel, S., Janvier, P., Barboux, P. et al. (1995). Synthesis, structure and reactivity of some functionalized zinc and copper(II) phosphonates. *Inorg. Chem.* 34 (1): 148–156.

43 Riou, D., Roubeau, O., and Férey, G. (1998). Composite microporous compounds. Part I: synthesis and structure determination of two new vanadium alkyldiphosphonates (MIL-2 and MIL-3) with three-dimensional open frameworks. *Microporous Mesoporous Mater.* 23 (1–2): 23–31.

44 Li, H., Eddaoudi, M., O'Keeffe, M., and Yaghi, O.M. (1999). Design and synthesis of an exceptionally stable and highly porous metal-organic framework. *Nature* 402 (6759): 276–279.

45 Li, D., Xu, H.-Q., Jiao, L., and Jiang, H.-L. (2019). Metal-organic frameworks for catalysis: state of the art, challenges, and opportunities. *EnergyChem* 1 (1): 100005.

46 Zhou, H.-C.J. and Kitagawa, S. (2014). Metal–organic frameworks (MOFs). *Chem. Soc. Rev.* 43 (16): 5415–5418.

47 Li, J.-R., Kuppler, R.J., and Zhou, H.-C. (2009). Selective gas adsorption and separation in metal–organic frameworks. *Chem. Soc. Rev.* 38 (5): 1477–1504.

48 Corma, A., García, H., and Llabrés i Xamena, F.X. (2010). Engineering metal organic frameworks for heterogeneous catalysis. *Chem. Rev.* 110 (8): 4606–4655.

49 Cui, Y.; Yue, Y.; Qian, G.; Chen, B., Luminescent functional metal–organic frameworks. *Chem. Rev.*2012,112 (2), 1126–1162.

50 Fujita, M., Kwon, Y.J., Washizu, S., and Ogura, K. (1994). Preparation, clathration ability, and catalysis of a two-dimensional square network material composed of cadmium(II) and 4,4'-bipyridine. *J. Am. Chem. Soc.* 116 (3): 1151–1152.

51 Liu, H.-J. and Dong, B. (2021). Recent advances and prospects of MXene-based materials for electrocatalysis and energy storage. *Mater. Today Phys.* 20: 100469.

52 Lim, K.R.G., Handoko, A.D., Nemani, S.K. et al. (2020). Rational design of two-dimensional transition metal carbide/nitride (MXene) hybrids and nanocomposites for catalytic energy storage and conversion. *ACS Nano* 14 (9): 10834–10864.

53 Barsoum, M.W. and Radovic, M. (2011). Elastic and mechanical properties of the MAX phases. *Annu. Rev. Mater. Res.* 41 (1): 195–227.

54 Li, Y., Ye, M., Tang, R. et al. (2020). Pressure-induced isostructural phase transition in Ti_3AlC_2: experimental and theoretical investigation. *Phys. Chem. Chem. Phys.* 22 (23): 13136–13142.

55 Peng, J., Chen, X., Ong, W.-J. et al. (2019). Surface and heterointerface engineering of 2D MXenes and their nanocomposites: insights into electro- and photocatalysis. *Chem* 5 (1): 18–50.

56 Khazaei, M., Ranjbar, A., Arai, M. et al. (2017). Electronic properties and applications of MXenes: a theoretical review. *J. Mater. Chem. C* 5 (10): 2488–2503.

57 Li, Z. and Wu, Y. (2019). 2D early transition metal carbides (MXenes) for catalysis. *Small* 15 (29): 1804736.

58 Alhabeb, M., Maleski, K., Anasori, B. et al. (2017). Guidelines for synthesis and processing of two-dimensional titanium carbide ($Ti_3C_2T_x$ MXene). *Chem. Mater.* 29 (18): 7633–7644.

59 Persson, I., el Ghazaly, A., Tao, Q. et al. (2018). Tailoring structure, composition, and energy storage properties of MXenes from selective etching of In-plane, chemically ordered MAX phases. *Small* 14 (17): 1703676.

60 Huang, K., Li, Z., Lin, J. et al. (2018). Two-dimensional transition metal carbides and nitrides (MXenes) for biomedical applications. *Chem. Soc. Rev.* 47 (14): 5109–5124.

61 Halim, J., Palisaitis, J., Lu, J. et al. (2018). Synthesis of two-dimensional $Nb_{1.33}C$ (MXene) with randomly distributed vacancies by etching of the quaternary solid solution $(Nb_{2/3}Sc_{1/3})_2$ AlC MAX phase. *ACS Appl. Nano Mater.* 1 (6): 2455–2460.

62 Kong, F., He, X., Liu, Q. et al. (2018). Improving the electrochemical properties of MXene Ti_3C_2 multilayer for Li-ion batteries by vacuum calcination. *Electrochim. Acta* 265: 140–150.

63 Kayali, E., Vahid Mohammadi, A., Orangi, J., and Beidaghi, M. (2018). Controlling the dimensions of 2D MXenes for ultrahigh-rate pseudocapacitive energy storage. *ACS Appl. Mater. Interfaces* 10 (31): 25949–25954.

64 Papadopoulou, K., Chroneos, A., Parfitt, D., and Christopoulos, S.-R.G. (2020). A perspective on MXenes: their synthesis, properties, and recent applications. *J. Appl. Phys.* 128: 170902.
65 Ronchi, R.M., Arantes, J.T., and Santos, S.F. (2019). Synthesis, structure, properties and applications of MXenes: current status and perspectives. *Ceram. Int.* 45 (15): 18167–18188.
66 Bai, Y., Zhou, K., Srikanth, N. et al. (2016). Dependence of elastic and optical properties on surface terminated groups in two-dimensional MXene monolayers: a first-principles study. *RSC Adv.* 6 (42): 35731–35739.
67 Magnuson, M., Halim, J., and Näslund, L.-Å. (2018). Chemical bonding in carbide MXene nanosheets. *J. Electron. Spectrosc. Relat. Phenom.* 224: 27–32.
68 Zha, X.-H., Luo, K., Li, Q. et al. (2015). Role of the surface effect on the structural, electronic and mechanical properties of the carbide MXenes. *EPL (Europhysics Letters)* 111 (2): 26007.
69 Yorulmaz, U., Özden, A., Perkgöz, N.K. et al. (2016). Vibrational and mechanical properties of single layer MXene structures: a first-principles investigation. *Nanotechnology* 27 (33): 335702.
70 Enyashin, A.N. and Ivanovskii, A.L. (2013). Two-dimensional titanium carbonitrides and their hydroxylated derivatives: structural, electronic properties and stability of MXenes Ti_3C_2–xNx$(OH)_2$ from DFTB calculations. *J. Solid State Chem.* 207: 42–48.
71 Khazaei, M., Arai, M., Sasaki, T. et al. (2012). Novel electronic and magnetic properties of two-dimensional transition metal carbides and nitrides. *Adv. Funct. Mater.* 23 (17): 2185–2192.
72 Ivanovskii, A.L. and Enyashin, A.N. (2013). Graphene-like transition-metal nanocarbides and nanonitrides. *Russ. Chem. Rev.* 82 (8): 735–746.
73 Anasori, B., Xie, Y., Beidaghi, M. et al. (2015). Two-dimensional, ordered, double transition metals carbides (MXenes). *ACS Nano* 9 (10): 9507–9516.
74 Dong, L., Kumar, H., Anasori, B. et al. (2017). Rational design of two-dimensional metallic and semiconducting spintronic materials based on ordered double-transition-metal MXenes. *J. Phys. Chem. Lett.* 8 (2): 422–428.
75 Berdiyorov, G.R. (2015). Effect of surface functionalization on the electronic transport properties of Ti_3C_2 MXene. *EPL (Europhysics Letters)* 111 (6): 67002.
76 Sarikurt, S., Çakır, D., Keçeli, M., and Sevik, C. (2018). The influence of surface functionalization on thermal transport and thermoelectric properties of MXene monolayers. *Nanoscale* 10 (18): 8859–8868.
77 Ying, G., Dillon, A.D., Fafarman, A.T., and Barsoum, M.W. (2017). Transparent, conductive solution processed spincast 2D Ti_2CT_x (MXene) films. *Mater. Res. Lett.* 5 (6): 391–398.
78 Dillon, A.D., Ghidiu, M.J., Krick, A.L. et al. (2016). Highly conductive optical quality solution-processed films of 2D titanium carbide. *Adv. Funct. Mater.* 26 (23): 4162–4168.
79 Urbankowski, P., Anasori, B., Hantanasirisakul, K. et al. (2017). 2D molybdenum and vanadium nitrides synthesized by ammoniation of 2D transition metal carbides (MXenes). *Nanoscale* 9 (45): 17722–17730.

80 Si, C., Zhou, J., and Sun, Z. (2015). Half-metallic ferromagnetism and surface functionalization-induced metal–insulator transition in graphene-like two-dimensional Cr_2C crystals. *ACS Appl. Mater. Interfaces* 7 (31): 17510–17515.

81 Si, C., Jin, K.-H., Zhou, J. et al. (2016). Large-gap quantum spin hall state in MXenes: d-band topological order in a triangular lattice. *Nano Lett.* 16 (10): 6584–6591.

82 Wong, Z.M., Tan, T.L., Yang, S.-W., and Xu, G.Q. (2018). Enhancing the photocatalytic performance of MXenes via stoichiometry engineering of their electronic and optical properties. *ACS Appl. Mater. Interfaces* 10 (46): 39879–39889.

83 Balcı, E., Akkuş, Ü.Ö., and Berber, S. (2017). Band gap modification in doped MXene: Sc_2CF_2. *J. Mater. Chem. C* 5 (24): 5956–5961.

84 Lee, Y., Cho, S.B., and Chung, Y.-C. (2014). Tunable indirect to direct band gap transition of monolayer Sc_2CO_2 by the strain effect. *ACS Appl. Mater. Interfaces* 6 (16): 14724–14728.

85 Hong, L., Klie, R.F., and Öğüt, S. (2016). First-principles study of size- and edge-dependent properties of MXene nanoribbons. *Phys. Rev. B* 93 (11).

86 Balcı, E., Akkuş, Ü.Ö., and Berber, S. (2018). Controlling topological electronic structure of multifunctional MXene layer. *Appl. Phys. Lett.* 113 (8): 083107.

87 Chaudhuri, K., Wang, Z., Alhabeb, M. et al. (2019). Optical properties of MXenes. In: *2D Metal Carbides and Nitrides (MXenes)*, 327–346. Springer International Publishing.

88 Halim, J., Persson, I., Moon, E.J. et al. (2019). Electronic and optical characterization of 2D Ti_2C and Nb_2C (MXene) thin films. *J. Phys. Condens. Matter* 31 (16): 165301.

89 Li, R., Zhang, L., Shi, L., and Wang, P. (2017). MXene Ti_3C_2: an effective 2D light-to-heat conversion material. *ACS Nano* 11 (4): 3752–3759.

90 Jiang, X., Liu, S., Liang, W. et al. (2018). Broadband nonlinear photonics in few-layer MXene $Ti_3C_2T_x$ (T = F, O, or OH). *Laser Photonics Rev.* 12 (2): 1700229.

91 Jiao, L., Seow, J.Y.R., Skinner, W.S. et al. (2019). Metal–organic frameworks: structures and functional applications. *Mater. Today* 27: 43–68.

92 Furukawa, H., Cordova, K.E., O'Keeffe, M., and Yaghi, O.M. (2013). The chemistry and applications of metal-organic frameworks. *Science* 341 (6149).

93 Perry Iv, J.J., Perman, J.A., and Zaworotko, M.J. (2009). Design and synthesis of metal–organic frameworks using metal–organic polyhedra as supermolecular building blocks. *Chem. Soc. Rev.* 38 (5): 1400.

94 Tranchemontagne, D.J., Mendoza-Cortés, J.L., O'Keeffe, M., and Yaghi, O.M. (2009). Secondary building units, nets and bonding in the chemistry of metal–organic frameworks. *Chem. Soc. Rev.* 38 (5): 1257–1283.

95 Lu, W., Wei, Z., Gu, Z.-Y. et al. (2014). Tuning the structure and function of metal–organic frameworks via linker design. *Chem. Soc. Rev.* 43 (16): 5561–5593.

96 Deng, H., Grunder, S., Cordova Kyle, E. et al. (2012). Large-pore apertures in a series of metal-organic frameworks. *Science* 336 (6084): 1018–1023.

97 Burtch, N.C., Jasuja, H., and Walton, K.S. (2014). Water stability and adsorption in metal–organic frameworks. *Chem. Rev.* 114 (20): 10575–10612.

98 Howarth, A.J., Liu, Y., Li, P. et al. (2016). Chemical, thermal and mechanical stabilities of metal–organic frameworks. *Nat. Rev. Mater.* 1 (3).

99 Bosch, M., Zhang, M., and Zhou, H.-C. (2014). Increasing the stability of metal-organic frameworks. *Adv. Chem.* 2014: 1–8.

100 Férey, G., Mellot-Draznieks, C., Serre, C. et al. (2005). A chromium terephthalate-based solid with unusually large pore volumes and surface area. *Science* 309 (5743): 2040–2042.

101 Park, K.S., Ni, Z., Côté, A.P. et al. (2006). Exceptional chemical and thermal stability of zeolitic imidazolate frameworks. *Proc. Natl. Acad. Sci. U.S.A.* 103 (27): 10186.

102 Huang, X.-C., Lin, Y.-Y., Zhang, J.-P., and Chen, X.-M. (2006). Ligand-directed strategy for zeolite-type metal–organic frameworks: zinc(II) imidazolates with unusual zeolitic topologies. *Angew. Chem. Int. Ed.* 45 (10): 1557–1559.

103 Furukawa, H., Gándara, F., Zhang, Y.-B. et al. (2014). Water adsorption in porous metal–organic frameworks and related materials. *J. Am. Chem. Soc.* 136 (11): 4369–4381.

104 Cavka, J.H., Jakobsen, S., Olsbye, U. et al. (2008). A new zirconium inorganic building brick forming metal organic frameworks with exceptional stability. *J. Am. Chem. Soc.* 130 (42): 13850–13851.

105 Feng, D., Gu, Z.-Y., Li, J.-R. et al. (2012). Zirconium-metalloporphyrin PCN-222: mesoporous metal–organic frameworks with ultrahigh stability as biomimetic catalysts. *Angew. Chem. Int. Ed.* 51 (41): 10307–10310.

106 Loiseau, T., Serre, C., Huguenard, C. et al. (2004). A rationale for the large breathing of the porous aluminum terephthalate (MIL-53) upon hydration. *Chem. Eur. J.* 10 (6): 1373–1382.

107 Colombo, V., Galli, S., Choi, H.J. et al. (2011). High thermal and chemical stability in pyrazolate-bridged metal–organic frameworks with exposed metal sites. *Chem. Sci.* 2 (7): 1311–1319.

108 Nguyen, J.G. and Cohen, S.M. (2010). Moisture-resistant and superhydrophobic metal–organic frameworks obtained via postsynthetic modification. *J. Am. Chem. Soc.* 132 (13): 4560–4561.

109 Yang, C., Kaipa, U., Mather, Q.Z. et al. (2011). Fluorous metal–organic frameworks with superior adsorption and hydrophobic properties toward oil spill cleanup and hydrocarbon storage. *J. Am. Chem. Soc.* 133 (45): 18094–18097.

110 Xuan, W., Zhu, C., Liu, Y., and Cui, Y. (2012). Mesoporous metal–organic framework materials. *Chem. Soc. Rev.* 41 (5): 1677–1695.

111 Eddaoudi, M., Kim, J., Rosi, N. et al. (2002). Systematic design of pore size and functionality in isoreticular MOFs and their application in methane storage. *Science* 295 (5554): 469–472.

112 Senkovska, I. and Kaskel, S. (2014). Ultrahigh porosity in mesoporous MOFs: promises and limitations. *Chem. Commun.* 50 (54): 7089–7098.

113 Farha, O.K., Eryazici, I., Jeong, N.C. et al. (2012). Metal–organic framework materials with ultrahigh surface areas: is the sky the limit? *J. Am. Chem. Soc.* 134 (36): 15016–15021.

114 Jalil, A., Zafar Ilyas, S., Agathopoulos, S. et al. (2021). 2D $CoGeSe_3$ monolayer as a visible-light photocatalyst with high carrier mobility: theoretical prediction. *Appl. Surf. Sci.* 565: 150588.

115 Linsebigler, A.L., Lu, G., and Yates, J.T. (1995). Photocatalysis on TiO_2 surfaces: principles, mechanisms, and selected results. *Chem. Rev.* 95 (3): 735–758.

116 Qutub, N., Singh, P., Sabir, S. et al. (2022). Enhanced photocatalytic degradation of acid blue dye using CdS/TiO_2 nanocomposite. *Sci. Rep.* 12 (1): 1–18.

117 Sadeghfar, F.; Ghaedi, M.; Zalipour, Z., Chapter 4 – Advanced oxidation. In *Interface Science and Technology*, Ghaedi, M., Ed. Elsevier: 2021; Vol. 32, pp. 225–324.

118 Rauf, M. and Ashraf, S.S. (2009). Fundamental principles and application of heterogeneous photocatalytic degradation of dyes in solution. *Chem. Eng. J.* 151 (1–3): 10–18.

119 Krishnan, S., Rawindran, H., Sinnathambi, C., and Lim, J. (2017). Comparison of various advanced oxidation processes used in remediation of industrial wastewater laden with recalcitrant pollutants. In: *IOP Conference Series: Materials Science and Engineering*, 012089. IOP Publishing.

120 Coha, M., Farinelli, G., Tiraferri, A. et al. (2021). Advanced oxidation processes in the removal of organic substances from produced water: potential, configurations, and research needs. *Chem. Eng. J.* 414: 128668.

121 Jacobs, C.B., Vickrey, T.L., and Venton, B.J. (2011). Functional groups modulate the sensitivity and electron transfer kinetics of neurochemicals at carbon nanotube modified microelectrodes. *Analyst* 136 (17): 3557–3565.

122 Liu, Q., Tan, X., Wang, S. et al. (2019). MXene as a non-metal charge mediator in 2D layered CdS@ Ti_3C_2@ TiO_2 composites with superior Z-scheme visible light-driven photocatalytic activity. *Environ. Sci. Nano* 6 (10): 3158–3169.

123 Ibragimova, R., Erhart, P., Rinke, P., and Komsa, H.-P. (2021). Surface functionalization of 2D MXenes: trends in distribution, composition, and electronic properties. *J. Phys. Chem. Lett.* 12 (9): 2377–2384.

124 Im, J.K., Sohn, E.J., Kim, S. et al. (2021). Review of MXene-based nanocomposites for photocatalysis. *Chemosphere* 270: 129478.

125 Li, Y., Liu, Y., Wang, Z. et al. (2021). In-situ growth of Ti_3C_2@MIL-NH_2 composite for highly enhanced photocatalytic H_2 evolution. *Chem. Eng. J.* 411: 128446.

126 Zhao, J., Liu, L.-W., Liu, F. et al. (2019). Conductive Ti_3C_2 and MOF-derived CoS_x boosting the photocatalytic hydrogen production activity of TiO_2. *CrystEngComm* 21.

127 Tian, P., He, X., Zhao, L. et al. (2019). Ti_3C_2 nanosheets modified Zr-MOFs with Schottky junction for boosting photocatalytic HER performance. *Sol. Energy* 188: 750–759.

128 You, Z., Liao, Y., Li, X. et al. (2021). State-of-the-art recent progress in MXene-based photocatalysts: a comprehensive review. *Nanoscale* 13 (21): 9463–9504.

129 Hong, L. -f., Guo, R. -t., Yuan, Y. et al. (2020). Recent progress of two-dimensional MXenes in photocatalytic applications: a review. *Mater. Today Energy* 18: 100521.

130 Tian, P., He, X., Zhao, L. et al. (2018). Enhanced charge transfer for efficient photocatalytic H_2 evolution over UiO-66-NH_2 with annealed $Ti_3C_2T_x$ MXenes. *Int. J. Hydrogen Energy* 44.

131 Li, X., Bai, Y., Shi, X. et al. (2021). Applications of MXene ($Ti_3C_2T_x$) in photocatalysis: a review. *Mater. Adv.* 2 (5): 1570–1594.

132 Tunesi, M.M., Soomro, R.A., Han, X. et al. (2021). Application of MXenes in environmental remediation technologies. *Nano Convergence* 8 (1): 1–19.

133 Long, R., Yu, Z., Tan, Q. et al. (2021). Ti_3C_2 MXene/NH_2-MIL-88B (Fe): research on the adsorption kinetics and photocatalytic performance of an efficient integrated photocatalytic adsorbent. *Appl. Surf. Sci.* 570: 151244.

134 Li, N., Liu, J., Liu, J.J. et al. (2019). Adenine components in biomimetic metal-organic frameworks for efficient CO_2 photoconversion. *Angew. Chem. Int. Ed. Engl.* 58 (16): 5226–5231.

135 Liao, W.M., Zhang, J.H., Wang, Z. et al. (2018). Semiconductive amine-functionalized Co(II)-MOF for visible-light-driven hydrogen evolution and CO_2reduction. *Inorg. Chem.* 57 (18): 11436–11442.

136 Saeed, A., Chen, W., Shah, A.H. et al. (2021). Enhancement of photocatalytic CO_2 reduction for novel $Cd_{0.2}Zn_{0.8}S$@Ti_3C_2 (MXenes) nanocomposites. *J. CO_2 Util.* 47.

137 Chen, W., Han, B., Xie, Y. et al. (2020). Ultrathin Co-Co LDHs nanosheets assembled vertically on MXene: 3D nanoarrays for boosted visible-light-driven CO_2 reduction. *Chem. Eng. J.* 391: 123519.

138 Najam, T., Shah, S.S.A., Peng, L. et al. (2022). Synthesis and nano-engineering of MXenes for energy conversion and storage applications: recent advances and perspectives. *Coord. Chem. Rev.* 454.

139 Yu, L., Liu, B., Wang, Y. et al. (2021). Recent progress on MXene-derived material and its' application in energy and environment. *J. Power Sources* 490: 229250.

8

Titanium Carbide (TiC) MXene-Based Layered Double Hydroxide (LDH) Composites for Energy and Environment Applications

Azmat A. Khan[1], Muhammad Tahir[2], and Nazish Khan[3]

[1] *Baluchistan University of Information Technology, Engineering and Management Sciences (BUITEMS), Department of Chemical Engineering, Airport Road, Beleli, Quetta, Pakistan P.O. Box 87100*
[2] *UAE University, Chemical and Petroleum Engineering Department, P.O. Box 15551, Sheik Khalifa Bin Zayed Street, Asharej, Al Ain, United Arab Emirates*
[3] *The University of Agriculture Swat, Department of Plant Breeding and Genetics, Takhta Band, Mingora, P.O. Box 19040, Swat, Pakistan*

8.1 Introduction

Presently, the energy shortage and environmental pollution are key problems at the forefront of global consideration. The growing population and rapid expansion of industrialization have caused an acute energy crisis and environmental contamination. Therefore, the efficient advancement of innovative energy technologies and oxidation techniques by employing numerous nanomaterials as catalysts is required to lessen these problems [1]. Currently, several two-dimensional (2D) materials have been employed to improve the catalytic performance by virtue of their nontoxic behavior, high durability, and huge surface area. Among these 2D materials, the layered double hydroxides (LDH) and titanium carbide MXene (TiC MXene) having highly active sites with high density and high uniformity can work as ideal models to modify the important parameters for energy storage and environmental applications. LDHs have the general formula $[M^{2+}{}_{1-x}M^{3+}{}_{x}(OH)_2]_{x+}(A^{n-})_{x/n}\cdot mH_2O$, where M^{2+} are divalent ($M = Mg^{2+}$, Co^{2+}, Cu^{2+}, Fe^{2+}, Ni^{2+}, or Zn^{2+}) and M^{3+} ($M = Al^{3+}$, Cr^{3+}, Ga^{3+}, Fe^{3+}, In^{3+}, or Mn^{3+}) are trivalent metal cations; x is the molar ratio of $M^{3+}/(M^{2+} + M^{3+})$ and normally ranges from 0.2 to 0.33; the A^{n-} (e.g. $CO_3{}^{2-}$, Cl^-, $NO_3{}^-$, $SO_4{}^{2-}$) denotes the anions in the interlayer gallery of metal hydroxide layers.

LDHs have been widely utilized in energy storage and conversion application due to their unique properties. For example, the Fe-, Co-, and Ni-based LDHs have demonstrated efficient performance for the oxygen evolution reaction (OER) and hydrogen evolution reaction (HER); and the space of the layered structure can be used as storage sites for Li^+ in Li-ion batteries. However, the LDHs still endure poor performance owing to the shortage of active sites and effective electronic transport. Thus, the regulation of composition by tuning the components and structural changes such as exfoliation and obtaining different morphologies (nanosheets,

Titanium Carbide MXenes: Synthesis, Characterization, Energy, and Environmental Applications, First Edition. Edited by Muhammad Tahir.

flower-, and urchin-like structure) may result in improved performance. The tuning of composition and structural changes may result in the enhanced charge transfer, ionic diffusion and delivering more active sites to achieve higher performance. Moreover, the assembly of positively charged LDH nanosheets with negatively charged layered nanosheets such as carbon materials, graphitic carbon nitride (g-C_3N_4), metal oxides, and MXene can encourage the construction of hybrid nanosheets, attributing to the further enhanced charge transfer, optimized band structure, avoiding the aggregation of nanosheets, and exposure of catalytic active sites [2].

Among 2D materials, the TiC MXene material has excellent electrical conductivity, low cost, high flexibility, large surface area, plentiful surface functional groups, and excellent hydrophilicity [3]. Usually, TiC MXenes are attained by etching the bulk MAX phase, whereas M demonstrates the Ti metal, A is for Al, and X represents the C element. For example, $Ti_3C_2T_x$ (T_x is the surface termination group) is normally achieved by removing Al atom layers from Ti_3AlC_2 by employing strong acids such as hydrofluoric acid (HF). Therefore, not only TiC MXenes but the composites formed by TiC MXene have obtained significant research attention, holding huge consideration for application energy storage, conversion, and environmental applications. On the basis of excellent properties of LDH and TiC MXene, the formation of the LDH-based TiC MXene composite may result in high-performance LDH-based TiC MXene composite materials.

8.2 Basic Principles of Energy Storage and Conversion

The energy storage and conversion systems can be broadly classified into (i) electrochemical and (ii) photochemical systems. Both types have different working principles. The electrochemical systems mainly comprise supercapacitors, batteries, and fuel cells, while the photochemical systems involve CO_2 reduction and water splitting for the production of fuels.

8.2.1 Principle of Electrochemical Energy Storage and Conversion System

The electrochemical systems are either employed for energy storage or for conversion. The diverse range of electrochemical energy storage and conversion devices is demonstrated in Figure 8.1 based on their energy delivery time varying with the type of mechanism involved in energy storage or conversion systems [4]. The energy storage systems include capacitors and batteries, while fuel cells are used for energy conversion.

8.2.1.1 Supercapacitors

Supercapacitors are further divided into electrochemical double layer (EDLC) supercapacitors, pseudocapacitors, and hybrid capacitors based on the mechanism of energy storage. In EDLCs, the charge storage and discharge mechanism are

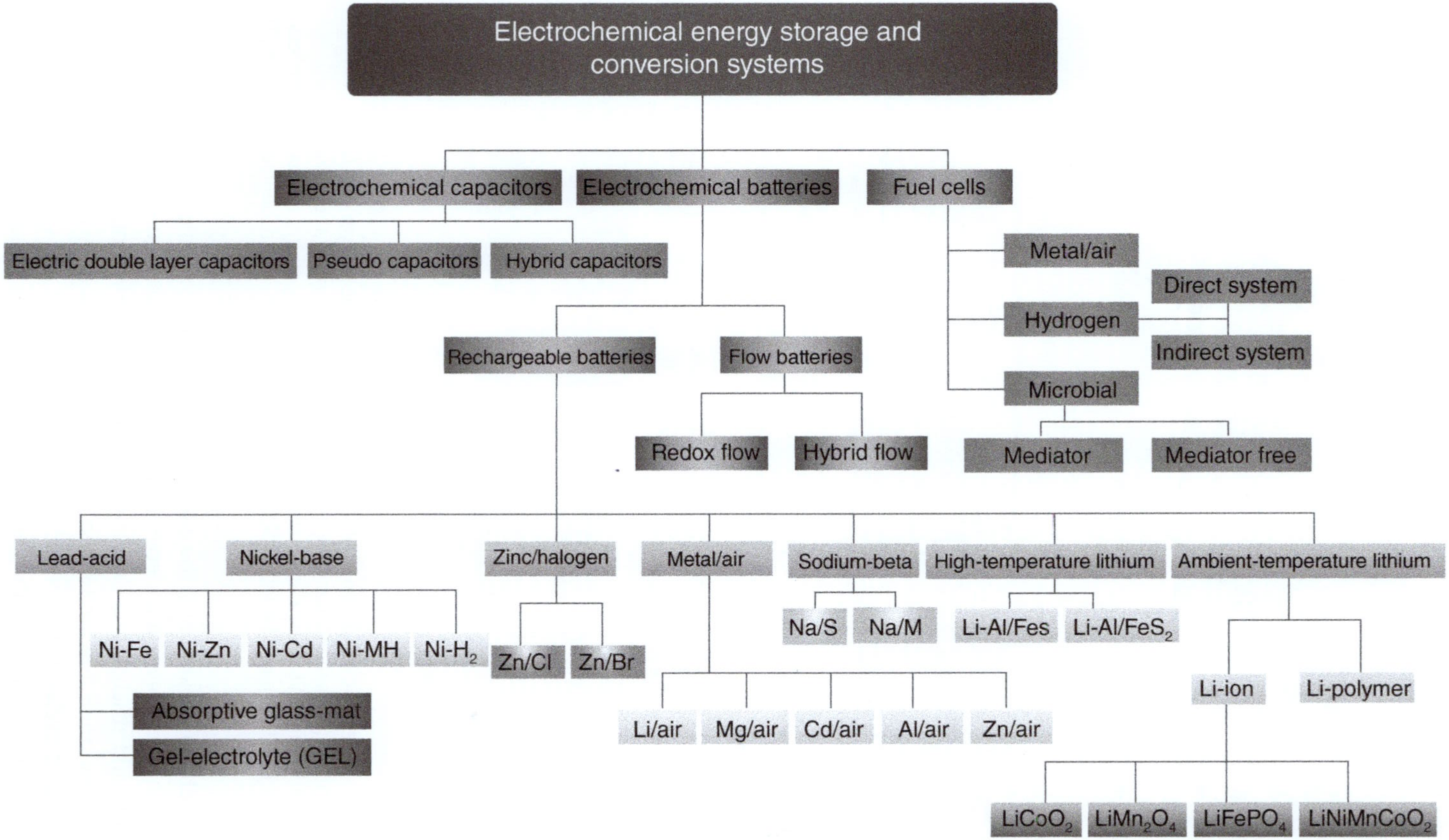

Figure 8.1 Classification of electrical energy storage and conversion devices. Source: Abbas et al. [4]/MDPI/CC BY 4.0.

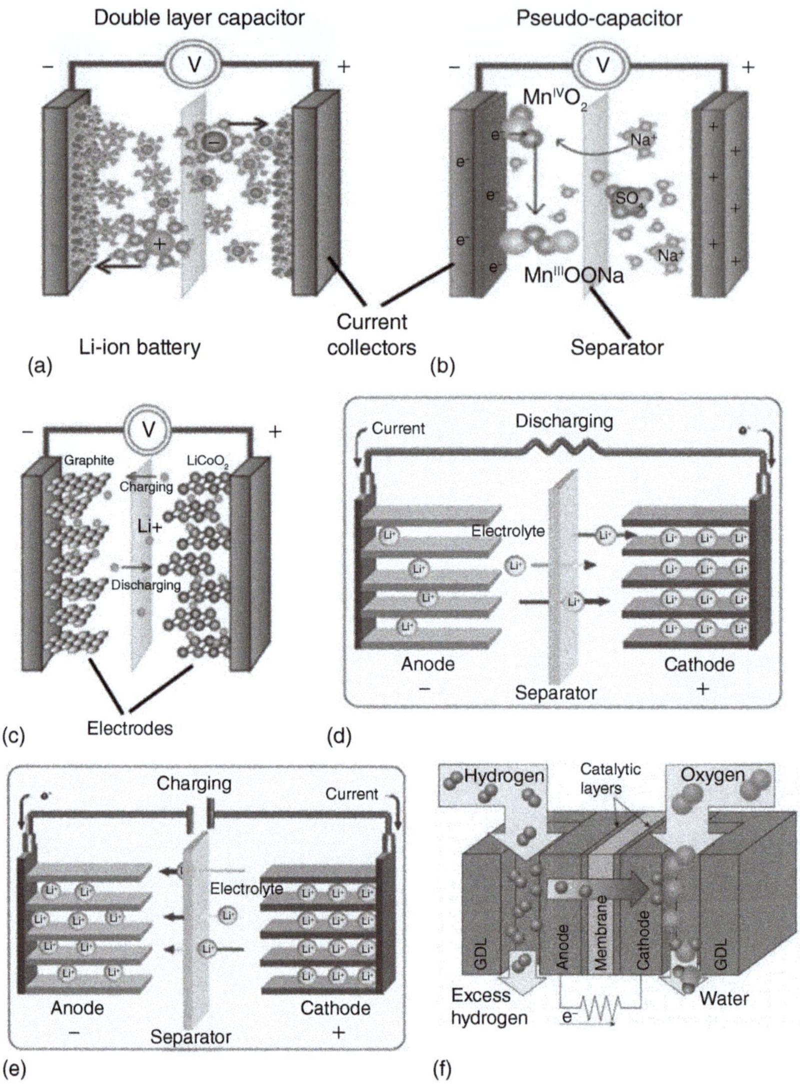

Figure 8.2 Schematic illustrations of (a) electrochemical double layer (EDLC) supercapacitors, (b) pseudocapacitors, (c) hybrid capacitors. Source: (a–c) Moussa et al. [5]/IOP Publishing, (d) battery during charging cycle, (e) battery during discharging process. Source: (d, e) Sashmitha and Rani [6]/Springer Nature, and (f) fuel cell. Source: Zakrisson [7].

based on charge separation at the nanoscale at the electrochemical interface formed between the electrolyte and electrode as demonstrated in Figure 8.2a. The process of charge and discharge does not involve chemical oxidation–reduction (redox) reactions and therefore are termed as nonfaradaic. The EDLCs possess comparatively longlife cycles as no chemical reaction is involved and only physical charge transfer occurs. Contrarily, the pseudocapacitors involve faradaic electrochemical reactions to produce charges that are passed across the double layer, resulting in faradaic current passing through the supercapacitor as demonstrated in Figure 8.2b. These processes happening in the supercapacitor improve the energy density and specific capacitance. However, the pseudocapacitors endure from a comparatively lower power density and lack cycling stability as compared to EDLCs. Hybrid supercapacitors combine mechanisms from both EDLCs and pseudocapacitors. The hybrid capacitors comprised non-faradaic electrodes with ion adsorption occurring on their surfaces and faradaic electrodes undergoing redox reactions. The non-faradaic electrodes work as cathodes, and the faradaic electrodes act as anodes, as demonstrated in Figure 8.2c. In the process of charging, the anions from the electrolyte are adsorbed on the cathode surface, while the cations take part in the redox reactions on the anode. On the other hand, during the process of discharging, the anions are desorbed from the surface of the cathode, and the cations de-intercalate/de-alloy from the anode. Thus, both anions and cations of the electrolyte take part in the energy storage mechanism of hybrid capacitors, where the anions are involved in the non-faradaic process and the cations take part in the faradaic process [8, 9].

8.2.1.2 Batteries

Batteries are devices that store chemical energy and convert it into electrical energy, whereas the chemical reactions in a battery contain electron transport from one electrode to another, through an external circuit that provides an electric current that can be used to do work. To balance the flow of electrons, charged ions also flow through an electrolyte solution that is in contact with both electrodes. Different electrodes and electrolytes produce different chemical reactions that affect the battery performance and energy storage and voltage. The battery contains a positive electrode (anode), a negative electrode (cathode), and liquid or solid electrolytes that store energy. During the discharge process, the electrochemical reactions occur at both the anode and cathode concurrently, whereas the circuit is completed with electron flow through an external circuit initiating from the anode and accumulated at the cathode as demonstrated in Figure 8.2d. In the charging process, reverse reactions take place, whereas the battery is recharged through an external potential applied at both electrodes [10, 11].

8.2.1.3 Fuel Cells

A fuel cell is a device that transforms the chemical energy of a fuel, usually hydrogen, and an oxidizing agent, normally oxygen, into electrical energy through redox reactions. The main components of a fuel cell are two electrodes isolated from each other with a membrane (either solid or liquid electrolyte) as demonstrated in Figure 8.2f. The fuel, such as H_2, is oxidized at the anode to produce electrons and protons during entering of the oxidant from the cathode. These electrons leave the anode through an external circuit to create electric current before arriving at the cathode. Finally, the protons diffused through the membrane recombine with the oxygen atoms and exhaust electrons at the anode to generate water as the only waste of the complete reaction. The electric current produced can then be utilized for different applications [12, 13].

8.2.2 Working Principle of Photochemical Processes

Photocatalysis is the process or activity that occurs by the interaction of a light source to the surface of a material (i.e. semiconductor materials). Photocatalysis comprises seven key steps that are categorized into four main processes as demonstrated in Figure 8.3a. These steps include (stage 1) light harvesting; (stage 2) charge carrier excitation; (stages 3, 4, and 5) charge separation and transfer, and (stages 6 and 7) surface electrocatalytic reactions. In photocatalytic water splitting, the holes react with H_2O to produce protons and O_2 with the help of the oxidation reaction. The protons then combine with the electrons to produce H_2 gas. The photocatalytic

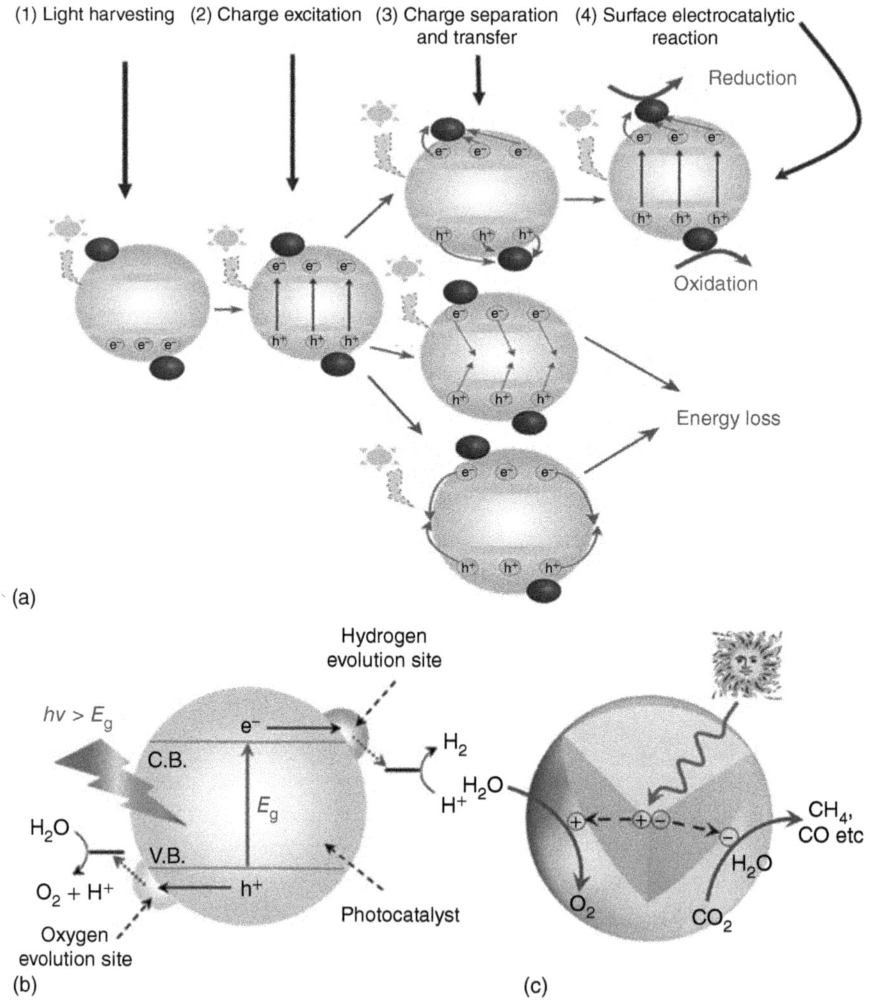

Figure 8.3 Schematic of the (a) steps involved in the photocatalytic process. Source: Khan and Tahir [14]/with permission of Elsevier, (b) photocatalytic water splitting. Source: Maeda and Domen [15]/with permission of American Chemical Society, and (c) photocatalytic CO_2 reduction. Source: Tu et al. [16]/John Wiley & Sons.

process of water splitting is demonstrated in Figure 8.3b. Similarly, in the process of photocatalytic CO_2 reduction, CO_2 is reduced by electrons, while H_2O is oxidized by holes to produce intermediates. These intermediates then combine to produce different types of gases such as CO, CH_4, and CH_3OH. The photocatalytic process of CO_2 reduction is demonstrated in Figure 8.3c [17].

8.3 Properties of TiC MXene

The electrical conductivity of TiC MXene is a crucial property that needs to be explored for application purpose. The electrical conductivity depends on optical and structural properties of TiC MXene that may change due to the variation in the layer spacing, surface functional groups, lateral sizes, delamination yield, and quantity of vacancies and defects.

8.3.1 Structural Properties

The structure of TiC MXenes comprises the intralayer skeleton, terminating groups on the surface of the skeleton, and the empty interlayer area as shown in Figure 8.4. The layers of Ti and C atoms are stacked to form the skeleton region, while the surface-terminating groups through hydrogen bonding and van der Waals forces between the atoms result in the interaction between the layers [18]. The TiC MXene has the ability to form different shapes such as zero-dimensional (0D) materials, (quantum dots, etc.), one-dimensional (1D) nanomaterials (nanowires, nanotubes, etc.), two-dimensional (2D) nanomaterials (layered and nanosheets), and three-dimensional (3D) nanomaterials. Among these structures, the 2D nanomaterials demonstrate significant variation in the surface, electron energy levels, and other features compared with 3D materials. The bonding strength between the interlayers depends on the orientation relative to the entire sheet, number, and distribution of the –OH groups. The presence of water molecules between the layers may also affect the hydrogen bonding. Furthermore, an enormous amount of terminal functional groups is unevenly distributed on the $Ti_3C_2T_x$ surface. These surface groups have the ability to form bonding directly with the exposed surface of $Ti_3C_2T_x$ achieved through top–down etching. The surface groups present

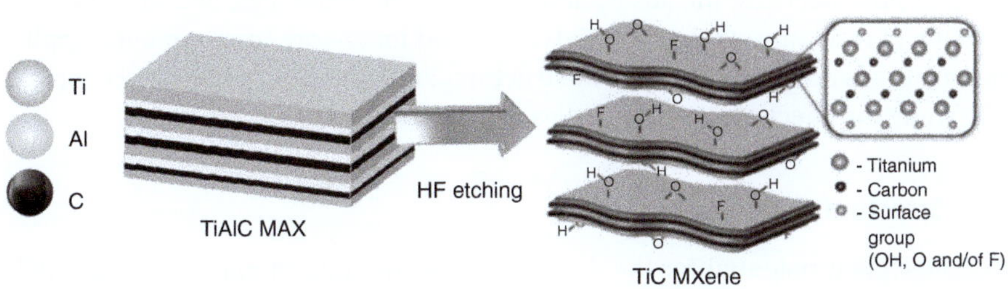

Figure 8.4 A typical representative structure of TiC MXene. Source: Adapted from [18].

have an effect on the properties of $Ti_3C_2T_x$ that are analyzed through techniques such as electron transmission electrons microscopy (TEM), neutron scattering, and neutron magnetic resonance (NMR) techniques. These techniques affirm the surface functional groups present on the surface of $Ti_3C_2T_x$, haphazardly dispersed with –O, –OH, and –F directly forming a bond with the exposed surface of the TiC MXene plane. The high specific surface area could be achieved on the TiC MXene for providing tremendous surface-active sites to enhance the performance in the energy storage and conversion applications. The transportation of ions occurs easily through the TiC MXene layers due to broad spacing, while the presence of –O, –OH groups on the surface delivers excellent electron conductivity.

8.3.2 Optical Properties

The TiC MXenes demonstrate outstanding optical properties, suggesting that several desirable applications could be accomplished by controlling the optical properties through structural and environment modifications. The TiC MXenes demonstrate excellent optical transparency and robust surface-sensitive optical property. Naturally, the optical response of TiC MXene materials is dependent on their structural and electronic properties such as band structures, topology, and surface functional groups. The Ti_3C_2 having a multilayered structure and metallic property retains no absorption edge [19]. The optical properties of TiC MXene are affected by the interband in-plane transitions having orbital hybridization. For instance, in the $Ti_3C_2T_x$ material, the change in optical properties is owing to the termination of T_x that improves optical properties. Berdiyorov et al. [20] obtained enhanced absorption of light irradiations due to the increased reflective property of F- and OH-terminated $Ti_3C_2T_2$ under UV–visible light. Similarly, the $Ti_3C_2T_2$ flakes demonstrated excellent absorption in the UV–visible region (300–500 nm) and in the visible region (700–800 nm).

8.4 Properties of LDH

LDH-based materials have obtained high consideration due to their structural and intrinsic optical properties. These properties can influence the electronic structure and active sites, decrease the adsorption energy of intermediates, and increase the adsorption of reactants. As the materials employed for energy environmental application materials are governed by their structural and optical properties, it is important to study the effect of these properties on the performance.

8.4.1 Structural Properties

LDHs show great potential for the design and construction of materials employed for energy and environmental application. The structural attributes such as crystal facets, geometry, defects, and morphology can be modified to improve application in energy storage and conversion [21]. The distinctive structure having specific chemical composition containing cations and anions provides good capability to provide and alter active sites [22]. Through the variation of the structure and composition of LDHs, the active sites' number and strength can be efficiently customized [23].

The decrease in the size or thickness to the nanometer scale of the LDH triggers the decrease of metal cation coordination number that directs the alteration in electronic properties. Essentially, the surface properties of the LDH materials become dominant when the LDH thickness is reduced to 1 nm level [23].

8.4.2 Optical Properties

The optical properties of LDH are strongly dependent on the layered structure and variation in chemical composition. The tuning of metal cations can change light absorption through changing to visible wavelength, resulting in enhanced optical properties. The intercalation of anionic species in the interlayer of LDH may also alter the optical properties that need to be given specific attention when fabricating the LDH photocatalyst. The increase in the visible light region and variation of band structures can enhance the optical response and charge carrier separation and transportation. The diffuse reflectance UV–vis spectroscopy is employed to measure the optical properties. LDH shows absorption bands in 200–300, 300–500, and 600–800 nm regions due to ligand-to-metal charge transfer (LMCT) and *d*–*d* transitions [24].

8.5 Structural and Optical Properties of LDH/TiC MXene

8.5.1 Structural Properties

The material structural and morphological studies of LDH/TiC MXene nanocomposites may be conducted by employing TEM, SEM, and energy-dispersive X-ray (EDX) measurements. The LDH/TiC MXene can be obtained through the formation of core-shell structure, whereas in the most cases, TiC MXenes act as a core and LDH as a shell. However, by reducing the size of TiC MXenes to a very small size such as TiC quantum dots, the LDH an act as a core. For instance, Zhao et fabricated a 2D/2D core–shell of NiAl-LDH/$Ti_3C_2T_x$, whereas the NiAl–LDH sheets were evenly dispersed on the multilayered $Ti_3C_2T_x$ surface, showing a close interaction between NiAl-LDH and $Ti_3C_2T_x$ as demonstrated in Figure 8.5a,b. The core shell structure formation was due to the confinement obtained from the ultrathin MXenes having a negative charge [25].

Similarly, Khan et al. [26] fabricated the $Ti_3C_2T_{A/R}$/$Co_2Al_{0.95}La_{0.05}$-LDH composite with $Co_2Al_{0.95}La_{0.05}$-LDH nanosheets dispersed on the surface and in the layers of $Ti_3C_2T_{A/R}$, as shown in Figure 8.5c. Contrarily, Shi et al. [28] formed a flower-like hierarchical structure of TiC/NiAl-LDH, whereas the NiAl-LDH showed flower-like hierarchical structure acting as a substrate stuffed by the Ti_3C_2 nanosheets. The 3D

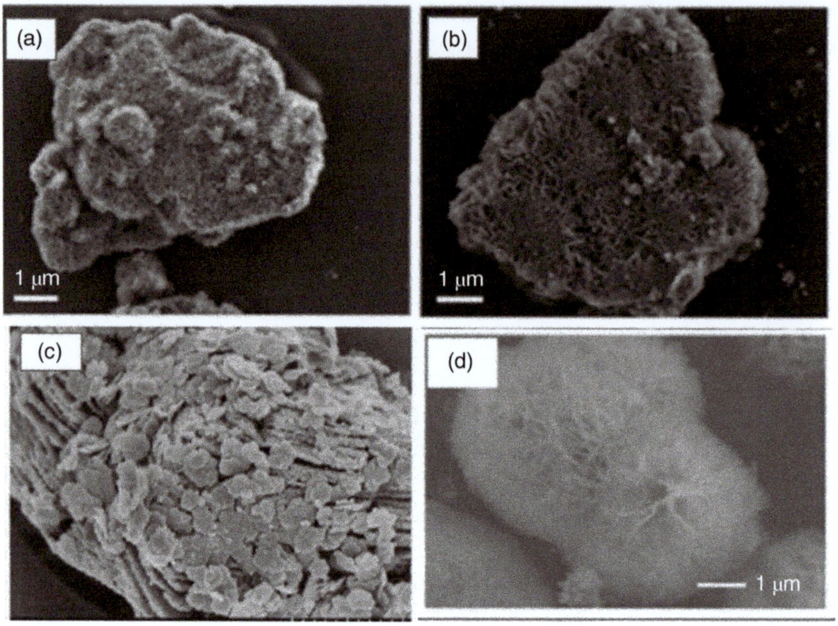

Figure 8.5 (a, b) 2D/2D Core–shell NiAl-LDH/$Ti_3C_2T_x$. Source: (a, b) Zhao et al. [25]/with permission from American Chemical Society, (c) $Co_2Al_{0.95}La_{0.05}$-LDH/$Ti_3C_2T_{A/R}$. Source: Ali Khan et al. [26]/with permission from American Chemical Society, (d) flower-like structure of NiCo-LDH/MXene. Source: Siwal et al. [27]/John Wiley & Sons, Inc./CC BY 4.0.

hierarchical LDH/TC hybrid photocatalyst with core–shell 2D/2D heterostructure was formed. Similarly, Siwal et al. [27] obtained NiCo-LDH flowers squeezed in between the MXene coatings, resulting in the formation of the NiCo-LDH/MXene composite demonstrated in Figure 8.5d.

8.5.2 Optical Properties

The construction of the 2D/2D heterojunction interface of TiC MXene with LDH has increasingly enticed attention, benefitting from the large contacted interfacial area for efficient charge migration to tune the optical and electronic properties. The photo-absorption capacity of the LDH/TiC MXene nanocomposite may be analyzed by employing UV–vis diffuse reflectance spectroscopy (DRS) and photoluminescence (PL) techniques. In the composites formed, their photo-absorption behavior demonstrate likeness to the LDH with minor increase in "he intensity of light irrad"ations absorption. Generally, a higher TiC MXene content corresponds to a stronger absorption band. Although the loading of TiC cannot strengthen the excitation of LDH, the charge migration may be enhanced [26, 28]. From PL spectroscopy, it can be determined that LDH as a solo material shows a high peak intensity as compared to LDH/TiC MXene. It can be believed that the reduction in peak intensity is due to the reduced recombination of LDH in the LDH/TiC MXene composite.

8.6 LDH-Based TiCMXene Composite Applications

8.6.1 Electrochemical Energy Storage Application

Recently, LDHs are being highly attractive candidates for their electrochemical energy storage and conversion applications. This is ascribed to the extraordinarily designed nanostructure through synthesis techniques, compositional tunability of components, and excellent morphology. Appealingly, layered LDH materials are largely functionalized by surface OH groups, aiding in better interaction with alternate high-throughput 2D materials forming robust nanocomposites. Therefore, hybridizing the corresponding properties of 2D LDH with highly conductive and surface-active 2D TiC MXene through synergetic hierarchical heterostructures is anticipated to resolve the problems of LDH. Therefore, employing LDH/TiC MXene composite potential material for energy storage and conversion, the focus has been given to synthesize heterostructures through strategies like electrostatic self-assembly and by in situ growth of LDH on MXene to form LDH/TiC MXene composite [29]. In some cases, LDH is used as a substrate that is doped with TiC MXene. Due to the excellent properties of LDH and TiC, MXenes and their interfacial interaction and electronic coupling may not only improve the structural stability, electrical conductivity, and electrolyte-accessibility but can also enhance the redox reaction kinetics [30].

8.6.2 Ni-Based LDH/TiC Composite

Ni-based LDH are auspicious materials for energy storage and conversion due to their environmental friendliness, low cost, and abundant reserves. Moreover, its high theoretical ability and excellent redox activity make it an excellent material for energy storage and conversion. However, the electrochemical and photochemical properties of LDH are limited by the low conductivity and self-agglomeration. The combination of LDH and MXene is expected to get materials with excellent electrochemical and photochemical properties [3]. Ni-based LDHs with TiC MXene have aroused substantial importance in energy storage due to their interchangeable anions and alterable interlayer spacing, improved electrical conductivity, and enormous amount of available active sites, causing exceptional activity and cycle stability. To increase the electrochemical activity of LDHs, substantial attempts have been dedicated to improve the structure and morphology of LDHs. For instance, NiAl-LDH platelets were homogeneously grown on MXene flake surface to fabricate a 3D porous structure that displayed extremely active sites for electrolyte penetration and decreased the change in volume of LDH during the process of charging and discharging. Ti_3C_2 MXene also assists in the conductive network formation, facilitating electron transport from the active material [31].

The assembly of 2D LDH and MXene with structural resemblance having complementary characteristics may improve the structural stability of the composite and offer the possibility with high energy density. The MXenes when reduced to the size of nanodots of $Ti_3C_2Cl_2$ MXene nanosheets (QD-$Ti_3C_2Cl_2$) and distributed on the surface of LDH can considerably reduce the interfacial resistance to accelerate the overall electron transfer and improve the electrochemical reaction kinetics. The formation of QD-$Ti_3C_2Cl_2$@NiAl-LDH energy storage activity has demonstrated capacitance of the value 2010.8 F g^{-1} at current density, energy density, and power density of 1.0 A g^{-1}, 100.5 Wh kg^{-1}, and 299.8 W kg^{-1}, respectively. Furthermore, after cycling for 10,000 cycles at 1.0 A g^{-1}, a capacitance retention of 94.1% is achieved that surpassed the previously worked 2D hybrid supercapacitor [32]. Likewise, a supercapacitor electrode was formed with MXene@PDDA/NiAl-LDH nanocomposite by employing the electrostatic assembly method. The hybrid was fabricated by employing exfoliated Ti_3C_2 and PDDA/NiAl-LDHs layers for the formation of 3D interconnected architectures. The architectures fabricated demonstrated a strong contact to proficiently attain the benefit of greater electron conductivity from Ti_3C_2 MXene and outstanding pseudocapacitance activity of PDDA/NiAl-LDHs, therefore proficiently hastening the transport of electrons or ions and improving the energy storage. Consequently, the MXene@PDDA/NiAl-LDH nanocomposite exhibited exceptional specific capacitance of 1825.8 F g^{-1} at a current density of 1.0 A g^{-1}, resulting in improved rate capacity after 100 cycles as demonstrated in Figure 8.6a. Also, the nanohybrid electrodes demonstrated extended cycling stability with very little degradation of 0.90% after 5000 cycles at a current density of 5.0 A g^{-1} as demonstrated in Figure 8.6b [33]. Moreover, the MXene@PDDA/NiAl-LDH electrode delivered a maximum energy density of 91.3 Wh kg^{-1} at a power density of 300 W kg^{-1} as demonstrated in Figure 8.6c. The manganese (Mn) element has shown the capability to enhance the electrochemical performance of Ni in the LDH; thus, NiMn-LDH has been employed as an exceptional for the supercapacitor. Thus, the hybridization of Ni–Mn LDH covering the MXene surface can better improve the electrochemical activity. The synergistic effect of electronic transport and structural coupling between NiMn-LDH and MXene can enhance the electrical conductivity and structural stability, respectively [34].

By varying the Ni^{2+} to Fe^{3+} molar ratio, the structure and morphology of NiFe-LDHs have the ability to be improved, with the structure offering a large specific surface area and greater number of active sites that can improve the diffusion of ions from the electrolyte to the surface of active materials. The obtained NiFe-LDHs may then demonstrate extraordinary OER performance in alkaline electrolytes because of the synergistic effect of multiple metal centers, bulk redox activity, and encouraging kinetics in the layered structure with bigger basal spacing. However, the utilization of OER performance for NiFe-LDHs remains insufficient owing to the inherently inadequate conductivity, lower electrochemical active surface area (ECSA), and a strong aggregation. In this scenario, Qiu et al. employed the electrochemical deposition method to form mesoporous NiFe-LDH nanosheets on Ti_3C_2 MXene scaffolded by nickel foam (NF), resulting in the formation of Ni-Fe-LDH/MXene/NF. The 3D structured MXene demonstrated superior conductivity and hydrophilic properties, that supported the mass/charge transport across the catalyst and enhanced the OER and HER kinetics. The NiFe-LDH/MXene/NF

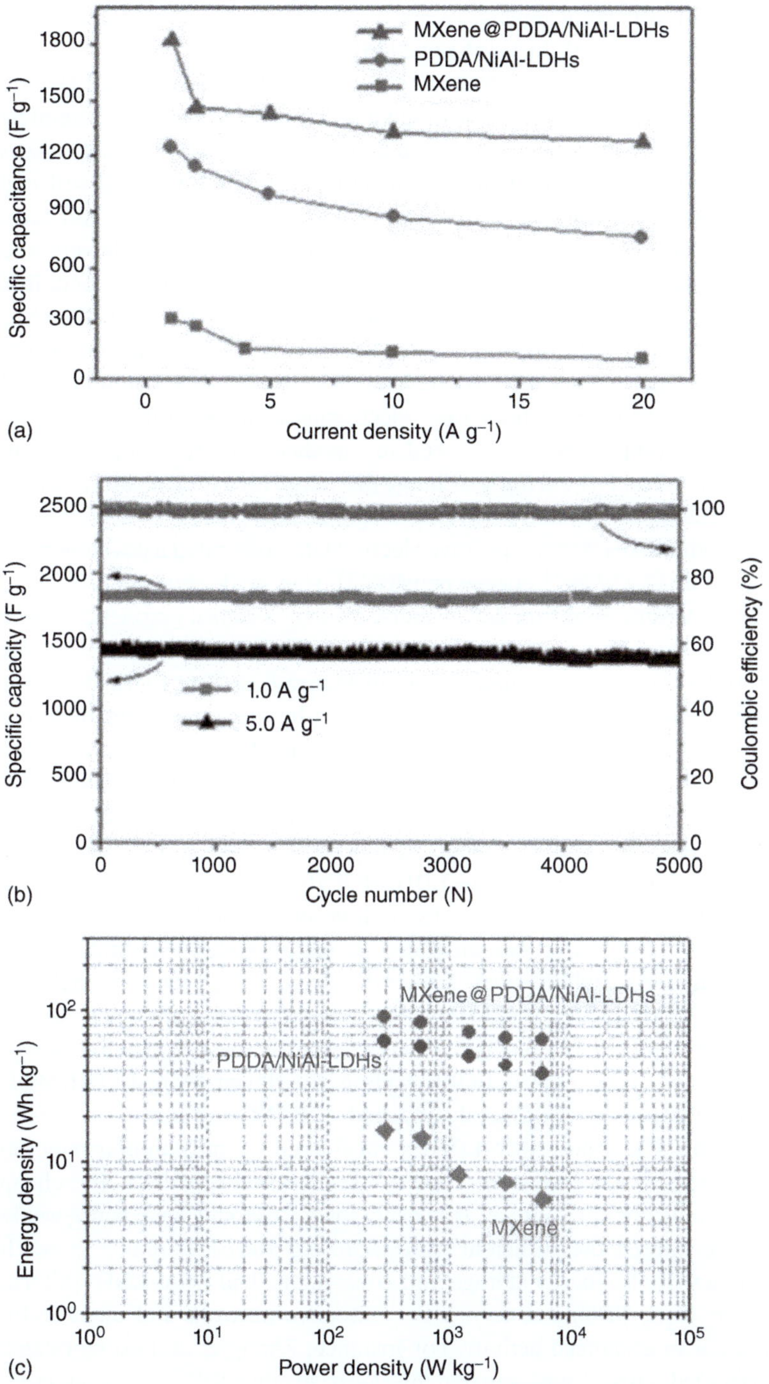

Figure 8.6 (a) Specific capacitances of theMXene, PDDA/NiAl-LDHs, s and MXene@PDDA/NiAl-LDHs. (b) Long cycling performance of the MXene@NiAl-LDH electrode at 1.0 and 5.0 A g^{-1} for 5000 cycles. (c) Gravimetric energy and power density for MXene, PDDA/NiAl-LDH, and MXene@PDDA/NiAl-LDH electrode.

achieved a maximum current density of 500 mA cm^{-2} at lower overpotential for OER 300 mV and HER 205 mV in 1.0 M KOH [35].

Likewise, NiFe-LDH/Ti_3C_2 nanohybrid with uniformly spread NiFe-LDH on Ti_3C_2 MXene nanosheets due to the synergism exhibited larger electrochemical performance for lithium storage. The reversible capacity of the heterostructure was 898.9 mAh g^{-1} at 0.1 A g^{-1}, and even at a current density of 1 A g^{-1}, it still had 726.1 mAh g^{-1}, without reduction occurring in capacity in lithium-ion half-cells. Furthermore, the lithium-ion capacitor demonstrated greater energy density and power [35]. Furthermore, a 3D porous NiFe-LDH/MXene electrode was acquired in another work that demonstrated superior specific capacitance of 720.2 F g^{-1} as compared to NiFe-LDH with 465 F g^{-1}. The NiFe-LDH/MXene nanocomposite-maintained capacitance of 86% after 1000 cycles 24% for NiFe-LDH demonstrated improved cycle stability. The improved electrochemical performance of the NiFe-LDH/MXene composites was ascribed to the sheet structure having exceptional stability for NiFe-LDH during the charge discharge time and network formed by the MXene causing accelerated electron transfer. Besides, the asymmetric supercapacitor of the NiFe-LDH/MXene positive electrode demonstrated a power density of 758.27 W kg^{-1} at 42.4 W kg^{-1} energy density [3]. Han et al. obtained hybrids of NiFe-LDH/Ti_3C_2 MXene quantum dots dispersed on an N-doped graphene surface (LDH/MQD/NG) through urea-assisted co-precipitation of Ni^{2+} and Fe^{2+} in the existence of MQDs and NG as demonstrated in Figure 8.6a. The LDH/MQD/NG hybrid comprised flower-like LDH/MQDs combined by the nanosheets anchored onto the surface of NG, showing a 3D hierarchical porous structure as demonstrated in Figure 8.6b. The obtained LDH/MQD/NG composite showed greater OER activity than pristine LDH/NG. A lower 1.50 V_{RHE} potential was required for the LDH/MQD/NG hybrids as compared to LDH/NG (1.52 V_{RHE}) to deliver the same 10 mA cm^{-2} current density. The LDH/MQD/NG hybrid electrode showed faster OER kinetics through a smaller Tafel slope of 57 mV dec^{-1} as compared to 60 mV dec^{-1} for LDH/NG as demonstrated in Figure 8.6c. In addition to the OER activity, long-term stability was achieved for the LDH/MQD/NG hybrids, as indicated by the slight decay (7.7%) of the current density after 10 hours of continuous testing and almost negligible degradation in the polarization curves after 5000 CV cycles. Upon repeated OER cycling, the flower-shaped LDH/MQD hybrids remain uniformly and intimately attached to the NG surface.

Among the LDH materials, NiCo-LDH is considered an outstanding material ascribed to its alterable ion exchange property, excellent redox activity, and cheap and environmentally friendly nature. However, NiCo-LDH still have some weaknesses that have limited the sufficient utilization, e.g. they are unable to reach the maximum activity in energy storage and conversion. The NiCo-LDH uniform arrangement on Ti_3C_2 MXene to form the conductive network and open architecture can result in enhanced activity. For instance, Zhang et al. [36] fabricated ultra-thin, wrinkled, curved with greater interlayer spacing NiCo-LDHs on Ti_3C_2 MXene as an anode material for Li-ion batteries. The design lowered the degradation of the composite structure and achieved superior electrical conductivity, reduced diffusion barrier, and improved discharge capacity. The discharge capacity is 562 mAh g^{-1} at 5.0 A g^{-1} for 800 cycles without degradation. Moreover, the rate capability and cycle performance were improved as compared to NiCo-LDH (~100 mAh g^{-1}) [36].

A trimetallic NiCoFe-LDH grown on nickel foam has demonstrated outstanding electrochemical properties as an active electrode in supercapacitors and as a catalyst material in the OER. Yet, the capacitance achieved still remains smaller, and therefore further improvement is necessary [37]. In this scenario, the NiCoFe-LDH coupled with the Ti_3C_2 MXene composite showed an outstanding specific capacity of 1990 F g^{-1} in 1 M KOH at of 1 A g^{-1} and maintained a specific capacity of 84.8% after 5000 cycles of charging and discharging [38].

It can be determined that Ni-based LDH and MXene composite structures deliver excellent performance in the field of energy storage and conversion with performance and improved stability.

8.6.3 Co-Based LDH TiC Composite

Co-based LDHs with multiple redox states are beneficial to establish electron transmission in the field of energy storage and conversion; however, the efficiency is still low and therefore needed to be improved. In this scenario, the Co-based LDHs may be combined with TiC MXene to attain improvement in performance of the composites for energy storage and conversion. For instance, delaminated $Ti_3C_2T_x$/CoAl-LDH as the cathode of the alkaline hybrid battery through electrostatic interaction is obtained. The $Ti_3C_2T_x$/CoAl-LDH composite formed a heterogeneous electrode that improved the kinetics of the Faraday reaction. Obtaining benefit from quick electron/ion transportation and coupled structure, the $Ti_3C_2T_x$/CoAl-LDH electrode exhibited 106 mAh g^{-1} specific capacity at a current density of 0.5 A g^{-1} with exceptional rate capacity and exceptional stability [39]. In recent times, Niu et al. [40] attained self-assembly of monolayered electrostatic orderly Ti_3C_2 MXene and CoAl-LDH with edge-enriched nanosheets as demonstrated in Figure 8.7a. The CoAl-LDH/Ti_3C_2 MXene hybrid with excellent assembly and invincible interfacial interaction between Ti_3C_2 MXene and CoAl-LDH leads to extraordinary electrical conductivity, available electrolyte ion surface area, and increased number of exposed electroactive sites, leading to intense electrochemical activity. The MXene/CoAl-LDH-80% (ML-80) demonstrated superior volumetric capacitance of 2472 C cm^{-3} with high rate capacity of 70.6% at current density of 20 A g^{-1} [39, 40].Likewise, Hao et al. [41] achieved CoFe-LDH/MXene nanohybrid for OER activity. Owing to the synergistic effect of the Ti_3C_2 MXene and CoFe-LDH, the as-attained nanohybrid offered an efficient OER activity having high stability with 319 mV overpotential as compared to other substrates, comprising CoFe-LDH/GO ($\eta_{j=10}$ = 365 mV), CoFe-LDH/C nanotube ($\eta_{j=10}$ = 372 mV), CoFe-LDH/C powder ($\eta_{j=10}$ = 387 mV) 10 mA cm^{-2} current density

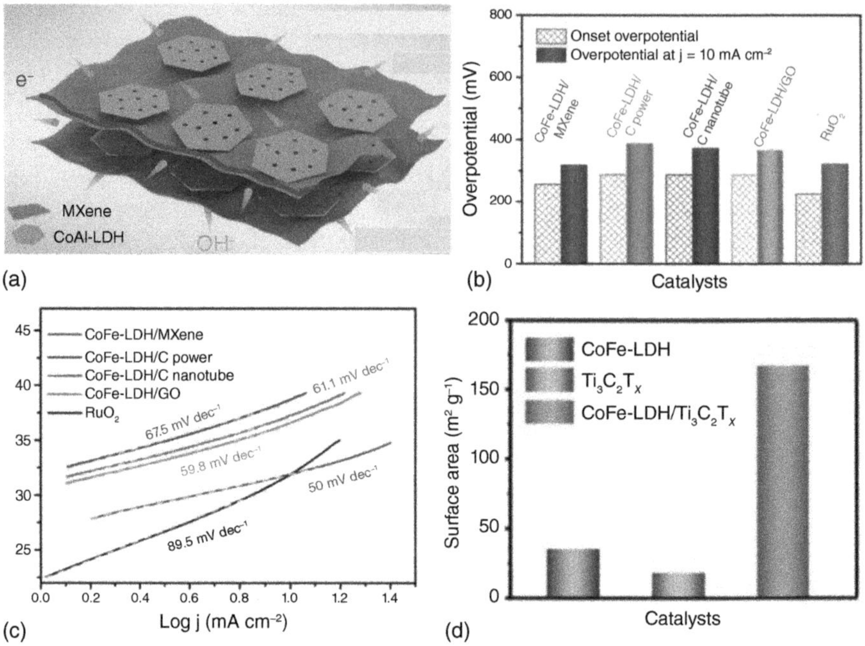

Figure 8.7 (a) Schematic representation of the Ti_3C_2 MXene/CoAl-LDH composite. Source: Niu et al. [40]/with permission of Elsevier; (b) comparison of catalysts in onset overpotential and overpotential to obtain 10 mA cm^{-2} ($\eta_{j=10}$) current density; (c) Tafel plots of RuO_2, CoFe-LDH/C powder, CoFe-LDH/C nanotube, and CoFe-LDH/GO and CoFe-LDH/MXene catalysts. Source: Hao et al. [41]/with permission of Elsevier; (d) N_2 adsorption–desorption of CoFe-LDH, $Ti_3C_2T_x$, and of CoFe-LDH/$Ti_3C_2T_x$ composites. Source: Lei et al. [42]/with permission of Elsevier.

as exhibited in Figure 8.7b. Furthermore, the Tafel slope achieved 50 mV dec^{-1}, which was very low in comparison to CoFe-LDH/C nanotube (61.1 mV dec^{-1}), CoFe-LDH/GO (59.8 mV dec^{-1}) and CoFe-LDH/C powder (67.5 mV dec^{-1}) as revealed in Figure 8.7c. In a similar way, CoFe-LDH has also been composited with MXene to construct a self-supporting film electrode of CoFe-LDH/$Ti_3C_2T_x$, used as the anode material for the storage of Cl^- ions. The as-attained nanocomposite of CoFe-LDH/$Ti_3C_2T_x$ showed a hierarchical micro-mesoporous structure that is determined through the N_2 adsorption–desorption isotherms with enhanced surface area as established in Figure 8.7d that hastened ion diffusion, resulting in enhanced desalination rate. The desalination capacity touched the value of 149.25 ± 6.17 mg g^{-1} at a current density of 50 mA g^{-1} and desalination rate of 3.86 mg g^{-1} min^{-1} at a current density of 100 mA g^{-1} [42].

In another work, CoNi-LDH/$Ti_3C_2T_x$ showed outstanding OER activity with high stability in basic media, which demonstrated a lower overpotential of 257.4 mV to obtain 100 mA cm^{-2} current density and 68 mV dec^{-1} Tafel slope. It was observed that transfer of electrons from CoNi-LDH to $Ti_3C_2T_x$ performed a crucial function in enhancing OER performance by decreasing the binding strength of intermediates [43].

8.7 Photocatalytic CO_2 Reduction Application

Photocatalytic reduction of CO_2 into fuels such as CO, CH_4, and CH_3OH is an auspicious attempt to store solar energy and mitigate greenhouse gas emissions simultaneously. The LDH material, a low-cost and easily available clay, lamellar structure, with exceptional semiconducting and adsorption properties, has drawn stimulated extensive interests in photocatalytic CO_2 reduction due to its superior properties such as layered structure, high porosity, surface area and high number of exposed surface-active sites, and facile synthesis. However, the pristine LDHs show poor photocatalytic performance due to the quick recombination of electron–hole pairs and need to do much work in improving the performance. To overcome these problems, the hybridization of LDH with MXene could competently amalgamate multifunctionalities, reduce shortcomings, and confer properties, which make efficient charge separation, resulting in exceptional photocatalytic performance. Especially, combining TiC with LDH can remarkably improve the migration and separation of charge carriers due to the highly conductive nature of TiC MXene. The enhancement in properties of the LDH was achieved by hybridizing LDH with TiC MXene.

8.7.1 Ni-Based LDH/TiC Composite

The Ni-based LDH is a member of LDH materials and has been frequently used for photocatalytic reactions. However, poor conductivity and lower separation efficiency for the photoinduced charge carriers is needed to be improved, that seriously restricts their useful activity. In this scenario, the TiC MXene is deemed as a high-efficiency cocatalyst for photocatalytic CO_2reduction due to the cheap price, excellent electrical conductivity, and hydrophilic surface nature. The hybridization of Ni-based LDH with TiC MXene may result in the formation of composite has found widespread application in the field of CO_2 reduction to fuels.

For instance, Shi et al. [28] constructed a 3D hierarchical NiAl-LDH/Ti_3C_2 nanohybrid through in situ loading of Ti_3C_2 nanosheets on the NiAl-LDH scaffold as shown in Figure 8.8a. The fabrication of the 2D/2D interface can be achieved by altering the Ti_3C_2 ratio and the precursors utilized for NiAl-LDH. The 3D structure with larger specific surface area, enhanced adsorption, and utilization of photons. The strong interaction between Ti_3C_2 and NiAl-LDH with numerous interfaces effectively improved the separation of the electron-holes in the NiAl-LDH sample. The excellent photocatalytic CO production rate on NiAl-LDH/Ti_3C_2 reached $11.82\,\mu mol\,g^{-1}\,h^{-1}$ with 92% selectivity and excellent stability as demonstrated in Figure 8.8b. The enhancement in photocatalytic activity was owing to the coupling with the high quantity of bare oxidation–reduction active sites, superb CO_2 capture, activation, and conversion.

Likewise, Zhao et al. [25] constructed a core/shell NiAl-LDH/$Ti_3C_2T_x$ nanosheet structure (NiAl-LDH/TNS) for photocatalytic reduction of CO_2 with the presence of ruthenium photosensitizers under visible light irradiations. The hybridization of 2D NiAl-LDH with 2D $Ti_3C_2T_x$ nanosheets established a Schottky junction that

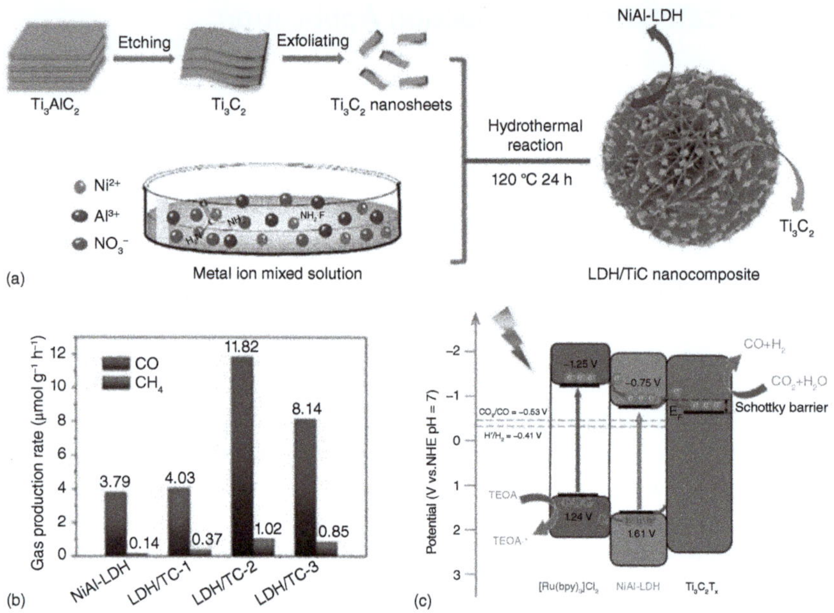

Figure 8.8 (a) Schematic of the 3D NiAl-LDH/Ti_3C_2 hybrid formation (b) CO and CH_4 production over NiAl-LDH and LDH/TiC MXene catalyst samples Source: Shi et al. [28]/with permission of American Chemical Society. (c) Schematic of charge transfer over the NiAl-LDH/TNS composite with $[Ru(bpy)_3]Cl_2$ under visible light. Source: Zhao et al. [25]/with permission of American Chemical Society.

quelled the recombination of electrons and facilitated the charge transfer as shown in Figure 8.8c. The NiAl-LDH/TNS demonstrated enhanced CO_2 photoreduction as compared to NiAl-LDH, $Ti_3C_2T_x$, and NiAl-LDH/multilayered $Ti_3C_2T_x$.

8.7.2 Co-Based TiC Composite

Among the LDHs, Co-based LDHs have found widespread application for CO_2 photoreduction due to high photocatalytic activity and thermal stability. For instance, through an in situ MOF-derived approach, Chen et al. [44] utilized divalent and trivalent Co to fabricate Co–Co LDH/MXene. The method transformed ZIF-67 MOFs into 2D Co–Co LDH distributed vertically on $Ti_3C_2T_x$ nanosheets, forming Co–Co LDH/$Ti_3C_2T_x$ hybrids (Co–Co LDH/TNS) as demonstrated in Figure 8.9a. Subsequently, the achieved Co–Co LDH/TNS hybrids were utilized with the presence of ruthenium (Ru) as photosensitizers for photocatalytic CO_2 reduction. The obtained hierarchical Co–Co LDH/TNS resulted in improved photoactivity toward visible-light reduction of CO_2. The CO was obtained as a main product with a 2.2-fold yield of 1.25×10^4 μmol h^{-1} g^{-1} as compared to Co–Co LDH nanosheets, revealed in Figure 8.9b. The increase in activity was due to enhanced electron transfer because of the synergetic effects of distinctive hierarchical structures and excellent conductive Ti_3C_2 MXene. Also, Khan et al. [26] fabricated CoAlLa-LDH rich of oxygen vacancies that was combined with $Ti_3C_2T_x$ MXene. First, $Ti_3C_2T_x$ was synthesized through etching, and simultaneously a CoAlLa-LDH was synthesized. The CoAlLa-LDH

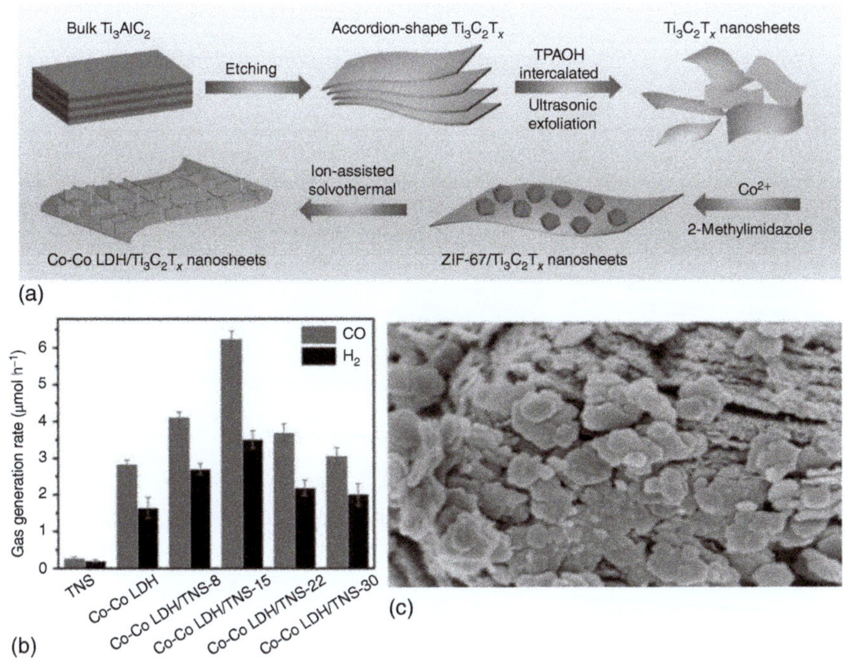

Figure 8.9 (a) Schematic of the synthesis procedure of CoCo-LDH/TNS nanosheets. (b) Photocatalytic performance of CoCo-LDH and CoCo-LDH/TNS hybrids with several TiC nanosheet content. Source: Chen et al. [44]/with permission of Elsevier. (c) FESEM of CoAlLa-LDH/Ti_3C_2 Source: Ali et al. [26]/with permission from American Chemical Society.

was spread on $Ti_3C_2T_x$ because of the electrostatic attraction to achieve the 2D/2D $Ti_3C_2T_x$/CoAlLa-LDH photocatalyst, as demonstrated in FESEM in Figure 8.9c. The hybridization of the composite causes effective charge transfer to improve the separation proficiency, resulting in superior photon consumption of light irradiations. The obtained composite resulted in effective photocatalytic production of CO and CH_4 under light irradiations. In photocatalytic bireforming of methane (BRM) and dry reforming of methane (DRM), CO and H_2, a potential syngas was obtained. The BRM demonstrated encouraging results due to effective attachment of the reactants over the defective rich active sites of the nanohybrid obtained.

8.8 Photocatalytic Degradation Application

LDH-based TiC MXene composites have found very little application in the field of photodegradation. Only few researchers have employed LDH TiC MXene for degradation of pharmaceuticals in water for wastewater reclamation. For instance, Wang et al. [45] fabricated a 3D CoAl-LDH/Ti_3C_2 nanohybrid in situ grown CoAl-LDH on Ti_3C_2 nanosheets. The optimized Ti_3C_2 directs the formation of 2D lotus leaf like nanoplates on 3D marigold-like structure as exhibited in Figure 8.10a and SEM in Figure 8.10b that improved the adsorption capability and photon utilization of the photocatalyst by the enhanced specific surface area. The 2D/2D structure improved

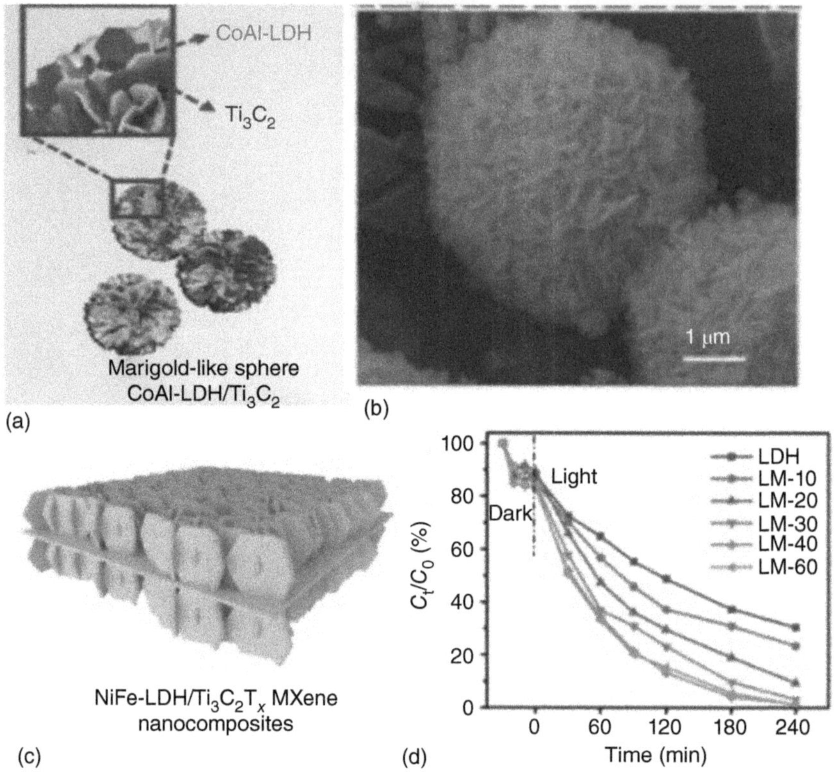

Figure 8.10 (a) Schematic of the 3D hierarchical structure of CoAl-LDH/Ti_3C_2 (b) SEM of CoAl-LDH/Ti_3C_2 Source: (a, b) Wang et al. [45]/with permission from Elsevier. (c) Schematic of NiFe-LDH/$Ti_3C_2T_x$ MXene nanocomposites. (d) Photocatalytic efficiency of the NiFe-LDH/$Ti_3C_2T_x$ MXene hybrids with different MXene loadings. Source: (c, d) Ma et al. [46]/with permission of Elsevier.

the transfer and separation of electron–hole pairs from CoAl-LDH to Ti_3C_2. As a result, CoAl-LDH/Ti_3C_2 hybrid showed a degradation efficiency of 85%, 56%, and 70% for tetracycline hydrochloride, chloramphenicol, and terramycin chemicals, respectively, that are 3.1-, 3-, and 2.1-fold larger than CoAl-LDH.

Similarly, the 2D/2D oxygen vacancy rich NiFe-LDH/$Ti_3C_2T_x$ MXene nanosheet hybrid served as an effective photocatalyst for the remediation of norfloxacin (NOR) from water exhibited in Figure 8.10c. With the increase of the quantity of $Ti_3C_2T_x$ MXene loading from 10 to 40 mg to the NiFe-LDH, the enhancement in the degradation of NOR was observed. Typically, the NiFe-LDH sample with loading of 40 mg (LM-40) shown in Figure 8.10d exhibited excellent photocatalytic activity of more than 98% NOR elimination. The hybridization of MXene with LDH lengthened the absorption of light irradiations in the visible region and lessened the photogenerated charge recombination. It can be observed that the LDH-based TiC nanocomposites have the ability to perform excellently for the reclamation of the aquatic environment by eradicating hazardous materials [46].

8.9 Conclusions and Future Recommendations

In this chapter, the synthesis procedure of LDH, TiC MXene, and LDH/TiC MXene materials is discussed. The addition of TiC MXene to the LDH proficiently lessens the restrictions of pure LDHs. The outstanding structural and morphological properties such as layered structures of TiC and LDH cause the formation of a hierarchal structure with improved charge carrier transport. The hybridization of TiC and LDHs caused the amalgamation of the optical properties, leading to decreased recombination with improved photocatalytic activity. The addition of TiC MXene improves the electrolyte ion transfer of composites, stabilizes LDH structure, and prevents aggregation of LDH. LDH/TiC MXene has found widespread application in the field of CO_2 reduction, resulting in production of potential fuels showing the effect on yield and selectivity toward different potential fuels. The assembly of LDH/TiC MXene nanomaterials for photocatalytic CO_2 reduction is an emerging field that is needed to be studied. Separately, both TiC MXene and LDH are exceptional 2D structural materials. Still, the fabrication of TiC and LDH involves acute environments; therefore, inventive techniques with lesser harsh environments such as normal pH are essential for the fabrication of LDH/TiC MXene materials. Furthermore, the proficiency of the CO_2 photoreduction is low and is required to be enhanced. The application of LDH/TiC MXene for photocatalytic degradation and H_2 production has been studied very seldom and demands investigation. Further research on more-flexible LDH/TiC MXene composites would lead to new research directions. Also, the LDH/TiC MXene is required to be examined to analyze the photodegradation of different pollutants.

Acknowledgment

This work was supported by United Arab Emirates University (UAEU) under the research fund no. 12N097.

References

1 Biswal, L., Nayak, S., and Parida, K. (2021). Recent progress on strategies for the preparation of 2D/2D MXene/g-C_3N_4 nanocomposites for photocatalytic energy and environmental applications. *Catal. Sci. Technol.* 11 (4): 1222–1248.

2 Chen, C., Tao, L., Du, S. et al. (2020). Advanced exfoliation strategies for layered double hydroxides and applications in energy conversion and storage. *Adv. Funct. Mater.* 30 (14): 1909832.

3 Zhou, H., Wu, F., Fang, L. et al. (2020). Layered NiFe-LDH/MXene nanocomposite electrode for high-performance supercapacitor. *Int. J. Hydrogen Energy* 45 (23): 13080–13089.

4 Abbas, Q., Mirzaeian, M., Hunt, M.R.C. et al. (2020). Current state and future prospects for electrochemical energy storage and conversion systems. *Energies* 13 (21): 5847.

5 Moussa, M., El-Kady, M.F., Zhao, Z. et al. (2016). Recent progress and performance evaluation for polyaniline/graphene nanocomposites as supercapacitor electrodes. *Nanotechnology* 27 (44): 442001.

6 Sashmitha, K. and Rani, M.U. (2022). A comprehensive review of polymer electrolyte for lithium-ion battery. *Polym. Bull.* 80: 89–135.

7 Zakrisson, E. (2011). The effect of start/stop strategy on PEM fuel cell degradation characteristics. Master thesis. Chalmers University of Technology.

8 Theerthagiri, J., Lee, S.J., Shanmugam, P., and Choi, M.Y. (2020). Chapter 1: Basic principles in energy conversion and storage. In: *Nanostructured, Functional, and Flexible Materials for Energy Conversion and Storage Systems* (ed. A. Pandikumar and P. Rameshkumar), 1–14. Elsevier.

9 González, A., Goikolea, E., Barrena, J.A., and Mysyk, R. (2016). Review on supercapacitors: technologies and materials. *Renewable Sustainable Energy Rev.* 58: 1189–1206.

10 Liang, Y., Zhao, C.-Z., Yuan, H. et al. (2019). A review of rechargeable batteries for portable electronic devices. *InfoMat* 1 (1): 6–32.

11 Kong, L., Tang, C., Peng, H.J. et al. (2020). Advanced energy materials for flexible batteries in energy storage: a review. *SmartMat* 1 (1): 1007.

12 Li, C., Liu, Y., Xu, B., and Ma, Z. (2019). Finite time thermodynamic optimization of an irreversible proton exchange membrane fuel cell for vehicle use. *Processes* 7 (7): 419.

13 Haile, S.M. (2003). Materials for fuel cells. *Mater. Today* 6 (3): 24–29.

14 Khan, A.A. and Tahir, M. (2019). Recent advancements in engineering approach towards design of photo-reactors for selective photocatalytic CO_2 reduction to renewable fuels. *J. CO_2 Util.* 29: 205–239.

15 Maeda, K. and Domen, K. (2010). Photocatalytic water splitting: recent progress and future challenges. *J. Phys. Chem. Lett.* 1 (18): 2655–2661.

16 Tu, W., Zhou, Y., and Zou, Z. (2014). Photocatalytic conversion of CO_2 into renewable hydrocarbon fuels: state-of-the-art accomplishment, challenges, and prospects. *Adv. Mater.* 26 (27): 4607–4626.

17 Wen, J., Xie, J., Chen, X., and Li, X. (2017). A review on g-C_3N_4-based photocatalysts. *Appl. Surf. Sci.* 391: 72–123.

18 Li, X., Bai, Y., Shi, X. et al. (2021). Applications of MXene ($Ti_3C_2T_x$) in photocatalysis: a review. *Mater. Adv.* 2: 1570–1594.

19 Mohanty, B., Giri, L., and Jena, B.K. (2021). MXene-derived quantum dots for energy conversion and storage applications. *Energy Fuel* 35 (18): 14304–14324.

20 Berdiyorov, G. (2016). Optical properties of functionalized $Ti_3C_2T_2$ (T = F, O, OH) MXene: first-principles calculations. *AIP Adv.* 6 (5): 055105.

21 Li, C., Wei, M., Evans, D.G., and Duan, X. (2014). Layered double hydroxide-based nanomaterials as highly efficient catalysts and adsorbents. *Small* 10 (22): 4469–4486.

22 Zhao, Y., Chen, P., Zhang, B. et al. (2012). Highly dispersed TiO_6 units in a layered double hydroxide for water splitting. *Chem. Eur. J.* 18 (38): 11949–11958.

23 Xu, M. and Wei, M. (2018). Layered double hydroxide-based catalysts: recent advances in preparation, structure, and applications. *Adv. Funct. Mater.* 28 (47): 1802943.

24 Tonda, S., Kumar, S., Bhardwaj, M. et al. (2018). g-C_3N_4/NiAl-LDH 2D/2D hybrid heterojunction for high-performance photocatalytic reduction of CO_2 into renewable fuels. *ACS Appl. Mater. Interfaces* 10 (3): 2667–2678.

25 Zhao, S., Pan, D., Liang, Q. et al. (2021). Ultrathin NiAl-layered double hydroxides grown on 2D $Ti_3C_2T_x$ MXene to construct core–shell heterostructures for enhanced photocatalytic CO_2 reduction. *J. Phys. Chem. C* 125 (19): 10207–10218.

26 Ali Khan, A. and Tahir, M. (2021). Construction of an S-scheme heterojunction with oxygen-vacancy-rich trimetallic CoAlLa-LDH anchored on titania-sandwiched Ti_3C_2 multilayers for boosting photocatalytic CO_2 reduction under visible light. *Ind. Eng. Chem. Res.* 60 (45): 16201–16223.

27 Siwal, S.S., Kaur, H., Chauhan, G., and Thakur, V.K. (2023). MXene-based nanomaterials for biomedical applications: healthier substitute materials for the future. *Adv. Biomed. Res.* 3 (1): 2200123.

28 Shi, Q., Zhang, X., Yang, Y. et al. (2021). 3D hierarchical architecture collaborating with 2D/2D interface interaction in NiAl-LDH/Ti_3C_2 nanocomposite for efficient and selective photoconversion of CO_2. *J. Energy Chem.* 59: 9–18.

29 Long, X., Wang, Z., Xiao, S. et al. (2016). Transition metal based layered double hydroxides tailored for energy conversion and storage. *Mater. Today* 19 (4): 213–226.

30 Zhang, R., Dong, J., Zhang, W. et al. (2022). Synergistically coupling of 3D FeNi-LDH arrays with $Ti_3C_2T_x$-MXene nanosheets toward superior symmetric supercapacitor. *Nano Energy* 91: 106633.

31 Wang, Y., Dou, H., Wang, J. et al. (2016). Three-dimensional porous MXene/layered double hydroxide composite for high performance supercapacitors. *J. Power Sources* 327: 221–228.

32 Zhao, Z., Wu, X., Luo, C. et al. (2022). Rational design of $Ti_3C_2Cl_2$ MXenes nanodots-interspersed MXene@NiAl-layered double hydroxides for enhanced pseudocapacitor storage. *J. Colloid Interface Sci.* 609: 393–402.

33 Zhao, Z. and Wu, X. (2020). Electrostatic-assembled MXene@NiAl-LDHs electrodes with 3D interconnected networks architectures for high-performance pseudocapacitor storage. *Adv. Mater. Interfaces* 7 (19): 2000831.

34 Wang, W., Jiang, D., Chen, X. et al. (2020). A sandwich-like nano-micro LDH-MXene-LDH for high-performance supercapacitors. *Appl. Surf. Sci.* 515: 145982.

35 Yu, M., Wang, Z., Liu, J. et al. (2019). A hierarchically porous and hydrophilic 3D nickel–iron/MXene electrode for accelerating oxygen and hydrogen evolution at high current densities. *Nano Energy* 63: 103880.

36 Zhang, R., Xue, Z., Qin, J. et al. (2020). NiCo-LDH/Ti_3C_2 MXene hybrid materials for lithium ion battery with high-rate capability and long cycle life. *J. Energy Chem.* 50: 143–153.

37 Shabangoli, Y., Rahmanifar, M.S., El-Kady, M.F. et al. (2018). An integrated electrochemical device based on earth-abundant metals for both energy storage and conversion. *Energy Storage Mater.* 11: 282–293.

38 Chen, J., Ren, Y., Zhang, H. et al. (2021). Ni-Co-Fe layered double hydroxide coated on Ti_3C_2 MXene for high-performance asymmetric supercapacitor. *Appl. Surf. Sci.* 562: 150116.

39 Dong, X., Zhang, Y., Ding, B. et al. (2018). Layer-by-layer self-assembled two-dimensional MXene/layered double hydroxide composites as cathode for alkaline hybrid batteries. *J. Power Sources* 390: 208–214.

40 Niu, H., Yang, X., Wang, Q. et al. (2020). Electrostatic self-assembly of MXene and edge-rich CoAl layered double hydroxide on molecular-scale with superhigh volumetric performances. *J. Energy Chem.* 46: 105–113.

41 Hao, C., Wu, Y., An, Y. et al. (2019). Interface-coupling of CoFe-LDH on MXene as high-performance oxygen evolution catalyst. *Mater. Today Energy* 12: 453–462.

42 Lei, J., Xiong, Y., Yu, F., and Ma, J. (2022). Flexible self-supporting CoFe-LDH/MXene film as a chloride ions storage electrode in capacitive deionization. *Chem. Eng. J.* 437: 135381.

43 Hu, L., Li, M., Wei, X. et al. (2020). Modulating interfacial electronic structure of CoNi LDH nanosheets with $Ti_3C_2T_x$ MXene for enhancing water oxidation catalysis. *Chem. Eng. J.* 398: 125605.

44 Chen, W., Han, B., Xie, Y. et al. (2020). Ultrathin Co–Co LDHs nanosheets assembled vertically on MXene: 3D nanoarrays for boosted visible-light-driven CO_2 reduction. *Chem. Eng. J.* 391: 123519.

45 Wang, T., Yang, Y., Deng, Q. et al. (2021). In-situ construction of 3D marigold-like CoAl-LDH/Ti_3C_2 heterosystem collaborating with 2D/2D interface for efficient photodegradation of multiple antibiotics. *Appl. Surf. Sci.* 569: 151084.

46 Ma, Y., Xu, D., Chen, W. et al. (2022). Oxygen-vacancy-embedded 2D/2D NiFe-LDH/MXene Schottky heterojunction for boosted photodegradation of norfloxacin. *Appl. Surf. Sci.* 572: 151432.

9

Titanium Carbide MXene-Based Perovskites Composites for Energy and Environment Applications

Mohamed Madi[2], Muhammad Tahir[1], and Naveen Kumar[3]

[1] *UAE University, Chemical and Petroleum Engineering Department, P.O. Box 15551, Sheik Khalifa Bin Zayed Street, Asharej, Al Ain, United Arab Emirates*
[2] *Universiti Teknologi Malaysia, School of Chemical and Energy Engineering, Faculty of Engineering, 81310, UTM, Johor Bahru, Johor, Malaysia*
[3] *SRM Institute of Science and Technology, Department of Physics and Nanotechnology, Kattankulathur, Chennai, 603203, Tamil Nadu, India*

9.1 Introduction

Two fundamental stumbling barriers to human social progress, global warming and energy scarcity, are progressively causing people to experience crisis [1, 2]. CO_2 is a major contributor to climate change and global warming produced by fossil-fuel burning as well as several other human activities [3]. This is why the safe elimination and utilization of CO_2 is critical for the preservation of the Earth's natural resources. Using photocatalysis to remove CO_2 from the atmosphere might meet present energy needs and alleviate environmental concerns via converting CO_2 to chemicals and solar fuels [4]. Furthermore, the photocatalytic conversion of CO_2 by the use of light irradiation, a renewable and beneficial method, has garnered significant interest because it is conducted at normal pressure and temperature and without the need for substantial energy input. More importantly, photocatalysis is a low-energy process that utilizes an infinite source of light energy [5]. Moreover, secondary pollution caused by photocatalytic materials is also extremely low [6].

High selectivity and faster reaction rates can be achieved by fine-tuning CO_2 conversion processes through the careful selection of catalyst materials. Perovskites are an especially interesting class of photocatalysts since they can be synthesized under benign circumstances and a broad range of catalytic properties can be controlled. Perovskites have broad-spectrum activity, enabling their use for a variety of purposes, including the production of value-added compounds for expanding chemical feedstocks and lowering atmospheric CO_2 emissions [7]. Perovskites, on the other hand, still suffer from their high electron–hole recombination and limited specific surface area [8]. As a result, a variety of techniques have been devised to increase the photocatalytic efficiency of perovskite. These strategies include the deposition of noble metals [9], adjustment of bandgap [10], and modulation of morphology [11].

Titanium Carbide MXenes: Synthesis, Characterization, Energy, and Environmental Applications, First Edition. Edited by Muhammad Tahir.

However, most of the strategies are costly and inefficient, rendering this process unaffordable [12].

Recently, low-cost structural materials are considered one of the best solutions that have been studied as co-catalysts for trapping and transporting photoinduced charge carriers with increased CO_2 conversion efficiencies under the visible light regions for the application of photocatalysis. In this regard, the broad family of MXenes, because of their exceptional conductive and photothermal properties, is one of the fastest-growing research domains among all existing 2D materials. Moreover, they are useful in a variety of industries due to their great proficiency in converting light to heat energy, ranging from energy storage materials to co-catalysts for semiconductors [13]. The use of MXene-based semiconductors as a mediator or heterojunction during the CO_2 reduction process has been accepted as a viable strategy for promoting charge separation and enhancing the photostability of the final product. More interestingly, it is possible to increase light absorption with improved photoactivity and stability by combining MXene multilayer materials with perovskite [14]. As a result, it was anticipated that the creation of a heterojunction with perovskite would enhance the photocatalytic efficiency of MXene and perovskite. As a result, the overall photocatalytic activity and charge separation is enhanced and the recombination rate is reduced as well.

This chapter will go through the fundamentals and mechanism of MXenes as a co-catalyst for the photoconversion of CO_2 into solar fuels. Furthermore, this chapter covers the combination of Ti_3C_2 with perovskites for improved photocatalytic efficiency. The applications of MXene-based perovskite into solar fuels have also been reviewed. Finally, we discuss the persistent obstacles and future directions for the development of MXenes as CO_2 reduction catalysts.

9.2 Properties and Application of Perovskite

Since the discovery of calcium titanate ($CaTiO_3$) by a Russian scientist in 1839, the term "perovskite" has been used to refer to a group of minerals with structural similarities to $CaTiO_3$ and a formula of ABX_3. After Miyasaka revealed solar cells made of perovskite materials in 2012, research on perovskite's use in solar energy conversion quickly increased. The general formula for perovskite is ANX_3, where the B site represents transition metals such as gold (Au), niobium (Nb), cobalt (Co), and ruthenium, and the A site contains monovalent and divalent cations with larger radii, such as cesium (Cs), barium (Ba), rubidium (Rb), serium (Sr), sodium (Na), and lanthanum (La) (Ru). The nonmetal elements bromine (Br), nitrogen (N), oxygen (O), and chlorine (Cl) are found in the X site of perovskites, which are divided into oxide perovskites, halide perovskites, and their derivatives.

The outer p orbitals of the B-site, which form the conduction band (CB), and the orbitals of the X-site, which make the valence band (VB), are where perovskites get their semiconductor-like characteristics. The atoms at the A site do not impact the energy band, but they do contribute to the stabilization of the perovskite lattice. On the other hand, it is possible to tune the bandgap and edges by partially replacing the atoms at the B site and changing the ions at the X site. This may result in perovskites having strong catalytic capabilities and good light utilization [15]. Perovskite oxides exhibit a greater adsorption coefficient than other oxide materials because of the ability of perovskite to contain a lot of electrons in the VB states. Another structural characteristic of these oxides is their ability to provide electron hole pairs with extended diffusion lengths. Based on defects in their lattice and tolerance factor, perovskites exhibit a wide range of compositional and structural diversity. Because the bulk and surface components of these materials are flexible, it is possible to produce perovskites with the necessary energy bands and surface areas. This favors the excitation of electrons for water splitting [16].

On account of their high catalytic activity, electronic conductivity, and O_2 vacancies, electrocatalysis is the predominant field of catalysis employing perovskite oxides as catalysts. However, a few perovskites have also been shown to generate solar fuels through photo catalysis. To present, the majority of perovskite materials published in the scientific literature have demonstrated multiple technological uses, excellent stability, and structural property changes. It is somewhat remarkable that the structure of perovskites can be adjusted to suit specific applications; for instance, the distorted form and reduced symmetry of some perovskites are significant due to their intriguing electric and magnetic properties.

9.3 Principle of Photocatalysis Using MXene/Perovskite Composite

MXene material use in photocatalysis was initiated in 2011 when Ti_3C_2 was exfoliated from Ti_3AlC_2 [17]. MXenes are a new type of two-dimensional layered materials composed of transition metal carbides, nitrides, and carbonitrides. In nature, there is no direct precursor for the MXene material; hence, the material is normally formed during the MAX phase. The general formula for the MAX phase is $M_{n+1}AX_n$, where n might be 1, 2, or 3. X in the formulation represents the C or N element, as well as the early transition metal element, IIIA or IVA [18]. Figure 9.1 demonstrates comprehensive structural perspectives of MXene and its composition. Layer A is less stable than layers M–X in the MAX structure. Regarding the layer structure of Ti_3AlC_2, it was possible to etch the Al layer in the solutions while leaving the Ti_3C_2. A huge number of MXenes may be synthesized using the same approach, thanks to the adaptability of the reactant precursors. More than 30 different kinds of MXenes have been discovered experimentally so far, and theoretical predictions indicate that there will be many more.

The bare MXenes cannot be employed as a photocatalyst since they are metallic materials and lack semiconductor characteristics [20]. Typically, metallic MXenes are utilized as excellent co-catalysts because of their excellent electrical conductivity, the amount of the interlayer space of the Mxene material to adsorb gas molecules, their lower Fermi level, and their termination groups [21]. Since MXenes are conductors and serve as co-catalysts, the mechanism of action of an MXene-based photocatalytic system is through accelerated charge separation and suppression of carrier

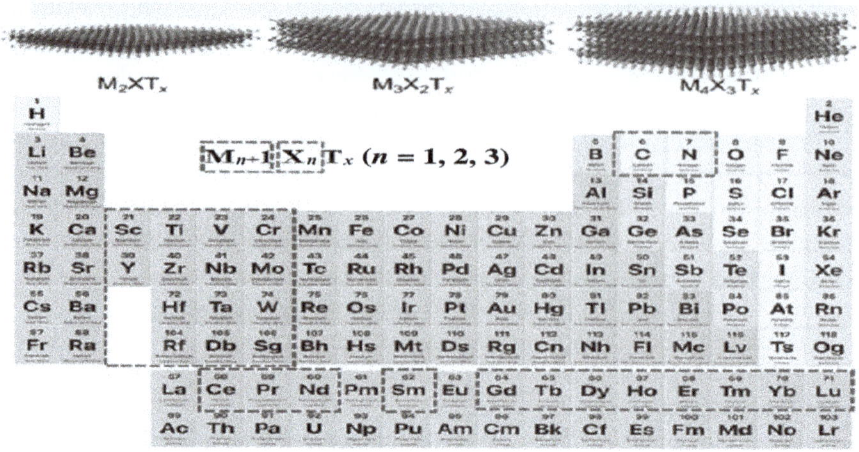

Figure 9.1 General structural classification of MXene with the compositional formula in which M stands for transitional metal elements, X consists of C and N, while T stands for surface termination groups. Source: Reproduced with permission from reference [19]. Copyright 2021 ACS Applied Energy Materials.

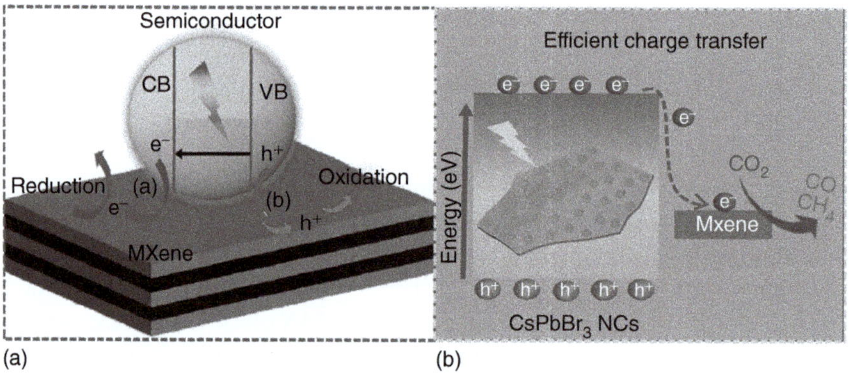

Figure 9.2 (a) Schematic of the working mechanism of MXenes applied in photocatalysis. Source: Sun et al. [22]/Springer Nature / CC BY 4.0, (b) diagram of the $CsPbBr_3$ NCs/MXene-n heterostructures suitable for photocatalytic CO_2 reduction. Source: Reproduced with permission from reference [23]. Copyright 2019 The Journal of Physical Chemistry Letters.

recombination, as shown in Figure 9.2a [24]. The photocatalysts absorb visible light, and photogenerated electrons are excited to the CB, while holes are left in the VB. The excited charge carriers are transferred to MXenes at the interface mainly because of the higher potential of MXenes. Electrons transfer to MXenes without recombination and react on the MXene surface to generate H_2 by reducing H^+ [25] and CH_4 and CO by reducing CO_2 [26], as shown in Figure 9.2b. The holes transfer to Mxenes and react to produce $OH^•$ that can be utilized for degradation of organics; electrons can also produce $OH^•$ for photocatalytic CO_2 reduction [27]. The charge transfer process from the photocatalyst to MXenes improves electron–hole pair separation and suppresses charge recombination in photo catalysts, thus enhancing the photoactivity.

Low-dimensional materials like MXene with perovskite can lead to enhanced optoelectronic capabilities, where quick separation and transfer of photogenerated electrons occur to result in enhanced photoreceptors and photocatalytic CO_2 reduction through photoresponse enhancement. Therefore, an efficient charge transfer is made possible, for example, by the relative energy level alignment between the $CsPbBr_3$ perovskite and MXene nanosheets [23]. The conduction band offset (1.5 eV) between the two components can effectively promote the separation of photogenerated excitons in $CsPbBr_3$ NCs and the transport of electrons to MXene nanosheets, as demonstrated in Figure 9.2b. These nanocomposites have also shown strong photocatalytic activity for selective reduction of CO_2 to CO and CH_4 under visible light. However, the design principles for MXene-based photocatalysts are still in their infancy because of a lack of understanding of the surface chemistry and active sites involved in photocatalytic processes. Thus, researchers must perform additional studies and development in terms of reducing charge recombination, boosting charge transfer rate, and enhancing photocatalytic conversion of CO_2 to value-added chemicals and fuels in order to overcome these obstacles.

9.4 Applications of TiC MXene-Based Perovskite Composite for CO_2 Reduction

Photocatalytic CO_2 reduction and finally conversion to hydrocarbon fuels under solar illumination mirrors the natural photosynthesis method and is one of the sustainable sources for addressing the energy requirements and mitigating the environmental effects of global warming. Many studies have been done to prove that adding MXenes as cocatalysts to photocatalytic CO_2 reduction is effective. In photocatalytic CO_2 reduction applications, nanocomposites based on Ti_3C_2 have shown excellent performance. These composites were made by combining several semiconductors, such as perovskites that are compatible with Ti_3C_2, to produce a composite with enhanced performance that is highly interconnected. Metal halide perovskite with a sufficient energy band structure is a viable photocatalyst for photocatalytic CO_2 reduction, with excellent visible-light responses. $FAPbBr_3/Ti_3C_2$ composite with a Schottky heterojunction for photocatalytic CO_2 reduction was found to have an impressive photocatalytic performance, according to Que. and colleagues. The production of CO for $FAPbBr_3/Ti_3C_2$ increased significantly, with the greatest yield of 283 441 $\mu mol\, g^{-1}\, h^{-1}$, which was 1.84 times that of $FAPbBr_3$. This excellent performance can be attributed to the tightly anchored $FAPbBr_3$ quantum dots on $Ti_3C_2T_x$ nanosheets via the Schottky heterojunction, as well as $Ti_3C_2T_x$ MXene nanosheets serving as electron acceptors and providing catalytic sites for boosting photocatalytic CO_2 reduction [28].

It was also previously reported that $Cs_2AgBiBr_6$ NCs were able to self-assemble on MXene nanosheets via mutual Coulomb electrostatic interaction for the first time [29]. The existence of MXene nanosheets decreases the exciton binding energy of $Cs_2AgBiBr_6$ NCs while simultaneously promoting the formation of free charge carriers. Due to suppressed carrier recombination, the ultrafast photogenerated electron transfer from $Cs_2AgBiBr_6$ to MXene with a time scale of 1.1 ps significantly increases the charge carrier lifetime by twofold. Using a simple solution growth method, astudy has demonstrated for the first time the successful fabrication of a $MAPbBr_3$ perovskite/MXene heterostructure by depositing $MAPbBr_3$ perovskite nanocrystals on thin layers of $Ti_3C_2T_x$ MXene nanosheets [30]. An ideal mixed-solvent ratio not only assures the remarkable

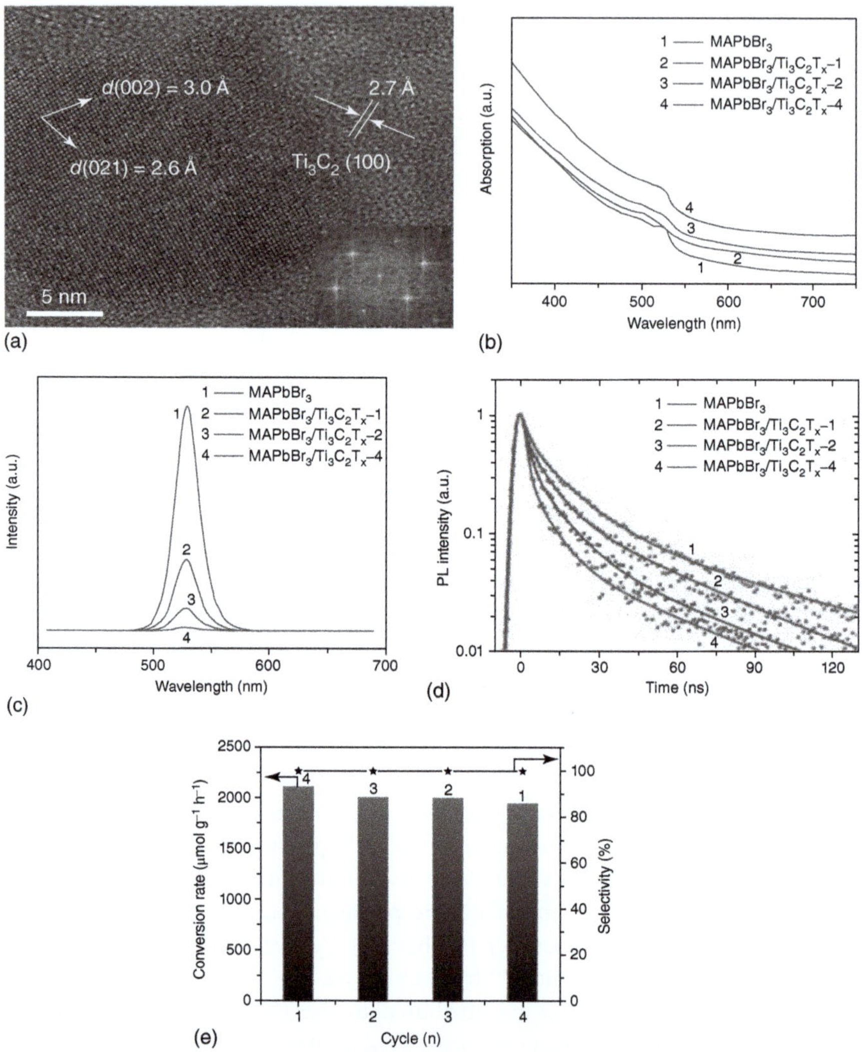

Figure 9.3 (a) HRTEM images of heterostructures of $MAPbBr_3/Ti_3C_2T_{x-1}$, and (b) UV–vis absorption of $MAPbBr_3/Ti_3C_2T_x$ heterostructures, and (c) PL spectra of heterostructures of $MAPbBr_3/Ti_3C_2T_{x-1}$, and (d) TRPL spectra of the $MAPbBr_3$ NCs and $MAPbBr_3/Ti_3C_2$ $MAPbBr_3/Ti_3C_2T_x$ heterostructures with different content of $Ti_3C_2T_x$ nanosheets. Source: (a–d) Zhang et al. [30]/John Wiley & Sons, and (e) cycle stability by $Cs_3Bi_2Br_9$/MXene-7.5 under visible-light illumination. Source: Li et al. [31]/with permission of American Chemical Society.

dispersity of $Ti_3C_2T_x$ nanosheets but also possesses a low solvent polarity for the good crystallization of perovskite nanocrystals, enabling the growth of perovskite nanocrystals on $Ti_3C_2T_x$ nanosheets, as shown by HRTEM images in Figure 9.3a. It is also shown in Figure 9.3b,c how $Ti_3C_2T_x$ nanosheets differ in amount between $MAPbBr_3$ and $Ti_3C_2T_x$ heterostructures, as well as how their UV–vis and PL spectra are enhanced compared to those of $MAPbBr_3$ NCs. Time-resolved photoluminescence (TRPL) spectroscopy, as shown in Figure 9.3d, reveals the interfacial energy transfer behavior within $MAPbBr_3/Ti_3C_2T_x$ heterostructures. This could provide a beneficial route for the use of halide perovskite/2D material heterostructures, inspiring applications like photocatalysis. All-inorganic, bismuth-based, lead-free perovskite heterostructures can be fabricated in situ on monolayer $Ti_3C_2T_x$ MXene nanosheets, according to a separate study [31]. On MXene nanosheets, ligand-free $Cs_3Bi_2Br_9$ nanocrystals with uniform crystal size and high crystallinity were uniformly distributed, and an intimate interface contact with a strong interaction was formed. As depicted in Figure 9.3e, the incorporation of monolayer MXene nanosheets significantly improved the stability of $Cs_3Bi_2Br_9$ nanocrystals.

In general, it is expected that functionalized MXene will become a unique and adaptable photocatalysis material. To advance this application, more knowledge of the detailed bonding properties among pure MXene surface species, the regulation rules between surface metal atoms and surface functional groups, the interfacial interaction of MXenes and composite phases, and other related topics is required. This information will allow for a greater comprehension of the photocatalytic process and a greater comprehension of techniques for optimizing the efficiency of MXene-based CO_2 photocatalysis. The continual advancements in MXene synthesis, modification, characterization, and mechanism analysis are building the framework for CO_2 pollution management and energy conversion, despite the fact that there are still significant obstacles to overcome before large-scale applications can be realized.

9.5 Applications of TiC MXene-Based Perovskite Composite for Degradation

For photocatalytic degradation of pollutants over perovskite nanostructures, it is crucial that the photocatalyst efficiently absorbs light of proper wavelength, resulting in the formation of charge carriers. These charge carriers will transfer to the surface of the perovskite nanoparticles, where they will interact with the suitable electron donors/acceptors to produce the reactive oxygen species responsible for the oxidative degradation of pollutants. The photocatalyst's redox potentials govern the formation of these reactive species/radicals, primarily ·OH. In certain cases, photogenerated electrons or holes could directly partake in the redox removal of the pollutant species. It is not just about how much light is absorbed by the photocatalyst but also how well the photogenerated charge carriers can be separated from each other and the band potentials that are playing an important role in the degradation of pollutants. Furthermore, because photocatalysis is a surface process, the adsorption capability of the perovskite photocatalyst toward the pollutants is critical to ensuring adequate interaction for the degradation process. Various photocatalyst design strategies are being investigated in order to synthesize perovskite-based materials with improved catalytic properties while keeping these important photocatalyst aspects in mind. Such techniques include fine-tuning the photocatalyst's morphology, size, textural properties, and crystal structure [32], doping the perovskite with metal or non-metal ions and their respective combinations [33], NP deposition of noble metals [34], combining perovskite substances with carbon nanomaterials [35], and heterojunction configuration with additional semiconductors [36]. The ultimate objective of these strategies is to enhance the overall catalytic efficiency of perovskite materials.

The carbon-based 2D layered structures of recently reported well-synthesized 2D layered materials containing transition-metal elements of carbide and nitride, also known as MXenes, have garnered considerable attention [37]. MXenes could indeed enhance photocatalytic activity by boosting charge carrier transfer, acting as a robust support, limiting the photocatalyst size, and enhancing reactant adsorption. Moreover, surface terminations could indeed help modify the bandgap and work function of MXenes, so whether the efficiency of MXenes as cocatalysts allowed to serve as electron or hole reservoirs in a photocatalytic reaction is dependent on the surface-terminated groups [38]. As shown in Figure 9.4a, water treatment can be improved by using $BiFeO_3$-based nanostructures, such as Ti_3C_2-MXene/$BiFeO_3$ [39]. Ti_3C_2-MXene/ $BiFeO_3$ has been shown to degrade both colored and colorless organic pollutants in water. Under optimal conditions, 100% removal of Congo Red and acetophenone was accomplished in 42 and 150 minutes, respectively, in the presence of visible light [39]. Moreover, absorption spectra of $BiFeO_3$/MXene demonstrate the degradation efficiency of organic

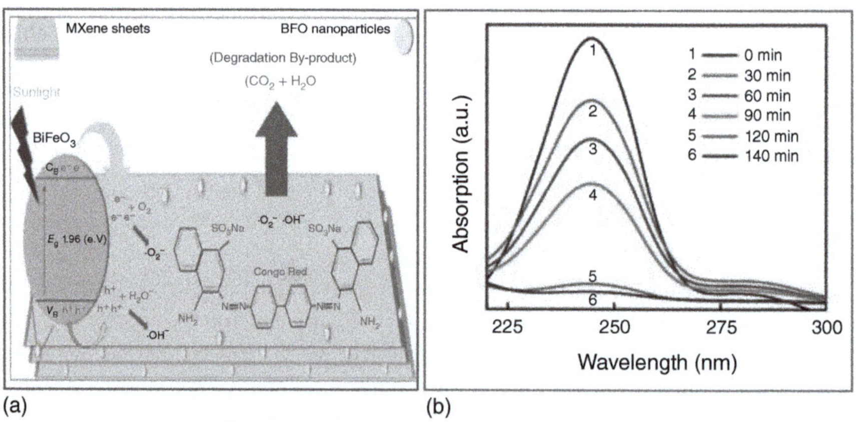

Figure 9.4 (a) Degrading organic dye molecules using $BiFeO_3$/MXene nanohybrids, and (b) biodegradation efficiency of organic pollutants in solution as shown by $BiFeO_3$/MXene absorbance spectra over time. Source: Iqbal et al. [39]/American Chemical Society. (c) Ion concentrations in seawater and wastewater before and after solar evaporation were measured. Source: Reproduced with permission from Fan et al. [40]. Copyright 2021 Applied Catalysis B: Environmental.

pollutants in solution at various times as shown in Figure 9.4b. This research demonstrated the composite material's ability to deal with various types of pollutants, which is essential for the sustainable use of such photocatalysts in water and wastewater treatment. In another study, inspired by the transpiration and natural photosynthesis of plants, the bifunctional characteristics of mixed-dimensional $Ti_3C_2T_x$(MXene)/$La_{0.5}Sr_{0.5}CoO_3$ (LSC) nano heterostructures to degrade pollutants via photochemical reactions and produce fresh water through the photothermal effect were investigated. MXene/LSC hybrid hydrogels were able to achieve highly efficient solar water evaporation due to their high solar absorption, rapid water transfer, and outstanding thermal insulation.

9.6 Prospects and Challenges

The current work mentioned in this chapter has revealed the outstanding potential of MXene composited with perovskite for improving photocatalytic degradation of pollutants, and photocatalytic CO_2 efficiently as well. However, research on MXene as a cocatalyst for the photocatalytic process is still in its infancy and a long way from practical application. Furthermore, more studies are required to fully investigate MXene's capabilities. There are still a lot of issues to be worked out before MXene can be used as an efficient co-catalyst for utilizing sunlight. To solve these issues, more study and MXene development must be undertaken by researchers in terms of synthesis techniques, etching agents, and storage conditions.

The photocatalytic process with MXene as a co-catalyst has been widely used; however, the mechanism remains a mystery. Theoretical simulations and calculations (e.g. first-principle density functional theory) are effective to understand the promotion effect of MXene on the photocatalytic reduction performance of CO_2. Until now, there are only a handful of pertinent reports. MXene was observed to compete with the conventional catalyst in terms of light absorption. Photocatalyst activity can be improved by combining several modification methods in order to achieve extremely efficient visible-light driven by MXene reactions. There are different methods available for improving the photocatalytic process, including modification, metal/nonmetal doping, vacancy engineering, and molecular engineering.

The preparation of MXene materials on a large scale remains a significant difficulty, and experimental proof of their enhancing effect in CO_2 photocatalysis as an example has been limited to a small number of materials, most notably Ti_3C_2 [41]. Therefore, the search for a technology that is flexible enough to synthesize MXene materials with precisely regulated compositions, characteristics, and surface functions must thus be pursued with vigor [42]. There have been recent reports of successful large-scale production of MXenes, which is likely to have a significant impact on the development of the field [43]. It would also be possible to boost photocatalytic activity even further by combining MXenes with active photocatalysts,the shapes and characteristics of which can be precisely controlled.

9.7 Conclusions

The Ti_3C_2 MXene has been the most extensively researched of the MXene family. Since MXenes have high electronic conductivity, a high specific surface area, a good band structure, a strong interfacial contact, and multiple hydrophilic groups on their surface, they can serve as co-catalysts in photocatalytic reactions. Perovskite semiconductors are also ideal candidates for different photocatalytic processes due to their outstanding properties such as regulated structure and morphologies, flexible band structure, regulated valence state, flexible oxidation state, and visible light sensitivity. As a result, MXene combined with perovskite is an excellent approach for increasing photocatalytic activity by extending the lifespan of photogenerated charge carriers by spatial separation of holes and electrons in the interfacial region. It is also a good idea to use MXenes as a support for light-harvesting and photothermal effect. It is also easy to enhance activation on MXene materials because of the abundance of adjustable surface functional groups, which are the active sites for photocatalytic processes. It is also easy to enhance CO_2 adsorption and activation on MXene materials because of the abundance of adjustable surface functional groups, which are the active sites for CO_2 reduction processes.

In CO_2 photocatalysis, selectivity is just as critical as activity and stability, if not more so. Therefore, MXene co-catalysts may play a critical role in determining product selectivity, which will require additional in-depth studies of the surface chemistry of CO_2 reduction processes on MXene-based photocatalysts. When combined with improvements in material preparation, photocatalytic CO_2 conversions using MXene-based perovskite have a promising future as a source of renewable solar fuel.

Acknowledgment

This work was supported by United Arab Emirates University (UAEU) under the research fund no. 12N097.

References

1 Qin, D., Zhou, Y., Wang, W. et al. (2020). Recent advances in two-dimensional nanomaterials for photocatalytic reduction of CO_2: insights into performance, theories and perspective. *J. Mater. Chem. A* 8 (37): 19156–19195.

2 Madi, M., Tahir, M., and Tasleem, S. (2021). Advances in structural modification of perovskite semiconductors for visible light assisted photocatalytic CO_2 reduction to renewable solar fuels: a review. *J. Environ. Chem. Eng.* 9 (5): 106264.

3 Nguyen, T.P., Nguyen, D.M.T., Le, H.K. et al. (2020). MXenes: applications in electrocatalytic, photocatalytic hydrogen evolution reaction and CO_2 reduction. *J. Mol. Catal.* 486: 110850.

4 Ikreedeegh, R.R. and Tahir, M. (2021). A critical review in recent developments of metal-organic-frameworks (MOFs) with band engineering alteration for photocatalytic CO_2 reduction to solar fuels. *J. CO2 Util.* 43: 101381.
5 Zhou, Y., Wang, W., Zhang, C. et al. (2020). Sustainable hydrogen production by molybdenum carbide-based efficient photocatalysts: from properties to mechanism. *Adv. Colloid Interface Sci.* 279: 102144.
6 Lei, L., Huang, D., Zhang, C. et al. (2020). F dopants triggered active sites in bifunctional cobalt sulfide@nickel foam toward electrocatalytic overall water splitting in neutral and alkaline media: experiments and theoretical calculations. *J. Catal.* 385: 129–139.
7 Tabish, A., Varghese, A.M., Wahab, M.A., and Karanikolos, G.N. (2020). Perovskites in the energy grid and CO_2 conversion: current context and future directions. *Catalysts* 10 (1): 95.
8 Xiang, S., Zhang, Z., Gong, C. et al. (2018). $LaFeO_3$ nanoparticle-coupled TiO_2 nanotube array composite with enhanced visible light photocatalytic activity. *Mater. Lett.* 216: 1–4.
9 Sun, X., Liu, G., and Xu, X. (2018). Defect management and efficient photocatalytic water oxidation reaction over Mg modified $SrNbO_2N$. *J. Mater. Chem. A* 6 (23): 10947–10957.
10 Gao, Y., Yang, G., Dai, Y. et al. (2020). Electrodeposited Co-substituted $LaFeO_3$ for enhancing the photoelectrochemical activity of $BiVO_4$. *ACS Appl. Mater. Interfaces* 12 (15): 17364–17375.
11 Thirumalairajan, S., Girija, K., Mastelaro, V.R., and Ponpandian, N. (2014). Photocatalytic degradation of organic dyes under visible light irradiation by floral-like $LaFeO_3$ nanostructures comprised of nanosheet petals. *New J. Chem.* 38 (11): 5480–5490.
12 Tahir, M., Ali Khan, A., Tasleem, S. et al. (2021). Titanium carbide (Ti_3C_2) MXene as a promising co-catalyst for photocatalytic CO_2 conversion to energy-efficient fuels: areview. *Energy Fuels* 35 (13): 10374–10404.
13 Lin, H., Gao, S., Dai, C. et al. (2017). A two-dimensional biodegradable niobium carbide (MXene) for photothermal tumor eradication in NIR-I and NIR-II biowindows. *J. Am. Chem. Soc.* 139 (45): 16235–16247.
14 Tasleem, S. and Tahir, M. (2020). Current trends in strategies to improve photocatalytic performance of perovskites materials for solar to hydrogen production. *Renewable Sustainable Energy Rev.* 132: 110073.
15 Tanaka, H. and Misono, M. (2001). Advances in designing perovskite catalysts. *Curr. Opin. Solid State Mater. Sci.* 5 (5): 381–387.
16 Bian, H., Li, D., Yan, J., and Liu, S. (2021). Perovskite – a wonder catalyst for solar hydrogen production. *J. Energy Chem.* 57: 325–340.
17 Naguib, M., Kurtoglu, M., Presser, V. et al. (2011). Two-dimensional nanocrystals produced by exfoliation of Ti_3AlC_2. *Adv. Mater. Processes* 23 (37): 4248–4253.
18 Lin, Y., Li, Y., Cao, Y., and Wang, X. (2021). Two-dimensional MOFs: design &synthesis and applications. *Chemistry* 16 (21): 3281–3298.

19 Sherryna, A. and Tahir, M. (2021). Role of Ti_3C_2 MXene as prominent Schottky barriers in driving hydrogen production through photoinduced water splitting: acomprehensive review. *ACS Appl. Energy Mater.* 4 (11): 11982–12006.

20 Peng, J., Chen, X., Ong, W.-J. et al. (2019). Surface and heterointerface engineering of 2D MXenes and their nanocomposites: insights into electro- and photocatalysis. *Chem* 5 (1): 18–50.

21 Wang, B., Wang, M., Liu, F. et al. (2020). Ti_3C_2: an ideal co-catalyst? *Angew. Chem.* 132 (5): 1930–1934.

22 Sun, Y., Meng, X., Dall'Agnese, Y. et al. (2019). 2D MXenes as Co-catalysts in photocatalysis: synthetic methods. *Nano-Micro Lett.* 11 (1): 79.

23 Pan, A., Ma, X., Huang, S. et al. (2019). $CsPbBr_3$ perovskite nanocrystal grown on MXene nanosheets for enhanced photoelectric detection and photocatalytic CO_2 reduction. *J. Phys. Chem. Lett.* 10 (21): 6590–6597.

24 Ye, M., Wang, X., Liu, E. et al. (2018). Boosting the photocatalytic activity of P25 for carbon dioxide reduction by using a surface-alkalinized titanium carbide MXene as cocatalyst. *ChemSusChem* 11 (10): 1606–1611.

25 Sun, Y., Jin, D., Sun, Y. et al. (2018). Wang, g-C_3N_4/$Ti_3C_2T_x$ (MXenes) composite with oxidized surface groups for efficient photocatalytic hydrogen evolution. *J. Mater. Chem. A* 6 (19): 9124–9131.

26 Yang, C., Tan, Q., Li, Q. et al. (2020). 2D/2D Ti_3C_2 MXene/g-C_3N_4 nanosheets heterojunction for high efficient CO_2 reduction photocatalyst: dual effects of urea. *Appl. Catal., B* 268: 118738.

27 Peng, C., Wang, H., Yu, H., and Peng, F. (2017). (111) TiO_{2-x}/Ti_3C_2: synergy of active facets, interfacial charge transfer and Ti^{3+} doping for enhance photocatalytic activity. *Mater. Res. Bull.* 89: 16–25.

28 Que, M., Zhao, Y., Yang, Y. et al. (2021). Anchoring of formamidinium lead bromide quantum dots on Ti_3C_2 nanosheets for efficient photocatalytic reduction of CO_2. *ACS Appl. Mater. Interfaces* 13 (5): 6180–6187.

29 Zhang, Z., Wang, B., Zhao, H.-B. et al. (2022). Self-assembled lead-free double perovskite-MXene heterostructure with efficient charge separation for photocatalytic CO_2 reduction. *Appl. Catal., B* 312: 121358.

30 Zhang, Z., Li, Y., Liang, C. et al. (2020). In situ growth of $MAPbBr_3$ nanocrystals on few-layer MXene nanosheets with efficient energy transfer. *Small* 16 (17): 1905896.

31 Li, Q., Song, T., Zhang, Y. et al. (2021). Boosting photocatalytic activity and stability of lead-free $Cs_3Bi_2Br_9$ perovskite nanocrystals via in situ growth on monolayer 2D $Ti_3C_2T_x$ MXene for C–H bond oxidation. *ACS Appl. Mater. Interfaces* 13 (23): 27323–27333.

32 Salavati-Niasari, M., Soofivand, F., Sobhani-Nasab, A. et al. (2016). Synthesis, characterization, and morphological control of $ZnTiO_3$ nanoparticles through sol-gel processes and its photocatalyst application. *Adv. Powder Technol.* 27 (5): 2066–2075.

33 Huang, X..-j., Yan, X., Wu, H.-y. et al. (2016). Preparation of Zr-doped $CaTiO_3$ with enhanced charge separation efficiency and photocatalytic activity. *Trans. Nonferrous Met. Soc. China* 26 (2): 464–471.

34 Ahn, C.W., Borse, P.H., Kim, J.H. et al. (2018). Effective charge separation in site-isolated Pt-nanodot deposited $PbTiO_3$ nanotube arrays for enhanced photoelectrochemical water splitting. *Appl. Catal., B* 224: 804–809.

35 Zhao, Y., Zhang, X., Liu, J. et al. (2018). Graphene oxide modified nano-sized $BaTiO_3$ as photocatalyst. *Ceram. Int.* 44 (13): 15929–15934.

36 Li, S., Hu, S., Jiang, W. et al. (2018). Hierarchical architectures of bismuth molybdate nanosheets onto nickel titanate nanofibers: facile synthesis and efficient photocatalytic removal of tetracycline hydrochloride. *J. Colloid Interface Sci.* 521: 42–49.

37 Gogotsi, Y. (2015). Transition metal carbides go 2D. *Nat. Mater.* 14 (11): 1079–1080.

38 Liu, X., Chen, T., Xue, Y. et al. (2022). Nanoarchitectonics of MXene/semiconductor heterojunctions toward artificial photosynthesis via photocatalytic CO_2 reduction. *Coord. Chem. Rev.* 459: 214440.

39 Iqbal, M.A., Tariq, A., Zaheer, A. et al. (2019). Ti_3C_2-MXene/bismuth ferrite nanohybrids for efficient degradation of organic dyes and colorless pollutants. *ACS Omega* 4 (24): 20530–20539.

40 Fan, D., Lu, Y., Zhang, H. et al. (2021). Synergy of photocatalysis and photothermal effect in integrated 0D perovskite oxide/2D MXene heterostructures for simultaneous water purification and solar steam generation. *Appl. Catal., B* 295: 120285.

41 Low, J., Zhang, L., Tong, T. et al. (2018). TiO_2/MXene Ti_3C_2 composite with excellent photocatalytic CO_2 reduction activity. *J. Catal.* 361: 255–266.

42 Hong Ng, V.M., Huang, H., Zhou, K. et al. (2017). Recent progress in layered transition metal carbides and/or nitrides (MXenes) and their composites: synthesis and applications. *J. Mater. Chem. A* 5 (7): 3039–3068.

43 Shuck, C.E. and Gogotsi, Y. (2020). Taking MXenes from the lab to commercial products. *Chem. Eng. J.* 401: 125786.

10

Titanium Carbide (Ti_3C_2) Based MXenes for Energy Storage Applications

Ebrima Ceesay[3]*, Shamaila Fatima*[3]*, Muhammad Z. Iqbal*[1,2]*, and Syed Rizwan*[3]

[1] *United Arab Emirates University (UAEU), Department of Chemical and Petroleum Engineering, Sheikh Khalifa Bin Zayed Street, PO Box 15551, Al Ain, United Arab Emirates*
[2] *United Arab Emirates University, National Water and Energy Center (NWEC), Sheikh Khalifa Bin Zayed Street, PO Box 15551, Al Ain, United Arab Emirates*
[3] *National University of Sciences and Technology (NUST), School of Natural Sciences, Department of Physics, Physics Characterization and Simulations Lab (PCSL), H-12, Islamabad 54000, Pakistan*

10.1 Introduction

Energy, a fundamental ingredient for sustainable development of our society and economy, is one of the most critical topics of research in the twenty-first century. With limitations offered by fossil fuels, there is enormous research going on to establish efficient, clean, and reliable energy sources as a supplement and substitute for fossil fuel soon.

With the popularization of electric automobiles and handy electronic devices, development and improvement of electrochemical energy storage devices (EESD) are needed to meet the required power and energy density needed. Also, harnessing energy from renewable energy sources, such as solar, wind, and hydro energy, greatly depends on whether (tide, wind, and sun) and these conditions are not continuous due to environmental weathering conditions. Hence, highly performing energy storage devices are required to save the energy generated [1].

Due to their high efficiency, versatility, and flexibility, electrochemical energy storage stands out to be better among the various alternative energy storage technologies. Batteries and supercapacitors, among EESD, are at the forefront as the dominant power sources for handy electronic devices like cell phones and laptops and exhibit promising capability to power electric vehicles and electric grids [2, 3]. Both batteries and supercapacitors store energy through electrochemical processes, consisting of an electrolyte, cathode, anode, and a separator as their main component. The batteries deliver high energy density but show limited cycle life and power density, whereas the supercapacitors provide high power density and very prolonged cycling life [4].

A highly efficient energy storage device is expected to store energy in large quantities, in small space, and release energy rapidly when needed. However, bringing

Titanium Carbide MXenes: Synthesis, Characterization, Energy, and Environmental Applications, First Edition. Edited by Muhammad Tahir.

highly efficient energy storage devices to fruition is an ultimate challenge and to date, researchers are putting enormous effort to bridge the gaps with regards to energy and power density of the existing electrochemical storage devices [5, 6].

The one most important component in electrochemical energy storage devices such as supercapacitor and batteries is the active electrode materials, and a great deal of consideration has been put forward recently in the cost of active electrode material, sustainability, environmental implications, energy density, as well as synthesis methods [7–9].

Transition metal carbides, nitrides or carbonitrides, generally called MXenes have been at the forefront of 2D materials since the discovery in 2011 as energy storage materials. They share common properties with other 2D materials but with more desirable quality and characteristics such as metallic conductivity, strength, flexibility, hydrophilicity, electronics, magnetism, optics, thermoelectric behaviors, corrosion resistance, and tunable surface chemistry [9, 10]. Their intrinsic structure and properties make them receive increasing attention and advantages of being employed in particular as an electrode material for electrochemical energy storage devices [10, 11].

This chapter focused on energy storage requirements, classification of MXene, its synthesis (titanium and vanadium carbide), general characterization of MXenes, thermal and cycling stability, and its application in electrochemical energy storage. In addition, the electrochemical storage devices will be discussed followed by conclusion.

10.2 Requirements for Energy Storage

Energy storage refers to the acquisition and utilization of the energy produced formerly for later use to check imbalances between production and utilization of energy. Energy storage technologies allow us to store extra energy and release it when generation or demand is insufficient. To satisfy the demands of today's energy needs, devices that store energy like secondary batteries and electrochemical capacitors are being developed, although the effectiveness of these devices depends on their efficiency, inexpensiveness, and stability of the electrode materials being utilized. Hence, the main challenge is the discovery of efficient production methods to fabricate electrode materials that will aid in meeting future energy demands. To fulfil this, devices that store energy (secondary batteries and electrochemical capacitors in particular) should meet standards and needs such as high power and energy density, stability, durability, cost-effectiveness in production, and safe and environmentally friendly.

Researchers continue to investigate new electrode materials in pursuit of a novel solution for energy storage systems. By using MXenes to fabricate electrode for an electrochemical capacitors, they were demonstrated to have exceptional performance [12]. This new member of the two-dimensional nanomaterial is an exceptional candidate for energy storage owing to its large laminar-structured spacing, superior conductivity, rapid ion and molecule transport, hydrophilic nature, easily available structure, cycling and thermal stability, adjustable thickness, and a large surface-to-volume ratio [13, 14]. Considering the preceding information, MXenes appear to be the most auspicious electrode material for batteries and supercapacitors.

10.3 Classification of MXenes

MXenes are ceramics that make up one of the major groups of two-dimensional (2D) materials to date. They were discovered in 2011 at Drexel University, USA [12]. MXenes are derived from bulk crystal called MAX. The MAX phase refers to an enormous family of hexagonally stacked ternary carbides/nitrides [15]. They consist of M_2AX, M_3AX_2, M_4AX_3 and so on, where

M = transition elements like Ti, Nb, V, Mo, W, and Ta
A = Elements of groups III and IV such as Al, Si, and Ga
X = carbon C, nitrogen N, or both (carbonitrides C/N) [16, 17].

Through wet chemical etching using mostly acidic-, fluoride-containing etchants, MXenes are synthesized by cautiously eliminating the A element ($M_{n+1}AX_n$) in the parent MAX. This method of synthesizing MXenes results in the functionalization of surfaces with –O, –F, or –OH groups indicated as T_x in $M_{n+1}X_nT_x$.

There are two major types of transition-metal MXenes, namely, the single (mono) transition-metal MXenes and the double transition metal MXenes.

10.3.1 Mono Transition-Metal MXenes

They consist of a single transition metal. Using the generic formulation $M_{n+1}X_nT_x$, where n could be 1, 2, or 3, resulted three common structures M_2X (Ti_2C, V_2C, Nb_2C, Mo_2C) [18–20], M_3X_2 (Ti_3C_2, Zr_3C_2, Hf_3C_2) [21–23], and M_4X_3 (Nb_4C_3, Ta_4C_3, Ti_4N_3) [24–27]. MXene from M_2XT_x to $M_3X_2T_x$ and $M_4X_3T_x$ by changing n changes the thickness of the 2D sheets, which is in the range of 1 nm. These MXenes consist of three atomic configurations, ranging from M_2X to M_3X_2 and M_4X_3, enabling the discovery and modeling of materials based on specific requirements [28, 29]. In 2011, the first MXene Ti_3C_2 was derived from Ti_3AlC_2 [12].

In MXenes, the atoms are organized in a hexagonal pattern, with the “X” atoms stuffing in the octahedral spaces between the M atoms. When it comes to the M_2X phase, the stacking is AB (hexagonal) and ABC (face-centered cubic), but in M_3X_2 and M_4X_3, it is both AB and ABC. MXenes with a higher “n” index are less stable than their n counterparts because of this difference in stacking. However, Mo_2C is more stable than Mo_4C_3, which is less stable than Mo_3C_2, according to Ref. [30].

10.3.2 Double Transition n-metal (DTM) MXenes

Double transition-metal MXenes comprises two layers of transition metals represented by M′ and M″ as compared to the mono transition-metal MXene which is made of M layers of a single transition metal. Structurally, double transition-metal MXenes are further divided into two types: solid-solution MXenes and ordered double transition-metal MXenes as described below:

10.3.2.1 Solid-solution MXenes

The two metal atoms at the M site are arbitrarily distributed in solid-solution MXenes using $M'_{2-n}M''_nXT_x$ where $n = 0.4, 0.8, 1.2$ and 1.6 [31]. In 2021, niobium containing double-metal solid-solution MXenes ($Ti_{2-y}Nb_yT_x$ and $V_{2-y}Nb_yT_x$; $0 < y < 2$) were synthesized and investigated by W. Likui et al. thus opening doors and providing a guide for adjusting the electrochemical properties of MXenes by turning the M-site chemistry [32].

10.3.2.2 Ordered DTMs MXenes

Two distinct types of ordering exist in ordered DTMs (double transition metal), that is out-of-plane ordering and in-plane ordering, depending on the site of the M′ and M″ transition metal in the structure.

Out-of-Plane MXene (o-MXene): Structure and Composition Out-of-plane MXenes are represented by the formula $M'_2M''X_2T_x$ or $M'_2M''_2X_3T_x$, where M′ and M″ are distinct transition metals. The o-MXene structure consists of layers of transition metals

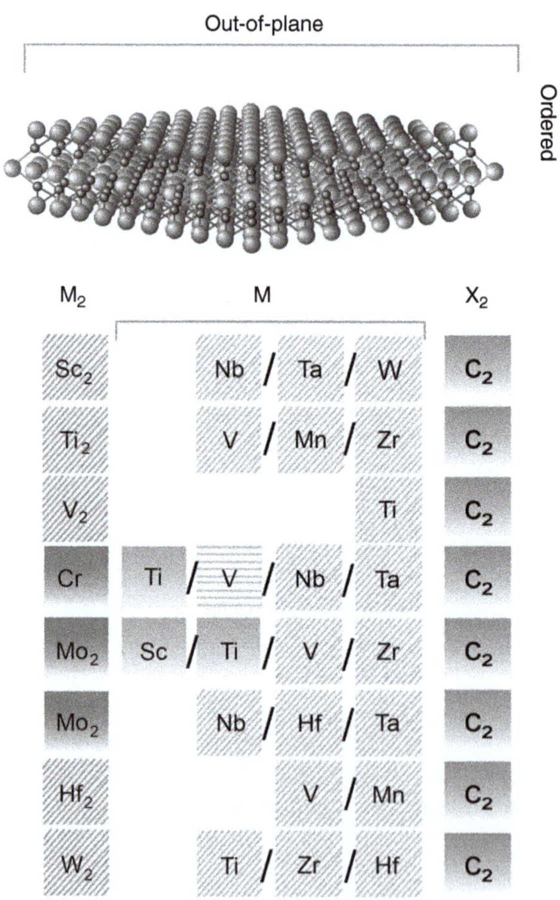

Figure 10.1 Structure of out-of-plane MXene. Source: Reproduced with permission from Iqbal et al. [33].

arranged in different atomic planes with the M″ transition metal packed between layers of M′ transition metals (Figure 10.1). Chemically ordered double transitional metal carbides in the form $M'_2M''_2C$ and $M'_2M''_2C_3$ such as Mo_2TiC_2, Cr_2TiC_2, $Mo_2Ti_2C_2$, $Mo_2Ti_2C_3$ and Mo_2ScC_2 [30, 34–39] were derived from the parent o-MAX phases . Liu et al. in 2014 reported the first o-MAX phase in the alloying of M elements in Cr_2TiAlC_2 [40]. o-MAX phases allow the existence of MXenes with surface layers like Cr (Cr_2TiAlC_2) which do not exist in 2D flakes like Cr_2Alc as the outer layer (Cr) will dissolve completely during etching [30].

In Plane MXene (i-MXene): Structure and Composition In-plane ordered MXenes are represented by the formula $M'_{4/3}M''_{2/3}XT_x$. The i-MXene structure consists of two different transition metals M′ and M″ situated in alternating sites in each M-layer of the atomic plane (Figure 10.2). The i-MXenes were discovered during the formation of o-MAX phase Mo_2ScAlC_2 by Refs. [41, 42]. This phase is mostly explained through a monoclinic ($C_{2/c}$) structure due to the presence of monoclinic distortion in the layers [43]. It has been noticed that for an i-MAX phase to be formed, the size of the M″ transition metal should be larger than M′ and the composition should be a 2 : 1 for M′:M″ [44]. Vacancy-ordered MXene from chemically ordered i-MAX phases $(Mo_{2/3}Sc_{1/3})_2AlC$, $(Mo_{2/3}Y_{1/3})_2AlC$, $(W_{2/3}Sc_{1/3})_2AlC$, and $(W_{2/3}Y_{1/3})_2AlC$ includes $Mo_{1.33}C$ and $W_{1.33}C$ MXene [45].

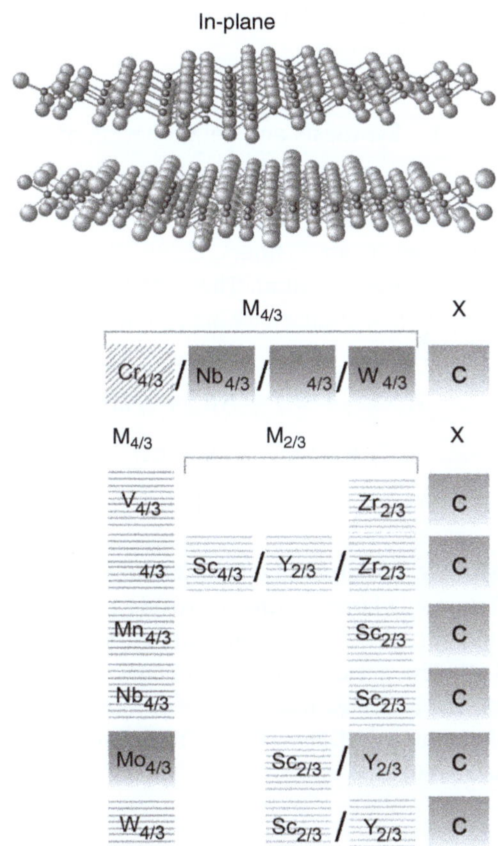

Figure 10.2 Structure of in-plane MXene. Source: Reproduced with permission from Iqbal et al. [33].

10.4 Synthesis of Titanium and Vanadium Carbide MXenes

MXene is produced by exfoliating the three-dimensional (3D) MAX phases. The route for MXene synthesis depends on several variables, including etching time, particle size of respective powder's MAX phase, temperature, and etchant used. Researchers modify these factors to increase yield, reduce etching time, and improve the quality of MXenes.

There are several types of MXenes synthesized using HF only as an etchant [12, 19, 21, 23, 30, 46–48] where so far, HF is the most effectively studied and researched etchant used for MXene synthesis. Other etchants including LiF, NaF, KF, and NH_4F in HCl (fluoride salts in HCl) [49–53], molten salt etchants [26, 54, 55], alkaline extraction [56, 57], halogenation[58, 59], and oxidant/ HF etchant were also used to synthesize MXenes ($Ti_3C_2T_x$) fromTi_3SiC_2 [60].

Delamination of multi-layer MXenes can be carried using the wet power (for a higher yield) or using the dried power. The delamination is required to produce MXene sheets which are easily dispersible in water and to increase the surface area of MXene particles. Each delamination protocol results in colloidal dispersions with varying yields and different properties when organic or inorganic intercalants, solvents, or mechanical force are used. Intercalants such as dimethyl sulfoxide (DMSO), tetrabutylammonium hydroxide (TBAOH), or tetramethylammonium hydroxide (TMAOH) can be used to delaminate multi-layer MXenes to few or single-layer MXene.

10.4.1 Synthesis and delamination of $Ti_3C_2T_x$

10.4.1.1 Synthesis of $Ti_3C_2T_x$

The Ti_3SiC_2 MAX powder is dispersed in H_2O_2/HF solution under magnetic stirring at 40 °C for 45 hours. Using DI water, the sediment is washed several times until the pH is neutral. This is followed by filtration and then dried in a vacuum oven [60]. A similar etching protocol is use for Ti_3AlC_2 MAX powder but in HF only [12, 14].

$$Ti_3AlC_2 + 3HF = AlF_3 + 3/2H_2 + Ti_3C_2 \quad (10.1)$$

$$Ti_3C_2 + 2H_2O = Ti_3C_2(OH)_2 + H_2 \quad (10.2)$$

$$Ti_3C_2 + 2HF = Ti_3C_2F_2 + H_2 \quad (10.3)$$

10.4.1.2 Delamination of the Multi-layered $Ti_3C_2T_x$ (ml-$Ti_3C_2T_x$)

Tetramethylammonium hydroxide (TMAOH) is added to multilayered $Ti_3C_2T_x$ and stirred for a moment, diluted in DI water, and allowed to stir for 24 hours [14]. The basic solution is brought to neutral by washing in DI water multiple times via centrifugation. This is followed by vacuum-assisted filtration to obtain delaminated MXene (Figure 10.3) and dried at ambient temperature in a vacuum oven [60].

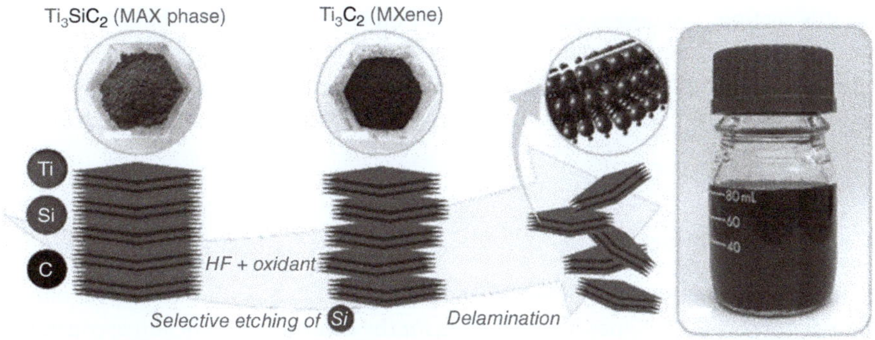

Figure 10.3 Showing the synthesis and delamination of ml-$Ti_3C_2T_x$. Source: Adapted with permission from Alhabeb et al. [60]. John Wiley & Sons.

10.4.2 Synthesis and Delamination of V_2CT_x

10.4.2.1 Synthesis of V_2CT_x

Through wet chemical etching, V_2AlC MAX powder is dispersed gradually in an aqueous solution of HF while stirring, and the reaction continues in ambient temperature for 90 hours. The resulting acidic suspensions are washed several times using deionized water to neutralize the pH. This is followed by vacuum-assisted filtration and later dried at ambient temperature in a vacuum oven to obtain multilayer ml-V_2CT_x [19, 61].

10.4.2.2 Delamination of V_2CT_x

Freshly etched or dried powder ml-V_2CT_x powder is added TMAOH solution and allow to stir for six hours at room temperature. Afterward, after being dispersed in DI water, the solution is centrifuged at 3500 rpm for 10 minutes, and the first supernatant sediment dispersed in DI water is stirred for few minutes and centrifuged to collect the few layers of V_2CT_x, followed by drying at room temperature in a vacuum oven to obtain d-V_2CT_x [18].

Other successfully synthesized MXenes include Nb_2CT_x [19], Ti_2NT_x [62],$Ta_4C_3T_x$ and Ti_3CNT_x [46], $Mo_2TiC_2T_x$, $Mo_2Ti_2C_3T_x$ and $Cr_2TiC_2T_x$ [30], $Mo_{4/3}CT_x$ [43], and $Zr_3C_2T_x$ [48].

10.5 Typical Characterization of MXenes

One of the distinctive superiorities that MXenes possess over nanomaterials is their usefulness in terms of chemical composition, surface chemistry after etching, and other properties that can be easily manipulated and modified for specific applications. Each stage in MXene synthesis and processing impacts their properties and to take full advantage of MXenes, advanced characterization techniques should be applied to examine the morphology, composition, structure, and other properties of the parent MAX phase and etched MXenes.

In this section, discussion will be centered on common MXene characterization approaches at various stages, using techniques such as Raman Spectroscopy (RAMAN), X-Ray Diffraction (XRD), X-Ray Photoelectron Spectroscopy (XPS), Scanning Electron Microscopy (SEM) and Energy Dispersive X-ray Spectroscopy (EDS).

10.5.1 X-ray Diffraction

XRD is used for phase determination and structural investigation of several nanomaterials, especially MXenes. It is also used for the determination of lattice constant, interplanar distance, size of the crystal using Scherer formula, the volume's unit cell, and several other parameters related to material structure using the crystallographic peaks (position, width, and intensity) that are present. XRD augments our knowledge of MXenes and how we can further customize their properties for diverse applications. Fundamentally, XRD is used to confirm whether MXene has been successfully synthesized from the MAX phase.

In general, using Miller indices (h, k, l), MXene indices are listed as (002), (004), (006), etc., for multilayer MXene since they are made from the parent MAX phases (M_3AX_2).However, for single flakes, the alignment becomes random and to get the indices, one should divide the c-lattice parameter by 2 to get the d-spacing of single-layer MXene [63].

The crystalline structure of the Ti_3AlC_2 MAX phase is shown in (Figure 10.4) by diffraction peaks at 9.6°, 38.8°, 41.5°, 48.3°, 52.1°, 56.2°, 59.9°, 65.3°, 70.2°, and 73.6°. For the etched (ex-$Ti_3C_2T_x$) and delaminated (d-$Ti_3C_2T_x$), there has been a considerable decrease in the intensity of the peak at 38.8° as compared to Ti_3AlC_2, indicating the removal of the A layer (Al or Si) during the etching process [64]. The

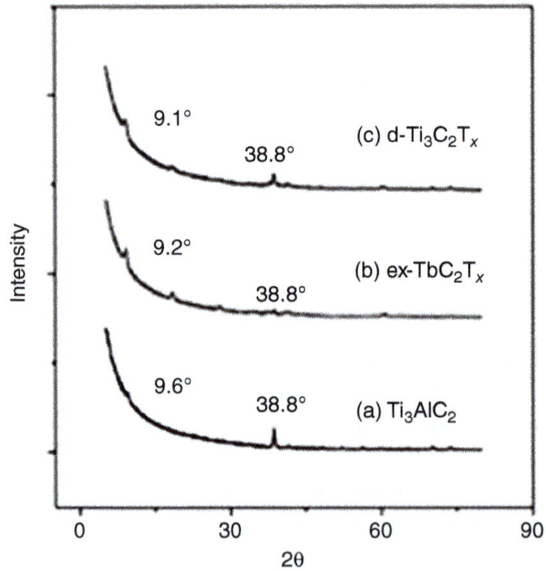

Figure 10.4 XRD of (a) Ti_3AlC_2, (b) exfoliated $Ti_3C_2T_x$ (ex-$Ti_3C_2T_x$), and (c) delaminated $Ti_3C_2T_x$ (d-$Ti_3C_2T_x$) [64]. Source: Reproduced with permission from Lim et al. [64].

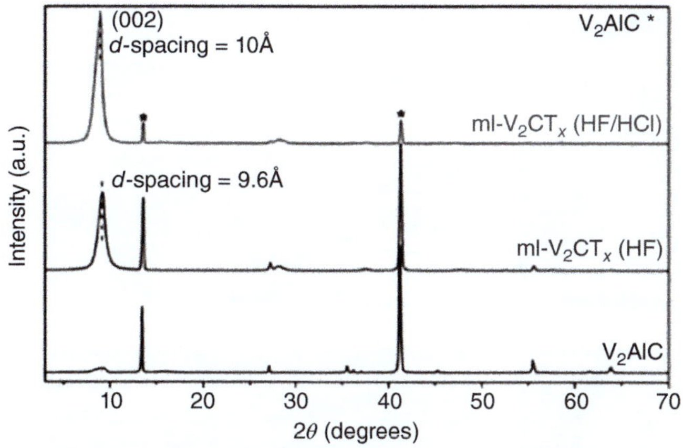

Figure 10.5 XRD patterns showing V_2AlC MAX and multilayer V_2CT_x via HF and HF/HCl etching. Source: With permission of Matthews et al. [18]/American Chemical Society.

basal plane peak in the d-$Ti_3C_2T_x$ nanosheets was found to have been moved to a lower 2θ angle (9.1°), shifting from 9.6° in the Ti_3AlC_2 MAX phase to which corresponds to a d-spacing of 9.68 Å. Also, the removal of Al from Ti_3AlC_2 and the addition of surface terminations (–F and –OH functional groups) created a lower peak shift of the basal plane.

XRD can also be used to check the effects of the etchant and etching time on synthesized MXene. In their work, Kyle Matthews et al. in 2022 [18], studied the impacts of the etching time of V_2AlC in hydrofluoric acid and HF/HCl solutions and ascertained the minimum time required for the most complete etching of V_2AlC into V_2CT_x without inherent oxidation. As shown in Figure 10.5, the HF/HCl and HF etching results to a shift of (002) peak to 8.98° and 9.22° from 13.44° in V_2AlC, respectively. This is attributed to the successful removal of Al from V_2AlC and subsequent expansion of the $d_{(002)}$-spacing from 6.6 to 10 Å and 9.6 Å. The peak analysis also indicates that the mixed etchant has better intensity of 002 peak and a greater d spacing showing Al from V_2AlC being completely etched.

However, other features like oxide nanoparticles that are usually outside the layers of MXenes due to severe etching and oxidation cannot be detected using XRD [63].

10.5.2 Scanning Electron Microscopy (SEM)/Energy-Dispersive X-ray Spectroscopy (EDS)

Using a focused beam of high energy electrons, SEM can disclose information regarding the MAX and MXene, like surface morphology, crystalline structure, and alignment of materials within the sample. Combined with EDS, SEM can also give elemental composition of the MAX and MXenes, thus making it an important characterization technique.

Ti_3C_2 MXene was produced by selective removal of (etching) Si from MAX powders (Ti_3SiC_2) utilizing HF and oxidants/HF etchants. SEM images of Ti_3SiC_2 (Figure 10.6a) show a closely compacted layered structure before as compared to

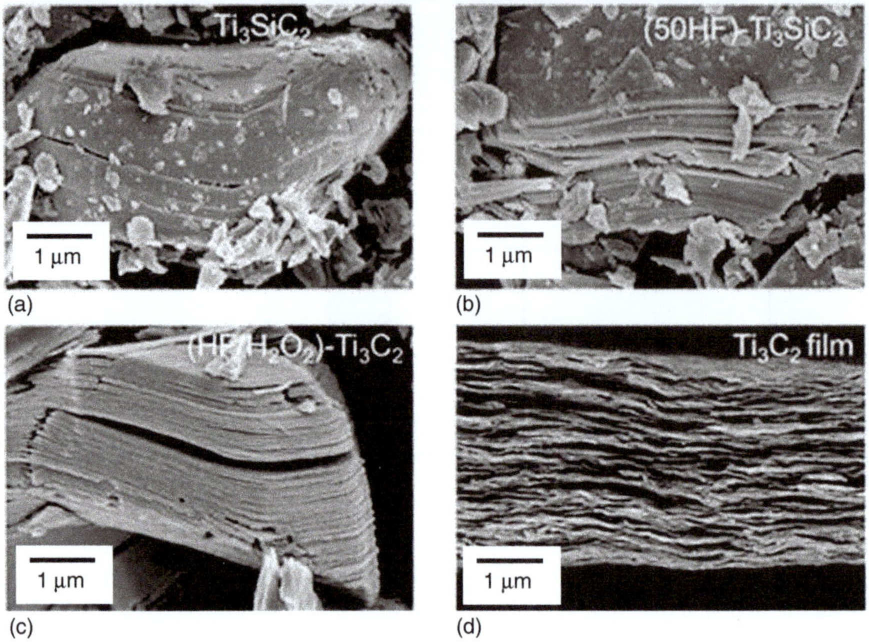

Figure 10.6 SEM images of the MAX precursor and MXene. (a)Ti_3SiC_2, (b) Ti_3C_2 etched in 50 wt% HF, (c) Ti_3SiC_2 etched using HF/H_2O_2 to produces Ti_3C_2 MXene, and (d) Ti_3C_2 delaminated using TMAOH. Source: Adapted with permission from Alhabeb et al. [60]. John Wiley & Sons.

(Figure 10.6b) after using HF (50 wt%) only. However, when HF/H_2O_2 was used, the layers expanded and further opened (Figure 10.6c) as compared to the stack layers observed in Figure 10.6a,b [60].

Recently, in their work (2022), A. Wangsheng et al. presented the synthesis of V_2C using a mixture of HCl and NaF as the etchant at varying etching time and temperatures. The SEM images point out that 90 °C reaction temperature and 96 hours reaction time are the most favorable conditions as shown in Figure 10.7d [65].

10.5.3 Raman Spectroscopy

Through the interaction of light with the chemical bonds contained in a material, Raman spectroscopy gives thorough information about the phase and polymorphs, chemical structure, crystallinity, and molecular interactions. The characterization technique has sensitivity that facilitates examining the surface groups and structure, when various synthesis procedures are used and can confirm MXene existence in a nanocomposite material [66–68].

The presence of V_2AlC (peak observed at 394 cm^{-1}) as shown in Figure 10.8 depicts the characteristic peaks of Al-V vibrations [70] and the broad peaks (657, 1339, 1704, and 2143 cm^{-1}) arises in MXene for the V-C vibrations [71]. The original V_2C peaks occur within 600–2100 cm^{-1}, while for MnO_2, peaks appear around 200 and 500 cm^{-1} to 600 cm^{-1}[72].

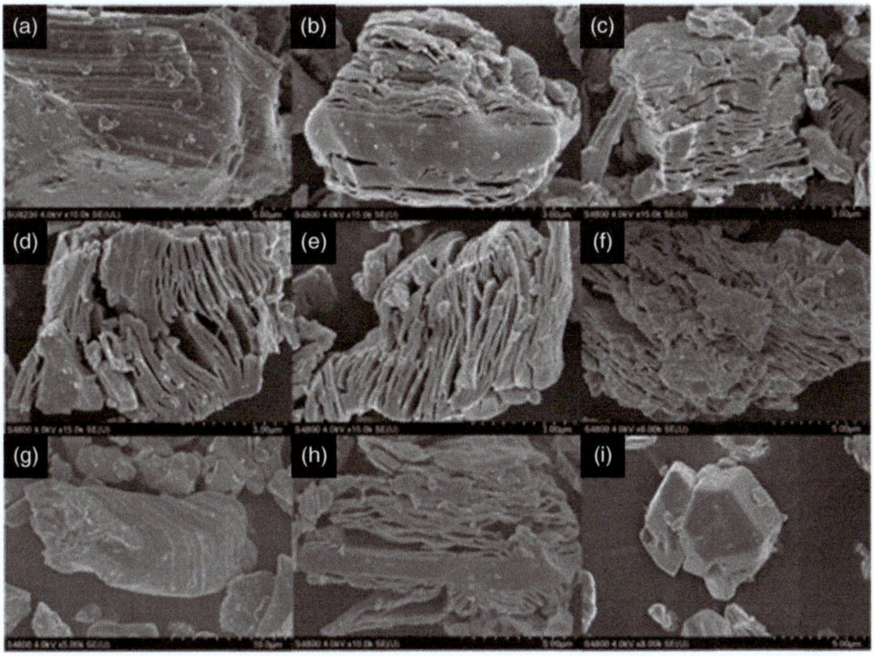

Figure 10.7 SEM image of (a) V_2AlC, (b–f) V_2C@24 h, V_2C@48 h, V_2C@72 h, V_2C@96 h, and V_2C@120 h, respectively. (g)V_2C@50 °C, (h) V_2C@70 °C, and (i) $Na_5Al_3F_{14}$. Source: Adapted with permission from Ai et al. [65]/MDPI/CC BY 4.0.

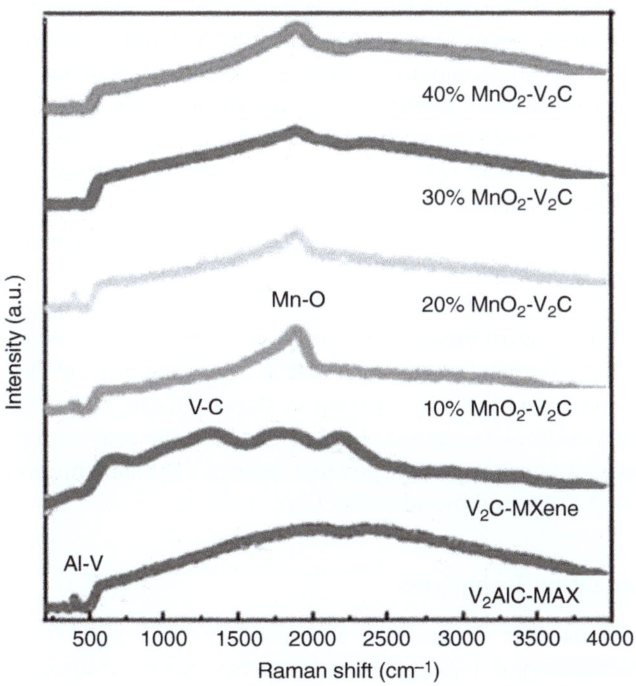

Figure 10.8 Raman spectroscopy showing characteristic peaks in V_2AlC, etched V_2C, and MnO_2-V_2C composite. Source: Fatima et al. [69]/MDPI/CC BY 4.0.

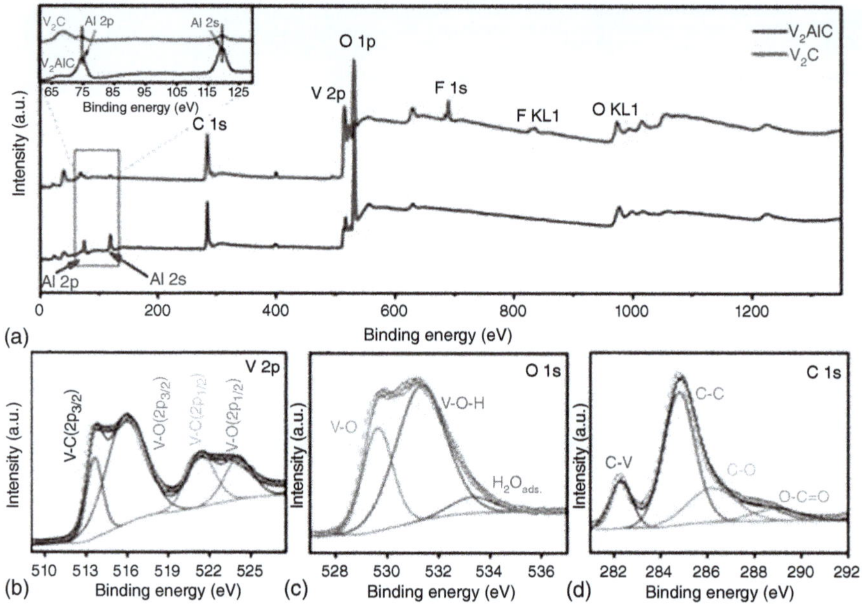

Figure 10.9 (a) The XPS spectra of V_2AlC and V_2C, (b) V 2p, (c) O 1s, (d) C 1s. Source: Ai et al. [65]/MDPI/CC BY 4.0.

10.5.4 X-ray Photoelectron Spectroscopy (XPS)

XPS is used for analyzing a material's external (surface) chemistry, elements present in it, and chemical and electronic state of the atoms within the material, especially MXenes. XPS is instrumental for surface analysis owing to its minimal penetration depth, capability to obtain information about the oxidation states, and chemical composition. Various papers presented the application of XPS techniques to investigate surface chemistry and composition [73, 74], intercalation process [74], surface groups, and thermal stability of materials [74, 75].

For MAX phase (V_2AlC) and the corresponding MXene (V_2C), XPS was employed to measure the composition of elements and termination species as shown in Figure 10.9a–d [65]. Nearly, a total vanishing of the Al 2p and Al 2s peaks in the XPS spectrum (Figure 10.9a) is attributed to the complete removal of the Al in the MAX phase [76]. In V_2C, the peaks are positioned at 513.6 and 521.2 eV for V—C bond and at 516.5 and 524.1 eV for V—O bond as shown Figure 10.9b [19]. Furthermore, the spectrum in the O 1s region (Figure 10.9c), and the peak fittings of C 1s (Figure 10.9d) are all consistent with previous reports [77] showing the V_2CT_x is attached to by V–O, O–H, and the adsorbed H_2O.

10.5.5 Other Characterization Techniques

Other techniques that are used to characterize MXenes include Atomic Force Microscopy – AFM (thickness and lateral dimension) [78], X-ray Absorption Spectroscopy – XAS (to analyze the fundamental electronic state of materials like oxidation state, bond length, and coordination) [20, 79], UV–vis Spectroscopy (optical properties and bandgap analysis) [18, 80], Terahertz Spectroscopy – THz (investigate electronic properties) [81], and Nuclear Magnetic Resonance – MMR(to measure the ratio between O, OH and F groups on surface of MXenes) [82]. The other characterization techniques are subject to specific applications of MXenes.

10.6 Electrochemical Energy Storage (EES) Devices

Batteries and electrochemical capacitors or supercapacitors are one of the leading EES technologies we have today. Basically, the underlying distinction between two is the storage mechanism as batteries store up energy in chemical reactants capable of producing charge. On the other hand, electrochemical capacitors store energy directly as charge [83].

Secondary batteries and supercapacitors can convert chemical and electric energy wholly and reversibly. Owing to their durability, the ability to store a large quantity of energy (energy density) and efficient released capacity, as well as minimal upkeep, much attention is focused on them as promising energy storage devices. Energy storage methods in ion batteries and supercapacitors include redox and non-redox reactions of the electrode material [84]. The battery and pseudo-capacitor operates on the basis of redox chemistry of specific electrode materials [85, 86].

A rudimentary feature common to batteries and supercapacitors is the electrode and electrolyte material. The chemistry and stability of electrode materials are the core attributes that define the overall performance for batteries and supercapacitors [87]. A suitable means to compare the working characteristics for batteries and supercapacitors indicated on the Ragone plot as shown in Figure 10.10.

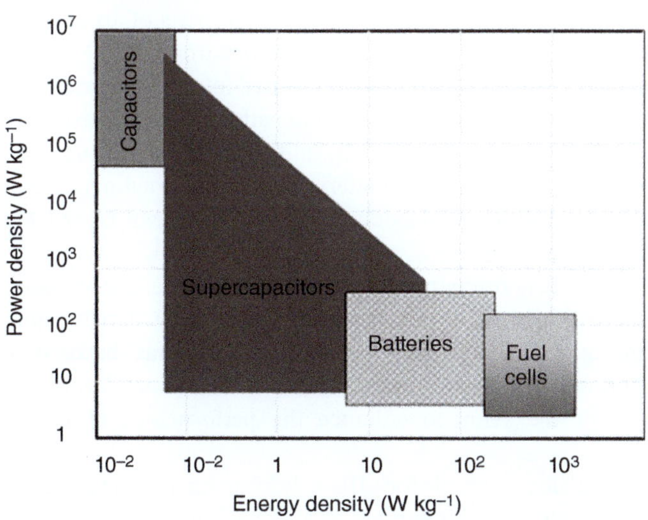

Figure 10.10 Ragone plot showing capacity in terms of power and energy density of EES devices.

10.6.1 MXene for Energy Storage

MXenes are good candidates in energy storage purposes according to their suitable properties such as (i) surface terminations, in particular O, produce surfaces redox-active transition metal oxide-like. (ii) The electrical conductivity resulting in fast electron transport at the electrochemically active sites, and (iii) the 2D layered structure and accessibility of cation intercalation resulting in swift ion transport. The kind of charge storage mechanism for both batteries and supercapacitors depends on the electrolyte and electrode material used.

This section will highlight the application of MXenes as the electrode material for the world's two most popular electrochemical energy storage devices, namely, batteries and supercapacitors.

10.6.2 Batteries

Batteries are the fundamental electrochemical energy storage (EES) device available in our times. They are used in electric grids, electronics, and automobiles and come in different size, shape, and capacity depending on the application. The most common among them is the lithium-ion battery. Others includes Lithium-sulfur, Potassium-ion, Calcium-ion, Aluminum-ion, and Sodium-ion.

10.6.2.1 Lithium-Ion Batteries (LIBs)

The lithium-ion battery is at this time the principal EES device and power supplier for electronic devices such as mobile phones, laptop, computers, and large-scale stationary energy storage systems and is utilized gradually in electric automobiles. They are currently recognized as the leaders in batteries in terms of the quantity of energy they can store (energy density), the number of production units, and they have the most in the world. In general, a rechargeable LIB consists of an organic electrolyte, a graphite anode, an $LiCoO_2$ cathode, and Li^+ ions from $LiCoO_2$ move to the anode through the electrolyte and lastly get intercalated into layered graphite, while the electrons are being moved from the anode to cathode through external circuit [88]. A cathode consisting of nickel, cobalt, and manganese (NMC_{811}) is also used for LIBs recently [89]. Other cathode materials with high gravimetric and volumetric capacity include transition metal oxides such as $LiNi_xCo_yMn_zO_2$ (NCM, with $x + y + z = 1$) [90].

Since the cathode material is a fundamental component for the energy density of the battery, with the critical market demand for energy density and cost-effectiveness, designing a cathode that can address this has become a highest priority [89]. Through extensive research efforts, numerous 2D materials have been in use over the years to enhance the performance of Li-ion batteries [91–95].

MXenes are deemed a promising electrode for LIBs as lithium ions (Li^+) predicted to diffuse swiftly and intercalated on the surface of MXenes, thereby enhancing substantial storage and steady-state capacity and outstanding high-rate capability [96, 97]. Ti_2CT_x was first used as an anode for lithium-ion batteries in 2012 by Naguib et al. At $C/25$ rate, it shows a constant capacity of 225 mAh g^{-1}. Also, after 120 cycles at $3C$ rate and 200 cycles at $10C$ rate, a steady cycling capacity of 80 and 70 mAh g^{-1} was observed, respectively [96]. Other MXenes and MXene-based materials used in LIBs include Nb_2CT_x [19, 98], V_2CT_x [19, 99], and Mo_2CT_x [100].

While lithium-ion batteries have remained the prime preference of energy devices for numerous applications, their potential in transportation and large-scale stationary energy storage systems has largely been hindered by the high risks of fire and explosion related to the use of flammable organic electrolytes [101]. Issues and concerns related to the access and cost of lithium metals have affected the application of LIBs in the future [102]. Also, the innate safety problem stemming from the formation of unnecessary metallic lithium dendrites during battery operation hinders the sustainable advancement of LIBs.

10.6.2.2 Other Batteries

Researchers continue to look for other alternatives, by exploring non-LIBs that are high performing and low cost as compared to the existing LIBs. Currently, the research on NLIBs primarily centers on the preparation of suitable cathode materials [103]. Some of the batteries that come after LIBs include Sodium-Ion Batteries [103–107], Lithium-Sulfur batteries [108–110], Potassium-Ion batteries [111, 112], Aluminum-Ion batteries [113, 114], Magnesium Air Battery [115] and Magnesium-Ion Batteries [103, 116], Zinc-Air batteries [117–119], and Aluminum-Air Batteries[120–122].

10.6.2.3 Application of MXenes as Electrode in Batteries

A variety of MXenes and MXene-based composites were use as the cathode material for batteries including $Ti_3C_2T_x$ [123], $Ti_3C_2T_x$/CNT [109], Nb_2CT_x, and V_2CT_x [19], albeit batteries so far have high energy density, but they lack the require power density and takes a longer time to recharge. For fast recharging, rapid power delivery, long life span, and stability, supercapacitors are used.

10.6.3 Supercapacitors (SCs)

Supercapacitors (electrochemical capacitors or ultracapacitors) are EES devices that store energy electrostatically (non-faradaic), surface redox reaction (faradaic), and electrodeposition of charge carrier ions on the porous carbon electrode surface [124]. It is an upgraded version of a typical capacitor but has low internal resistance and high capacitance, which opens the space for large power density [125].

SCs have desirable characteristics such as longer life span, superior specific power, and less hazard as compared to traditional capacitors, high energy fuels cells, and batteries and hence exhibits a promising solution among EES devices [126]. In addition, supercapacitors can function over a broad temperature range (such as 40–150 °C), can withstand severe environmental conditions, and are lightweight and take up less space, making these devices applicable in the fields of high-tech application, military, satellite, communications, space travel, and so on [127]. As shown Figure 10.10, SCs possess characteristics of both batteries and conventional capacitors. Like batteries, the electrode and electrolyte material use are the primary factors for the performance of SCs, as the transportation and intercalation of ions occurs between the electrode and electrolyte. Depending on the charge storage mechanism which is associated with the electrolyte and electrode used, SCs are categorized as Electrical Double Layer Capacitor (EDLC), Pseudo-capacitor, and Hybrid Capacitor [125].

10.6.3.1 Electrical Double Layer Capacitor (EDLC)

EDLC is the pioneer version of supercapacitors. The charge storage means predominantly rely on ion adsorption. Positive and negative charge ions are aligned to the opposite terminals when voltage is supplied. Due to electrostatic attraction, ions in the electrolyte move to the oppositely charged electrodes, followed by ion adsorption, leading to the creation of a double layer at the electrode and electrolyte interface (Figure 10.11a) [125].

EDLC has an exceptional specific capacitance and power density. However, owing to non-faradaic reactions as its method of storing charges, they have drawbacks in terms of energy density compared to pseudo-capacitors and hybrid capacitors.

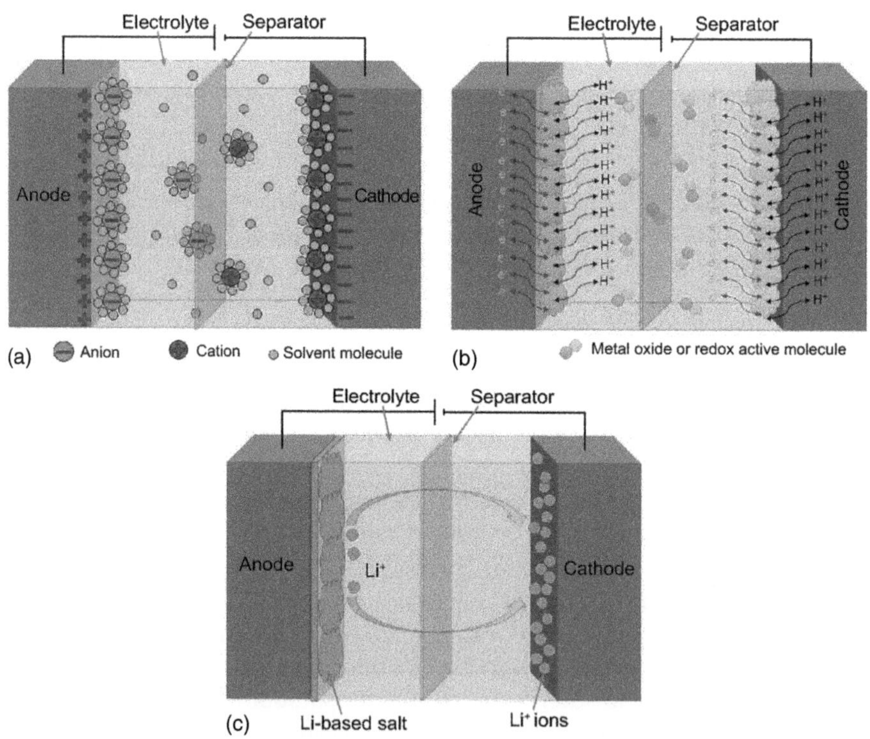

Figure 10.11 A representation of the types SCs (a) (EDLC), (b) pseudo-capacitor, and (c) HSC. Source: Chen et al. [128]/Oxford University Press/CC BY 4.0.

10.6.3.2 Pseudo-capacitor

Pseudo-capacitors store charges through the Faradaic process, involving oxidation and reduction reactions occurring within the electrolyte and at the electrode [126]. When a voltage is applied, electrons and holes are collected in each terminal, and the charged ions move toward the oppositely charged terminal (Figure 10.11b). The charge transfer happens between the electrode surface and the electrolyte reversibly, resulting in redox reactions occurring at the electrodes. Then, a Faraday current path is created in the bilayer, which contributes to the boost in energy density [125]. The downside of pseudo-capacitor includes electrodes degrading faster during charging and discharging and increasing internal resistance [127].

10.6.3.3 Hybrid Supercapacitor (HSC)

Hybrid supercapacitors use both storage mechanisms of the EDLC and pseudo-capacitor, using both faradaic and non-faradaic processes (Figure 10.11c). Hybrid supercapacitors typically have higher energy densities due to the Faraday reaction that happens at the anode made of the pseudo-capacitive electrode material. The cathode is usually made of activated carbon, which stores electrostatic energy in the bilayer on the surface of the electrode [127].

Current research in supercapacitors is aiming on the construction of hybrid supercapacitors because it overcomes the shortcomings of EDLC and pseudo-capacitor.

10.6.3.4 Application of MXenes as an Electrode in Supercapacitors

To improve the performance of SCs, a variety of electrode materials including MXene and its composites are designed and used.

At $10\,V\,s^{-1}$, a microporous $Ti_3C_2T_x$ film delivers up to $210\,F\,g^{-1}$, outperforming most well-known carbon supercapacitors. Also, MXene in hydrogel form were able to produce about $1500\,F\,cm^{-3}$, comparable to the performance of RuO_2 [129]. In their work, Qingmin Shan et al. synthesize and test the performance of V_2CT_x film in acidic (1 M H_2SO_4), alkaline (1 M KOH), and neutral (1 M $MgSO_4$) aqueous electrolyte at different scan rates. Greater than reported MXene electrodes [100, 130, 131], an excellent specific capacitance in 1 M H_2SO_4, 1 M $MgSO_4$, and 1 M KOH were 487, 225, and $184\,F\,g^{-1}$, respectively. Also, after 10,000 cycles, the capacitance retention was 83%, 94%, and 99% in the acidic, alkaline, and neutral electrolyte, respectively [132].

Other synthesized MXenes and MXene-based composites used as the electrode material for SCs include Nb_2CT_x [133], V_2CT_x [31, 132], N-doped $Ti_3C_2T_x$ [134], $V_2CT_x/Ti_3C_2T_x$ flexible films [135, 136], $Ti_3C_2T_x$/CNTs films [137], $Ti_3C_2T_x/Nb_2CT_x$ [138], $Mo_{1.33}CT_x/Ti_3C_2T_x$ [139], MnO_2-V_2C nanocomposite [69], $MnO_2/Ti_3C_2T_x$ [140], and Nb-doped $Ti_3C_2T_x$ [141].

The drawbacks of supercapacitors centered on the electrode and electrolyte use. Binder-free electrode and hybrid gel electrolyte can be explored to improve SC performance [125].

In summary, an ideal choice of the electrode material and electrolyte leads to excellent electrochemical performance of SCs.

10.7 Thermodynamic and Cycle Stability of MXenes

Electrochemical stability of the electrode material is a significant factor influencing the suitability of battery and supercapacitors in commercial applications. The electrode must be both thermal and cycling stable to a certain degree to be considered highly efficient for applications. Since MXenes and MXene-based composites are employed as the electrode material for batteries and supercapacitors, their stability is mostly investigated to determine their suitability.

10.7.1 Thermodynamic Stability of MXenes

Chemical termination of MXene surfaces and their interlayer chemistry are of essence to understanding and predicting their behavior, hence facilitating their preparation. These properties depend on their composition, processing, and operating conditions, which in turn impact their performance. The terminations, molecules, and ions attached on and within the layers of MXene during synthesis play a significant role in their stability. Delamination and intercalation of organic molecules and ions into the sheet structure can consequently affect MXene interlayer spacing and surface chemistry [142]. Their thermodynamic stability is probably related to functional groups but has not yet been investigated by quantitative experimental measurements.

Using calorimetric determination, the thermodynamic stability of the MAX phase and MXene were studied by S. Geetu et al. It was revealed that the enthalpy of MXene formation during wet chemical etching MAX is extremely exothermic, facilitating the removal of the A layer and indicating that LiF-HCl etching in aqueous solution yields highly stable MXene. The thermodynamic stability of MXene in contact with the aqueous solution indicates the significance of terminations/surface groups and lithium intercalation in the laminar structure [142].

Similarly, the thermal stability of $Ti_3C_2T_x$ was presented by L. Rui et al. using Differential Scanning Calorimetry (DSC) and Thermogravimetric Analysis (TGA). The $Ti_3C_2T_x$ MXene in a nitrogen atmosphere and temperature ranges from room temperature to 950 °C (Figure 10.12a) shown in three steps for changing weight. The

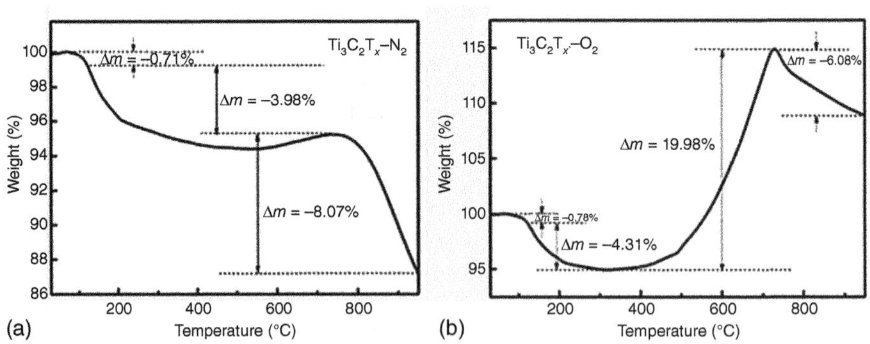

Figure 10.12 TG curves of $Ti_3C_2T_x$ in (a) nitrogen atmosphere and (b) in oxygen atmosphere. Source: Reproduced with permission from Liu and Li [143].

first is results from the release of free water that has been adsorbed to the $Ti_3C_2T_x$'s surface. The second stage results from the release of functional groups and bonding water from the $Ti_3C_2T_x$'s surface. The breakdown of $Ti_3C_2T_x$ MXene results in the third stage. Pure $Ti_3C_2T_x$ MXene decomposes at a temperature of around 785 °C. In Figure 10.12b, it is observed that the mass increases by 19.98% between 322 and 729 °C, which is caused by the oxidation of MXene. $Ti_3C_2T_x$ MXene has an oxidation temperature of around 322 °C and a decomposition temperature of approximately 729 °C [143].

The dynamic and structural stability, as well as the electronic and thermal stability of 2D MXene, was investigated by Aliakbari et al. through the plane wave method in quantum-Espresso [144, 145].Thermodynamic calculations revealed that enthalpy and heat capacity rise incessantly and steadily, while Debye temperature and stiffness reduce as pressure and n-index in the $Y_{n+1}C_n$ ($n = 1, 2,$ or 3) MXene structure increase. For entropy, it increases rapidly and constantly as temperature and n-index in MXenes increase.

10.7.2 Cycle Stability of MXenes

The stability of MXenes and its composites is evaluated in its application to batteries and Supercapacitor. On an electrochemical workstation, tests like Cyclic Voltammetry (CV) and Galvanostatic Charge Discharge (GCD) can be utilized to assess the stability of MXene electrodes.

When V_2CT_x and Nb_2CT_x were used in LIBs, both demonstrated exceptional capability of operating up to 10C cycling rates (Figure 10.13), maintaining a good reversible capacity of 210 and 170 mAh g^{-1}, respectively [19].

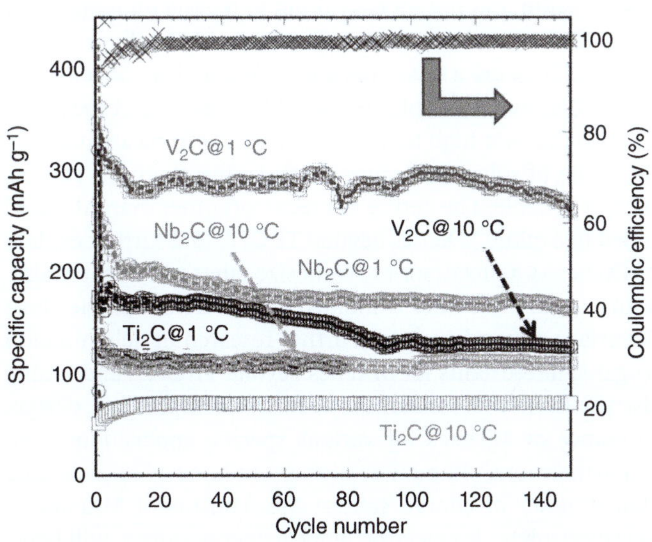

Figure 10.13 Specific capacity and coulombic efficiency of Nb_2CT_x and V_2CT_x electrodes at different cycle number and rates. Source: From Nahuib et al. [19]/with permission of American Chemical Society.

Similarly, as presented by Minjuan et al. in 2017, the cycling stability of Ti_3C_2 MXene was investigated at $5\,A\,g^{-1}$ for 6000 charge–discharge cycles and was able to maintain 70% of the capacity [146]. Yin and coworkers prepared and assessed $Ti_3C_2T_x$ and MnO_2-$Ti_3C_2T_x$ as the supercapacitor electrode. GCD at a current density of $5\,A\,g^{-1}$ was conducted for 5000 charge–discharge cycles, and $Ti_3C_2T_x$ was able to maintain 98% and 97% of its capacity and efficiency (coulombic), indicating its cycling stability [140].

A V_2CT_x film without a substrate was tested in 1 M H_2SO_4, 1 M $MgSO_4$, and 1 M KOH aqueous electrolytes as the electrode for a supercapacitor [132]. At $10\,A\,g^{-1}$, the capacitance it maintained after 10,000 cycles was 83% (in 1 M H_2SO_4), 94% (1 M KOH), and 99% (1 M $MgSO_4$), indicating the stability of V_2C as an SC electrode.

The cycling stability of other MXene electrodes for batteries [133, 147–149] and Supercapacitor [25, 150–152] was also presented in literatures.

10.8 Future Recommendations

Though encouraging data are continuously reported regarding MXenes for energy storage applications, a lot of questions still need to be addressed in order to proceed with mass-scale production of MXenes toward its practical applications. Therefore, presently, the research on MXenes is a long way from providing a practical solution to our everyday energy storage-related requirement. Increasing research in MXene synthesis is a light at the end of the tunnel, that most of the problems may be tackled by advancing the fundamental research on MXenes in the right direction [153].

One of the fundamental obstacles in MXene's mass-scale production is the choice of the etching solvent, which is HF at the moment. The fluorine-based solvents are extremely hazardous to health and not an easy thing to handle on large scale. Therefore, a big proportion of MXene research lies in finding a suitable fluoride-free solvent for MXene synthesis. There are a few reports on using highly concentrated alkaline solutions which can dissolve the Al phase from MAX; however, these methods are still not suitable since they use high temperatures. More efforts are urgently needed to overcome this issue of solvent selection to make the synthesis process non-toxic, safe, efficient, and scalable. The choice of precursors other than Al-Based MAX needs to be identified to explore MXenes beyond $Ti_3C_2T_x$ and further explore the possibility of using MXenes as a precursor to synthesize other novel materials.

The etched MXenes are usually not fully exfoliated and therefore single-sheet properties such as high surface area are limited. Further research in delamination procedures using non-organic intercalants for MXenes beyond $Ti_3C_2T_x$, and tuning MXene's surface chemistry with new intercalants is needed to comprehend, design, and improve the performance of MXenes for various specific applications even beyond energy storage. Furthermore, characterization procedures of MXene need standardization. A detailed study in charge storage mechanisms of MXenes in aqueous and nonaqueous electrolytes for electrochemical energy storage will boost the engineering of energy storage and open new doors for fundamentally altering MXene chemistry to improve its performance [154]. Molecular dynamic simulation and density functional theory may play a vital role in deciding the right mechanisms associated with energy storage and chemistry of MXenes. Finally, more efforts are needed to replace conventional electrode preparation with techniques such as printing, stamping, and centrifugal coating for highly compact MXene-based microelectrodes and scale up mass production [155].

With the aforementioned research directions, MXene as a reliable electrode for electrochemical energy storage devices will realize its true potential, and hence, it will help bringing it to the industrial-scale application.

10.9 Summary

In order to meet the current energy demands, energy storage devices should meet standards such as high power and energy density, stability, durability, cost-effectiveness in production, and safe and environmentally friendly. Electrochemical energy storage devices can certainly meet all these requirements by properly selecting the electrode materials. MXenes, owing to their unique characteristics, are potential candidates as electrode materials to improve the functioning of existing batteries and supercapacitors.

There are two major types of MXenes, namely, the mono-transition element MXenes (with only one transitional metal) and the double-transition element MXenes (with two transitional metals) which further include solid-solution and ordered MXenes. The synthesis and delamination protocol for $Ti_3C_2T_x$ and V_2CT_x depends on the MAX phase, the etchant/intercalant use, and the synthesis parameters such as reaction time, temperature, and stirring speed. So far, HF is the most famous etchant used in Al-based MAX to remove Al for producing MXenes.

Every single stage in synthesizing and tuning MXene impacts their functional and electronic properties, and to take benefit from these properties, a thorough understanding of those properties should be characterized with techniques such as X-ray Diffraction, Scanning Electron Microscopy, Raman Spectroscopy, and X-ray Photoelectron Spectroscopy (XPS) and investigate the morphology, composition, phase, and other properties of the parent MAX phase and etched MXenes.

Moreover, the two prominent electrochemical energy storage (batteries and supercapacitors), their types, storage mechanism, and limitations are highlighted. MXenes such as $Ti_3C_2T_x$, V_2CT_x, and Nb_2CT_x are more familiar MXenes studied as electrodes for both supercapacitors and batteries, exhibiting satisfactory performance in terms of unit energy and power they can store.

Acknowledgments

S. Rizwan thanks and acknowledges partial financial support from Higher Education Commission (HEC) of Pakistan (NRPU grant#14784). M.Z. Iqbal thanks and acknowledges partial financial support from UAEU through National Water and Energy Center (NWEC) (grant#31R166) and UAEU Program for Advanced Research (UPAR grant#12N002).

References

1 Liu, C., Li, F., Ma, L.P., and Cheng, H.M. (2010). Advanced materials for energy storage. *Adv. Mater.* 22 (8): E28–E62.

2 Armand, M. and Tarascon, J.-M. (2008). Building better batteries. *Nature* 451 (7179): 652–657.

3 Simon, P. and Gogotsi, Y. (2010). Materials for electrochemical capacitors. In: *Nanoscience and Technology: A Collection of Reviews from Nature Journals* (ed. P. Rodgers), 320–329. World Scientific.

4 Chen, P.-C., Shen, G., Shi, Y. et al. (2010). Preparation and characterization of flexible asymmetric supercapacitors based on transition-metal-oxide nanowire/single-walled carbon nanotube hybrid thin-film electrodes. *ACS Nano* 4 (8): 4403–4411.

5 Ji, L., Lin, Z., Alcoutlabi, M., and Zhang, X. (2011). Recent developments in nanostructured anode materials for rechargeable lithium-ion batteries. *Energy Environ. Sci.* 4 (8): 2682–2699.

6 Wang, B., Chen, J.S., Wu, H.B. et al. (2011). Quasiemulsion-templated formation of α-Fe_2O_3 hollow spheres with enhanced lithium storage properties. *J. Am. Chem. Soc.* 133 (43): 17146–17148.

7 Zhao, Y., Li, X., Yan, B. et al. (2016). Recent developments and understanding of novel mixed transition-metal oxides as anodes in lithium ion batteries. *Adv. Energy Mater.* 6 (8): 1502175.

8 Wang, H., Feng, H., and Li, J.J.S. (2014). Graphene and graphene-like layered transition metal dichalcogenides in energy conversion and storage. *Small* 10 (11): 2165–2181.

9 Su, D., Dou, S., and Wang, G. (2015). Ultrathin MoS_2 nanosheets as anode materials for sodium-ion batteries with superior performance. *Adv. Energy Mater.* 5 (6): 1401205.

10 Yu, H., Wang, Y., Jing, Y. et al. (2019). Surface modified MXene-based nanocomposites for electrochemical energy conversion and storage. *Small* 15 (25): 1901503.

11 Okubo, M., Sugahara, A., Kajiyama, S., and Yamada, A. (2018). MXene as a charge storage host. *Acc. Chem. Res.* 51 (3): 591–599.

12 Naguib, M., Kurtoglu, M., Presser, V. et al. (2011). Two-dimensional nanocrystals produced by exfoliation of Ti_3AlC_2. *Adv. Mater.* 23 (37): 4248–4253.

13 Mas-Balleste, R., Gomez-Navarro, C., Gomez-Herrero, J., and Zamora, F. (2011). 2D materials: to graphene and beyond. *Nanoscale* 3 (1): 20–30.

14 Alhabeb, M., Maleski, K., Anasori, B. et al. (2017). Guidelines for synthesis and processing of two-dimensional titanium carbide ($Ti_3C_2T_x$ MXene). *Chem. Mater.* 29 (18): 7633–7644.

15 Barsoum, M.W. (2013). *MAX Phases: Properties of Machinable Ternary Carbides and Nitrides*. Wiley.

16 Barsoum, M.W. (2000). The $M_{N+1}AX_N$ phases: a new class of solids: thermodynamically stable nanolaminates. 28 (1–4): 201–281.

17 Barsoum, M.W. and El-Raghy, T. (2001). The MAX phases: unique new carbide and nitride materials: ternary ceramics turn out to be surprisingly soft and machinable, yet also heat-tolerant, strong and lightweight. *Am. Sci.* 89 (4): 334–343.

18 Matthews, K., Zhang, T., Shuck, C.E. et al. (2021). Guidelines for synthesis and processing of chemically stable two-dimensional V_2CT_x MXene. *Chem. Mater.* 34 (2): 499–509.

19 Naguib, M., Halim, J., Lu, J. et al. (2013). New two-dimensional niobium and vanadium carbides as promising materials for Li-ion batteries. *J. Am. Chem. Soc.* 135 (43): 15966–15969.

20 Deeva, E.B., Kurlov, A., Abdala, P.M. et al. (2019). In situ XANES/XRD study of the structural stability of two-dimensional molybdenum carbide Mo_2CT_x: implications for the catalytic activity in the water–gas shift reaction. *Chem. Mater.* 31 (12): 4505–4513.

21 Li, Z., Wang, L., Sun, D. et al. (2015). Synthesis and thermal stability of two-dimensional carbide MXene Ti_3C_2. *Mater. Sci. Eng., B* 191: 33–40.

22 Naguib, M. (2017). Two-dimensional transition-metal carbides and carbonitrides. In: *Nanomaterials Handbook* (ed. Y. Gogotsi), 83–102. CRC Press.

23 Zhou, J., Zha, X., Zhou, X. et al. (2017). Synthesis and electrochemical properties of two-dimensional hafnium carbide. *ACS Nano* 11 (4): 3841–3850.

24 Ghidiu, M., Naguib, M., Shi, C. et al. (2014). Synthesis and characterization of two-dimensional Nb_4C_3 (MXene). *Chem. Commun.* 50 (67): 9517–9520.

25 Syamsai, R. and Grace, A.N. (2019). Ta_4C_3 MXene as supercapacitor electrodes. *J. Alloys Compd.* 792: 1230–1238.

26 Urbankowski, P., Anasori, B., Makaryan, T. et al. (2016). Synthesis of two-dimensional titanium nitride Ti_4N_3 (MXene). *Nanoscale* 8 (22): 11385–11391.

27 Yang, J., Naguib, M., Ghidiu, M. et al. (2016). Two-dimensional Nb-based M_4C_3 solid solutions (MXenes). *J. Am. Ceram. Soc.* 99 (2): 660–666.

28 Eklund, P., Rosen, J., and Persson, P.O.Å. (2017). Layered ternary $M_{n+1}AXn$ phases and their 2D derivative MXene: an overview from a thin-film perspective. *J. Phys. D: Appl. Phys.* 50 (11): 113001.

29 Högberg, H., Hultman, L., Emmerlich, J. et al. (2005). Growth and characterization of MAX-phase thin films. *Surf. Coat. Technol.* 193 (1–3): 6–10.

30 Anasori, B., Xie, Y., Beidaghi, M. et al. (2015). Two-dimensional, ordered, double transition metals carbides (MXenes). *ACS Nano* 9 (10): 9507–9516.

31 Guan, Y., Jiang, S., Cong, Y. et al. (2020). A hydrofluoric acid-free synthesis of 2D vanadium carbide (V_2C) MXene for supercapacitor electrodes. *2D Mater.* 7 (2): 025010.

32 Wang, L., Han, M., Shuck, C.E. et al. (2021). Adjustable electrochemical properties of solid-solution MXenes. *Nano Energy* 88: 106308.

33 Iqbal, M.Z., Khan, M.W., Siddique, S., and Aftab, S. (2022). MXenes: an exotic material for hybrid supercapacitors and rechargeable batteries. *J. Energy Storage* 56: 105914.

34 Meshkian, R., Tao, Q., Dahlqvist, M. et al. (2017). Theoretical stability and materials synthesis of a chemically ordered MAX phase, Mo_2ScAlC_2, and its two-dimensional derivate Mo_2ScC_2 MXene. *Acta Mater.* 125: 476–480.

35 Kim, H., Anasori, B., Gogotsi, Y., and Alshareef, H.N. (2017). Thermoelectric properties of two-dimensional molybdenum-based MXenes. *Chem. Mater.* 29 (15): 6472–6479.

36 Khazaei, M., Ranjbar, A., Arai, M., and Yunoki, S. (2016). Topological insulators in the ordered double transition metals $M_2'M''C_2$ MXenes (M′ = Mo, W; M″ = Ti, Zr, Hf). *Phys. Rev. B* 94 (12): 125152.

37 Si, C., You, J., Shi, W. et al. (2016). Quantum spin Hall phase in $Mo_2M_2C_3O_2$ (M = Ti, Zr, Hf) MXenes. *J. Mater. Chem. C* 4 (48): 11524–11529.

38 Khazaei, M., Ranjbar, A., Esfarjani, K. et al. (2018). Insights into exfoliation possibility of MAX phases to MXenes. *Chem. Phys.* 20 (13): 8579–8592.

39 Anasori, B., Dahlqvist, M., Halim, J. et al. (2015). Experimental and theoretical characterization of ordered MAX phases Mo_2TiAlC_2 and $Mo_2Ti_2AlC_3$. *J. Appl. Phys.* 118 (9): 094304.

40 Liu, Z., Wu, E., Wang, J. et al. (2014). Crystal structure and formation mechanism of $(Cr_{2/3}Ti_{1/3})_3AlC_2$ MAX phase. *Acta Mater.* 73: 186–193.

41 Tao, Q., Lu, J., Dahlqvist, M. et al. (2019). Atomically layered and ordered rare-earth i-MAX phases: a new class of magnetic quaternary compounds. *Chem. Mater.* 31 (7): 2476–2485.

42 Dahlqvist, M. and Rosen, J.J.N. (2022). The rise of MAX phase alloys-large-scale theoretical screening for prediction of chemical order and disorder. *Nanoscale* 14: 10958–10971.

43 Tao, Q., Dahlqvist, M., Lu, J. et al. (2017). Two-dimensional $Mo_{1.33}C$ MXene with divacancy ordering prepared from parent 3D laminate with in-plane chemical ordering. *Nat. Commun.* 8 (1): 1–7.

44 Dahlqvist, M., Lu, J., Meshkian, R. et al. (2017). Prediction and synthesis of a family of atomic laminate phases with Kagomé-like and in-plane chemical ordering. *Sci. Adv.* 3 (7): e1700642.

45 Mockute, A., Tao, Q., Dahlqvist, M. et al. (2019). Materials synthesis, neutron powder diffraction, and first-principles calculations of $(Mo_xSc_{1-x})_2$ AlC $i-$MAX phase used as parent material for MXene derivation. *Phys. Rev. Mater.* 3 (11): 113607.

46 Naguib, M., Mashtalir, O., Carle, J. et al. (2012). Two-dimensional transition metal carbides. *ACS Nano* 6 (2): 1322–1331.

47 Meshkian, R., Näslund, L.Å., Halim, J. et al. (2015). Synthesis of two-dimensional molybdenum carbide, Mo_2C, from the gallium based atomic laminate Mo_2Ga_2C. *Scr. Mater.* 108: 147–150.

48 Zhou, J., Zha, X., Chen, F.Y. et al. (2016). A two-dimensional zirconium carbide by selective etching of Al_3C_3 from nanolaminated $Zr_3Al_3C_5$. *Angew. Chem. Int. Ed.* 55 (16): 5008–5013.

49 Zhang, T., Pan, L., Tang, H. et al. (2017). Synthesis of two-dimensional $Ti_3C_2T_x$ MXene using HCl^+ LiF etchant: enhanced exfoliation and delamination. *J. Alloys Compd.* 695: 818–826.

50 Naguib, M., Unocic, R.R., Armstrong, B.L. et al. (2015). Large-scale delamination of multi-layers transition metal carbides and carbonitrides "MXenes". *Dalton Trans.* 44 (20): 9353–9358.

51 Halim, J., Lukatskaya, M.R., Cook, K.M. et al. (2014). Transparent conductive two-dimensional titanium carbide epitaxial thin films. *Chem. Mater.* 26 (7): 2374–2381.

52 Ling, Z., Ren, C.E., Zhao, M.Q. et al. (2014). Flexible and conductive MXene films and nanocomposites with high capacitance. *Proc. Natl. Acad. Sci.* 111 (47): 16676–16681.

53 Liu, F., Zhou, A., Chen, J. et al. (2017). Preparation of Ti_3C_2 and Ti_2C MXenes by fluoride salts etching and methane adsorptive properties. 416: 781–789.

54 Dong, H., Xiao, P., Jin, N. et al. (2021). Molten salt derived Nb_2CT_x MXene anode for Li-ion batteries. *ChemElectroChem* 8 (5): 957–962.

55 Li, M., Lu, J., Luo, K. et al. (2019). Element replacement approach by reaction with Lewis acidic molten salts to synthesize nanolaminated MAX phases and MXenes. *J. Am. Chem. Soc.* 141 (11): 4730–4737.

56 Li, T., Yao, L., Liu, Q. et al. (2018). Fluorine-free synthesis of high-purity $Ti_3C_2T_x$ (T = OH, O) via alkali treatment. *Angew. Chem. Int. Ed.* 57 (21): 6115–6119.

57 Yang, S., Zhang, P., Wang, F. et al. (2018). Fluoride-free synthesis of two-dimensional titanium carbide (MXene) using a binary aqueous system. *Angew. Chem.* 130 (47): 15717–15721.

58 Jawaid, A., Hassan, A., Neher, G. et al. (2021). Halogen etch of Ti_3AlC_2 MAX phase for MXene fabrication. *ACS Nano* 15 (2): 2771–2777.

59 Shi, H., Zhang, P., Liu, Z. et al. (2021). Ambient-stable two-dimensional titanium carbide (MXene) enabled by iodine etching. *Angew. Chem. Int. Ed.* 60 (16): 8689–8693.

60 Alhabeb, M., Maleski, K., Mathis, T.S. et al. (2018). Selective etching of silicon from Ti_3SiC_2 (MAX) to obtain 2D titanium carbide (MXene). *Angew. Chem.* 130 (19): 5542–5546.

61 Zahra, S.A. and Rizwan, S. (2022). MWCNT-modified MXene as cost-effective efficient bifunctional catalyst for overall water splitting. *RSC Adv.* 12 (14): 8405–8413.

62 Soundiraraju, B. and George, B.K. (2017). Two-dimensional titanium nitride (Ti_2N) MXene: synthesis, characterization, and potential application as surface-enhanced Raman scattering substrate. *ACS Nano* 11 (9): 8892–8900.

63 Shekhirev, M., Shuck, C.E., Sarycheva, A., and Gogotsi, Y. (2021). Characterization of MXenes at every step, from their precursors to single flakes and assembled films. *Prog. Mater Sci.* 120: 100757.

64 Lim, G.P., Soon, C.F., Morsin, M. et al. (2020). Synthesis, characterization and antifungal property of $Ti_3C_2T_x$ MXene nanosheets. *Ceram. Int.* 46 (12): 20306–20312.

65 Ai, W., Zhang, C., Xia, L. et al. (2022). Synthesis of high-quality two-dimensional V_2C MXene for supercapacitor application. *Energies* 15 (10): 3696.

66 Limbu, T.B., Chitara, B., Garcia Cervantes, M.Y. et al. (2020). Unravelling the thickness dependence and mechanism of surface-enhanced Raman scattering on $Ti_3C_2T_X$ MXene nanosheets. *J. Phys. Chem. C* 124 (32): 17772–17782.

67 Levitt, A.S., Alhabeb, M., Hatter, C.B. et al. (2019). Electrospun MXene/carbon nanofibers as supercapacitor electrodes. *J. Mater. Chem. A* 7 (1): 269–277.

68 Lioi, D.B., Neher, G., Heckler, J.E. et al. (2019). Electron-withdrawing effect of native terminal groups on the lattice structure of $Ti_3C_2T_x$ MXenes studied by resonance Raman scattering: implications for embedding MXenes in electronic composites. *ACS Appl. Nano Mater.* 2 (10): 6087–6091.

69 Fatima, M., Zahra, S.A., Khan, S.A. et al. (2021). Experimental and computational analysis of MnO_2@V_2C-MXene for enhanced energy storage. *Nanomaterials* 11 (7): 1707.

70 Chen, J., Chen, K., Tong, D. et al. (2015). CO_2 and temperature dual responsive "Smart" MXene phases. *Chem. Commun.* 51 (2): 314–317.

71 Spanier, J.E. et al. (2005). Vibrational behavior of the $M_{n+1}AX_n$ phases from first-order Raman scattering (M = Ti, V, Cr, A = Si, X = C, N). *Phys. Rev. B* 71 (1): 012103.

72 Widjaja, E. and Sampanthar, J.T. (2007). The detection of laser-induced structural change of MnO_2 using in situ Raman spectroscopy combined with self-modeling curve resolution technique. *Anal. Chim. Acta* 585 (2): 241–245.

73 Halim, J., Cook, K.M., Naguib, M. et al. (2016). X-ray photoelectron spectroscopy of select multi-layered transition metal carbides (MXenes). *Appl. Surf. Sci.* 362: 406–417.

74 Halim, J., Cook, K.M., Eklund, P. et al. (2019). XPS of cold pressed multilayered and freestanding delaminated 2D thin films of $Mo_2TiC_2T_z$ and $Mo_2Ti_2C_3T_z$ (MXenes). *Appl. Surf. Sci.* 494: 1138–1147.

75 Schultz, T., Frey, N.C., Hantanasirisakul, K. et al. (2019). Surface termination dependent work function and electronic properties of $Ti_3C_2T_x$ MXene. *Chem. Mater.* 31 (17): 6590–6597.

76 Wang, X., Lin, S., Tong, H. et al. (2019). Two-dimensional V_4C_3 MXene as high performance electrode materials for supercapacitors. *Electrochim. Acta* 307: 414–421.

77 Rakhi, R.B., Ahmed, B., Hedhili, M.N. et al. (2015). Effect of postetch annealing gas composition on the structural and electrochemical properties of Ti_2CT_x MXene electrodes for supercapacitor applications. *Chem. Mater.* 27 (15): 5314–5323.

78 Lipatov, A., Lu, H., Alhabeb, M. et al. (2018). Elastic properties of 2D $Ti_3C_2T_x$ MXene monolayers and bilayers. *Sci. Adv.* 4 (6): eaat0491.

79 Li, Z., Cui, Y., Wu, Z. et al. (2018). Reactive metal–support interactions at moderate temperature in two-dimensional niobium-carbide-supported platinum catalysts. *Nat. Catal.* 1 (5): 349–355.

80 Montazeri, K., Currie, M., Verger, L. et al. (2019). Beyond gold: spin-coated Ti_3C_2-based MXene photodetectors. *Adv. Mater.* 31 (43): 1903271.

81 Li, G., Kushnir, K., Dong, Y. et al. (2018). Equilibrium and non-equilibrium free carrier dynamics in 2D $Ti_3C_2T_x$ MXenes: THz spectroscopy study. *2D Mater.* 5 (3): 035043.

82 Harris, K.J., Bugnet, M., Naguib, M. et al. (2015). Direct measurement of surface termination groups and their connectivity in the 2D MXene V_2CT_x using NMR spectroscopy. *J. Phys. Chem. C* 119 (24): 13713–13720.

83 Goodenough, J.B., Abruna, H.D., and Buchanan, M.V. (2007). Basic Research Needs for Electrical Energy Storage. In: *Report of the Basic Energy Sciences Workshop on Electrical Energy Storage, April 2-4, 2007*. United States: N. p https://doi.org/10.2172/935429.

84 Simon, P., Gogotsi, Y., and Dunn, B.J.S. (2014). Where do batteries end and supercapacitors begin? *Science* 343 (6176): 1210–1211.

85 Liu, M., Lin, C., Gu, Y. et al. (2014). Oxygen reduction contributing to charge transfer during the first discharge of the CeO_2–$Bi_2Fe_4O_9$–Li battery: in situ X-ray diffraction and X-ray absorption near-edge structure investigation. *J. Phys. Chem. C* 118 (27): 14711–14722.

86 Mefford, J.T., Hardin, W.G., Dai, S. et al. (2014). Anion charge storage through oxygen intercalation in $LaMnO_3$ perovskite pseudocapacitor electrodes. *Nat. Mater.* 13 (7): 726–732.

87 Chen, K. and Xue, D. (2016). Materials chemistry toward electrochemical energy storage. *J. Mater. Chem. A* 4 (20): 7522–7537.

88 Dong, Y., Shi, H., and Wu, Z.S. (2020). Recent advances and promise of MXene-based nanostructures for high-performance metal ion batteries. 30 (47): 2000706.

89 Zhang, X., Qiu, Y., Cheng, F. et al. (2021). Realization of a high-voltage and high-rate nickel-rich NCM cathode material for LIBs by Co and Ti dual modification. *ACS Appl. Mater. Interfaces* 13 (15): 17707–17716.

90 Zhou, L., Zhang, K., Hu, Z. et al. (2018). Recent developments on and prospects for electrode materials with hierarchical structures for lithium-ion batteries. *Adv. Energy Mater.* 8 (6): 1701415.

91 Wang, G., Yang, J., Park, J. et al. (2008). Facile synthesis and characterization of graphene nanosheets. *J. Phys. Chem. C* 112 (22): 8192–8195.

92 Liu, H., Neal, A.T., Zhu, Z. et al. (2014). Phosphorene: an unexplored 2D semiconductor with a high hole mobility. *ACS Nano* 8 (4): 4033–4041.

93 Zhao, C., Gao, H., Chen, C., and Wu, H. (2015). Reduction of graphene oxide in Li-ion batteries. *J. Mater. Chem. A* 3 (36): 18360–18364.

94 Zou, F., Hu, X., Li, Z. et al. (2014). MOF-derived porous $ZnO/ZnFe_2O_4/C$ octahedra with hollow interiors for high-rate lithium-ion batteries. *Adv. Mater.* 26 (38): 6622–6628.

95 Pumera, M., Sofer, Z., and Ambrosi, A. (2014). Layered transition metal dichalcogenides for electrochemical energy generation and storage. *J. Mater. Chem. A* 2 (24): 8981–8987.

96 Naguib, M., Come, J., Dyatkin, B. et al. (2012). MXene: a promising transition metal carbide anode for lithium-ion batteries. *Electrochem. Commun.* 16 (1): 61–64.

97 Come, J., Naguib, M., Rozier, P. et al. (2012). A non-aqueous asymmetric cell with a Ti_2C-based two-dimensional negative electrode. *J. Electrochem. Soc.* 159 (8): A1368.

98 Mashtalir, O., Lukatskaya, M.R., Zhao, M.Q. et al. (2015). Amine-assisted delamination of Nb_2C MXene for Li-ion energy storage devices. 27 (23): 3501–3506.

99 Liu, F., Zhou, J., Wang, S. et al. (2017). Preparation of high-purity V_2C MXene and electrochemical properties as Li-ion batteries. *J. Electrochem. Soc.* 164 (4): A709.

100 Halim, J., Kota, S., Lukatskaya, M.R. et al. (2016). Synthesis and characterization of 2D molybdenum carbide (MXene). *Adv. Funct. Mater.* 26 (18): 3118–3127.

101 Wang, Y., Chao, D., Wang, Z. et al. (2021). An energetic CuS–Cu battery system based on CuS nanosheet arrays. *ACS Nano* 15 (3): 5420–5427.

102 Elia, G.A., Marquardt, K., Hoeppner, K. et al. (2016). An overview and future perspectives of aluminium batteries. *Adv. Mater.* 28: 7564–7579.

103 Aslam, M.K., Niu, Y., and Xu, M. (2021). MXenes for non-lithium-ion (Na, K, Ca, Mg, and Al) batteries and supercapacitors. *Adv. Energy Mater.* 11 (2): 2000681.

104 Palomares, V., Serras, P., Villaluenga, I. et al. (2012). Na-ion batteries, recent advances and present challenges to become low cost energy storage systems. *Energy Environ. Sci.* 5 (3): 5884–5901.

105 Wang, X., Kajiyama, S., Iinuma, H. et al. (2015). Pseudocapacitance of MXene nanosheets for high-power sodium-ion hybrid capacitors. *Nat. Commun.* 6 (1): 1–6.

106 Xie, X., Zhao, M.Q., Anasori, B. et al. (2016). Porous heterostructured MXene/carbon nanotube composite paper with high volumetric capacity for sodium-based energy storage devices. *Nano Energy* 26: 513–523.

107 Yang, C., Xin, S., Mai, L., and You, Y. (2021). Materials design for high-safety sodium-ion battery. *Adv. Energy Mater.* 11 (2): 2000974.

108 Liang, X. and Nazar, L.F. (2019). MXene materials as electrodes for lithium-sulfur batteries. In: *2D Metal Carbides and Nitrides (MXenes)* (ed. B. Anasori and Y. Gogotsi), 381–398. Springer.

109 Liang, X., Rangom, Y., Kwok, C.Y. et al. (2017). Interwoven MXene nanosheet/carbon-nanotube composites as Li–S cathode hosts. *Adv. Mater.* 29 (3): 1603040.

110 Zhao, M., Li, B.Q., Zhang, X.Q. et al. (2020). A perspective toward practical lithium–sulfur batteries. *ACS Cent. Sci.* 6 (7): 1095–1104.

111 Naguib, M., Adams, R.A., Zhao, Y. et al. (2017). Electrochemical performance of MXenes as K-ion battery anodes. *Chem. Commun.* 53 (51): 6883–6886.

112 Eftekhari, A. (2004). Potassium secondary cell based on Prussian blue cathode. *J. Power Sources* 126 (1–2): 221–228.

113 Vahid Mohammadi, A., Hadjikhani, A., Shahbazmohamadi, S., and Beidaghi, M. (2017). Two-dimensional vanadium carbide (MXene) as a high-capacity cathode material for rechargeable aluminum batteries. *ACS Nano* 11 (11): 11135–11144.

114 Huang, H., Cui, J., Liu, G. et al. (2019). Carbon-coated $MoSe_2$/MXene hybrid nanosheets for superior potassium storage. *ACS Nano* 13 (3): 3448–3456.

115 Zhang, L., Shao, Q., and Zhang, J. (2021). An overview of non-noble metal electrocatalysts and their associated air cathodes for Mg-air batteries. *Mater. Rep.: Energy* 1 (1): 100002.

116 Xu, M., Lei, S., Qi, J. et al. (2018). Opening magnesium storage capability of two-dimensional MXene by intercalation of cationic surfactant. *ACS Nano* 12 (4): 3733–3740.

117 Wang, S., Chen, S., Ma, L., and Zapien, J.A. (2021). Recent progress in cobalt-based carbon materials as oxygen electrocatalysts for zinc-air battery applications. *Mater. Today Energy* 20: 100659.

118 Liu, X., Zhang, G., Wang, L. et al. (2021). Structural design strategy and active site regulation of high-efficient bifunctional oxygen reaction electrocatalysts for Zn–air battery. *Small* 17 (48): 2006766.

119 Zhang, Y., Deng, Y.P., Wang, J. et al. (2021). Recent progress on flexible Zn-air batteries. *Energy Storage Mater.* 35: 538–549.

120 Katsoufis, P., Katsaiti, M., Mourelas, C. et al. (2020). Study of a thin film aluminum-air battery. *Energies* 13 (6): 1447.

121 Teabnamang, P., Kao-ian, W., Nguyen, M.T. et al. (2020). High-capacity dual-electrolyte aluminum–air battery with circulating methanol anolyte. *Energies* 13 (9): 2275.

122 Goel, P., Dobhal, D., and Sharma, R.C. (2020). Aluminum–air batteries: a viability review. *J. Energy Storage* 28: 101287.

123 Liang, X., Garsuch, A., and Nazar, L.F. (2015). Sulfur cathodes based on conductive MXene nanosheets for high-performance lithium–sulfur batteries. *Angew. Chem.* 127 (13): 3979–3983.

124 Lin, Z. and Simon, P. (2019). MXenes for supercapacitor application. In: *2D Metal Carbides and Nitrides (MXenes)* (ed. B. Anasori and Y. Gogotsi), 349–365. Springer.

125 Pershaanaa, M., Bashir, S., Ramesh, S. et al. (2022). Every bite of Supercap: a brief review on construction and enhancement of supercapacitor. *J. Energy Storage* 50: 104599.

126 Raghavendra, K.V.G., Vinoth, R., Zeb, K. et al. (2020). An intuitive review of supercapacitors with recent progress and novel device applications. *J. Energy Storage* 31: 101652.

127 Libich, J., Máca, J., Vondrák, J. et al. (2018). Supercapacitors: properties and applications. *J. Energy Storage* 17: 224–227.

128 Chen, X., Paul, R., and Dai, L. (2017). Carbon-based supercapacitors for efficient energy storage. *Natl. Sci. Rev.* 4 (3): 453–489.

129 Lukatskaya, M.R., Kota, S., Lin, Z. et al. (2017). Ultra-high-rate pseudocapacitive energy storage in two-dimensional transition metal carbides. *Nat. Energy* 2 (8): 1–6.

130 Ghidiu, M., Lukatskaya, M.R., Zhao, M.Q. et al. (2014). Conductive two-dimensional titanium carbide 'clay' with high volumetric capacitance. 516 (7529): 78–81.

131 Dall'Agnese, Y., Lukatskaya, M.R., Cook, K.M. et al. (2014). High capacitance of surface-modified 2D titanium carbide in acidic electrolyte. *Electrochem. Commun.* 48: 118–122.

132 Shan, Q., Mu, X., Alhabeb, M. et al. (2018). Two-dimensional vanadium carbide (V_2C) MXene as electrode for supercapacitors with aqueous electrolytes. *Electrochem. Commun.* 96: 103–107.

133 Shen, B., Liao, X., Zhang, X. et al. (2022). Synthesis of Nb_2C MXene-based 2D layered structure electrode material for high-performance battery-type supercapacitors. *Electrochim. Acta* 413: 140144.

134 Yang, C., Tang, Y., Tian, Y. et al. (2019). Methanol and diethanolamine assisted synthesis of flexible nitrogen-doped Ti_3C_2 (MXene) film for ultrahigh volumetric performance supercapacitor electrodes. 3 (1): 586–596.

135 Chen, K., Guan, Y., Cong, Y. et al. (2022). Vertically pillared $V_2CT_x/Ti_3C_2T_x$ flexible films for high-performance supercapacitors. *J. Alloys Compd.* 906: 164302.

136 VahidMohammadi, A., Liang, W., Mojtabavi, M. et al. (2021). 2D titanium and vanadium carbide MXene heterostructures for electrochemical energy storage. *Energy Storage Mater.* 41: 554–562.

137 Yang, L., Zheng, W., Zhang, P. et al. (2018). MXene/CNTs films prepared by electrophoretic deposition for supercapacitor electrodes. *J. Electroanal. Chem.* 830: 1–6.

138 Li, Z., Dall'Agnese, Y., Guo, J. et al. (2020). Flexible freestanding all-MXene hybrid films with enhanced capacitive performance for powering a flex sensor. *J. Mater. Chem. A* 8 (32): 16649–16660.

139 Etman, A.S., Halim, J., and Rosen, J. (2021). Mixed MXenes: Mo1. 33CTz and $Ti_3C_2T_z$ freestanding composite films for energy storage. *Nano Energy* 88: 106271.

140 Tang, Y., Zhu, J., Yang, C., and Wang, F. (2016). Enhanced supercapacitive performance of manganese oxides doped two-dimensional titanium carbide nanocomposite in alkaline electrolyte. *J. Alloys Compd.* 685: 194–201.

141 Fatima, M., Fatheema, J., Monir, N.B. et al. (2020). Nb-doped MXene with enhanced energy storage capacity and stability. *Front. Chem.* 8: 168.

142 Sharma, G., Naguib, M., Feng, D. et al. (2016). Calorimetric determination of thermodynamic stability of MAX and MXene phases. *J. Phys. Chem. C* 120 (49): 28131–28137.

143 Liu, R. and Li, W. (2018). High-thermal-stability and high-thermal-conductivity $Ti_3C_2T_x$ MXene/poly(vinyl alcohol) (PVA) composites. *ACS Omega* 3 (3): 2609–2617.

144 Perdew, J.P. and Wang, Y. (1992). Pair-distribution function and its coupling-constant average for the spin-polarized electron gas. *Phys. Rev. B* 46 (20): 12947.

145 Perdew, J.P., Burke, K., and Ernzerhof, M. (1996). Generalized gradient approximation made simple. *Phys. Rev. Lett.* 77 (18): 3865.

146 Cao, M., Wang, F., Wang, L. et al. (2017). Room temperature oxidation of Ti_3C_2 MXene for supercapacitor electrodes. *J. Electrochem. Soc.* 164 (14): A3933.

147 Li, X., Tang, Y., Liu, L. et al. (2022). Ti_3C_2 MXene with pillared structure for hybrid magnesium-lithium batteries cathode material with long cycle life and high rate capability. *J. Colloid Interface* 608: 2455–2462.

148 Xu, H., Zheng, R., Du, D. et al. (2022). V_2C MXene enriched with-O termination as high-efficiency electrocatalyst for lithium-oxygen battery. *Appl. Mater. Today* 27: 101464.

149 Li, Y.-M., Guo, Y.-L., and Jiao, Z.-Y. (2020). The effect of S-functionalized and vacancies on V_2C MXenes as anode materials for Na-ion and Li-ion batteries. *Curr. Appl Phys.* 20 (2): 310–319.

150 Venkateshalu, S., Cherusseri, J., Karnan, M. et al. (2020). New method for the synthesis of 2D vanadium nitride (MXene) and its application as a supercapacitor electrode. *ACS Omega* 5 (29): 17983–17992.

151 He, H., Xia, Q., Wang, B. et al. (2020). Two-dimensional vanadium carbide (V_2CT_x) MXene as supercapacitor electrode in seawater electrolyte. *Chin. Chem. Lett.* 31 (4): 984–987.

152 Tang, J., Mathis, T., Zhing, X. et al. (2021). Optimizing ion pathway in titanium carbide MXene for practical high-rate supercapacitor. *Adv. Energy Mater.* 11 (4): 2003025.

153 Das, P. and Wu, Z.-S. (2020). MXene for energy storage: present status and future perspectives. *J. Phys. Energy* 2 (3): 032004.

154 Gogotsi, Y. and Huang, Q. (2021). *MXenes: Two-Dimensional Building Blocks for Future Materials and Devices*, 5775–5780. ACS Publications.

155 Zhong, J., Sun, W., Wei, Q. et al. (2018). Efficient and scalable synthesis of highly aligned and compact two-dimensional nanosheet films with record performances. *Nat. Commun.* 9 (1): 1–11.

Index

Titanium Carbide MXenes: Synthesis, Characterization, Energy, and Environmental Applications, First Edition. Edited by Muhammad Tahir.